THE GEOLOGY OF DEVON

THE GEOLOGY OF DEVON

Edited by

E. M. Durrance

and

D. J. C. Laming

UNIVERSITY
of
EXETER
PRESS

First published in 1982 by
University of Exeter Press
Reed Hall, Streatham Drive
Exeter, Devon EX4 4QR
UK

Reprinted 1985 (Paperback)
Reprinted 1993, 1997, 1999

ISBN 0 85989 247 6

Cover illustration: Thatcher Rock, Torbay
Photograph: John Saunders

Printed and bound in Great Britain
by 4edge Ltd, Essex

Dedicated to the memory of

PROFESSOR SCOTT SIMPSON
MA *Cantab*. DrRerNat *Frankfurt* FGS
1915–1981

Professor of Geology
and Head of the Department of Geology
at the University of Exeter, 1959–75,
whose work greatly aided our understanding
of the geology of Devon.

Foreword

The county of Devon has some of the most varied and most visited geology in the British Isles. There can be few British geology students who, in the course of their training, have not spent at least one field course examining the cliffs or moors of the county. Devon is also known to geologists throughout the world as the type area of the Devonian System, established in 1839 by Sedgwick and Murchison to include Devon's oldest rocks.

Research on the geology of the county has a long history, in part connected with the exploration for economic minerals. In the last two decades there has been a dramatic increase, not only in the scope of investigations, but also in the number of researchers. New techniques, new observations and results, and new ideas are changing the interpretation of the geological history of the area.

In view of this diversity of interests, it is especially timely for the present volume to be published. It provides an opportunity for specialists to take stock of progress in their own field and, in so doing, to inform a wide readership. The aim has been not just to document progress but also to identify those gaps in knowledge that require further study.

Anyone who has experienced the difficulties of editing a multi-author publication will sympathise with the editors over their most difficult task for it requires a subtle blend of encouragement, goading and, above all, patience to achieve a finished product. The editors and authors are therefore to be congratulated on their success in working as a team and to be thanked for the resultant high quality of their labours. *The Geology of Devon* is a landmark in the understanding of our county.

J. W. Murray
Department of Geology
University of Exeter

Preface

Geological Field Work

It has often been remarked that geology is a subject best studied by actually looking at rocks, minerals and fossils, and their structures and relationships, in the field. Therefore, although this book mainly deals with descriptions from an interpretative viewpoint, at the end of each appropriate chapter a number of localities are listed which will serve to illustrate the main points dealt with in the text. The localities are mainly arranged in subject groupings, although some geographical subdivision is also present. Excursions to specific areas of Devon, to include visits to a number of sites of different character, may thus be constructed with the aid of the appropriate Ordnance Survey and Geological Survey maps, according to individual requirements. Excursion Guides to different parts of Devon are also published by the Geologists' Association:

> *Dartmoor: the North West Margin and Other Selected Areas*, by W. R. DEARMAN, 1962.
> *The Coast of South Devon and Dorset between Branscombe and Burton Bradstock*, by D. V. AGER and W. E. SMITH, 1965.
> *The Plymouth Area*, by D. M. Hobson, 1978.

Copies may be obtained from the Geologists' Association, Burlington House, Piccadilly, London.

Geological field work is an interesting and rewarding endeavour, but it should be borne in mind that in many cases rock exposures are situated on private land, and may be in hazardous localities or be hazardous themselves. The inclusion of a locality in this book, even where specific mention of the need for access permission or the presence of hazardous conditions is not made, does not imply that access is freely available and the localities are safe: it is the responsibility of the user of this book to obtain permission to enter private land and to avoid hazards of all kinds.

To help establish an acceptable framework of conduct, the Geologists' Association has published a Code for Geological Field Work which is reproduced (with permission) overleaf. It is hoped that users of this book will follow all recommendations given in the Code.

January 1982   E. M. DURRANCE and D. J. C. LAMING
Exeter

A Code for Geological Field Work

A Geological 'Code of Conduct' has become essential if opportunities for field work in the future are to be preserved. The rapid increase in field studies in recent years had tended to concentrate attention upon a limited number of localities, so that sheer collecting pressure is destroying the scientific value of irreplaceable sites. At the same time the volume of field work is causing concern to many site owners. Geologists must be seen to use the countryside with responsibility; to achieve this, the following general points should be observed.

1. Obey the Country Code, and observe local byelaws. Remember to shut gates and leave no litter.
2. Always seek prior permission before entering private land.
3. Don't interfere with machinery.
4. Don't litter fields or roads with rock fragments which might cause injury to livestock, or be a hazard to pedestrians or vehicles.
5. Avoid undue disturbance to wildlife. Plants and animals may inadvertently be displaced or destroyed by careless actions.
6. On coastal sections, consult the local Coastguard Service whenever possible, to learn of local hazards such as unstable cliffs, or tides which might jeopardise excursions possible at other times.
7. When working in mountainous or remote areas, follow the advice given in the pamphlet 'Mountain Safety', issued by the Central Council for Physical Education, and, in particular, inform someone of your intended route.
8. When exploring underground, be sure you have the proper equipment, and the necessary experience. Never go alone. Report to someone your departure, location, estimated time underground, and your actual return.
9. Don't take risks on insecure cliffs or rock faces. Take care not to dislodge rock, since other people may be below.
10. Be considerate. By your actions in collecting, do not render an exposure untidy or dangerous for those who follow you.

Collecting and Field Parties

1. Students should be encouraged to observe and record but not to hammer indiscriminately.
2. Keep collecting to a minimum. Avoid removing *in situ* fossils, rocks or minerals unless they are genuinely needed for serious study.
3. For teaching, the use of replicas is commended. The collecting of actual specimens should be restricted to those localities where there is a plentiful supply, or to scree, fallen blocks and waste tips.
4. Never collect from walls or buildings. Take care not to undermine fences, walls, bridges or other structures.

5. The leader of a field party is asked to ensure that the spirit of this Code is fulfilled, and to remind his party of the need for care and consideration at all times. He should remember that his supervisory role is of prime importance. He must be supported by adequate assistance in the field. This is particularly important on coastal sections, or over difficult terrain, where there might be a tendency for parties to become dispersed.

Health and Safety at Work Act

Since the introduction of this Act, safety measures are more strictly enforced on sites including quarries. Protective clothing, particularly safety helmets, must be worn by employees, so visitors are expected to observe the same precaution, often as a condition of entry. Suitable helmets are readily available, cheap to purchase, and should be part of the necessary equipment of all geologists. They must be worn at all times in quarries.

Visiting Quarries

1. An individual, or the leader of a party, should have obtained prior permission to visit.
2. The leader of a party should have made himself familiar with the current state of the quarry. He should have consulted with the Manager as to where visitors may go, and what local hazards should be avoided.
3. On each visit, both arrival and departure must be reported.
4. In the quarry, the wearing of safety hats and stout boots is recommended.
5. Keep clear of vehicles and machinery.
6. Be sure that blast warning procedures are understood.
7. Beware of rock falls. Quarry faces may be highly dangerous and liable to collapse without warning.
8. Beware of sludge lagoons.

Research Workers

1. No research worker has the special right to 'dig out' any site.
2. Excavations should be back-filled where necessary to avoid hazard to men and animals and to protect vulnerable outcrops from casual collecting.
3. Don't disfigure rock surfaces with numbers or symbols in brightly coloured paint.
4. Ensure that your research material and notebooks eventually become available for others by depositing them with an appropriate institution.
5. Take care that the publication of details does not lead to the destruction of vulnerable exposures. In these cases, do not give the precise location of such sites, unless this is essential to scientific argument. The details of such localities could be deposited in a national data centre for Geology.

Societies, Schools and Universities

1. Foster an interest in geological sites and their wise conservation. Remember that much may be done by collective effort to help clean up overgrown sites (with permission of the owner, and in consultation with the Nature Conservancy Council).
2. Create working groups for those amateurs who wish to do field work and collect, providing leadership to direct their studies.

3. Make contact with your local County Naturalists' Trust, Field Studies
   Centre, or Natural History Society, to ensure that there is coordination in
   attempts to conserve geological sites and retain access to them.

Reprinted by permission of the Geologists' Association. Further copies may be obtained from the Geologists' Association, Burlington House, Piccadilly, London.

Contributors

K. E. Beer, Institute of Geological Sciences, Exeter—Chapter Six: Metalliferous Mineralisation.

R. A. Cullingford, Department of Geography, University of Exeter—Chapter Eleven: The Quaternary.

J. Dangerfield, Institute of Geological Sciences, London—Chapter Ten: The Tertiary Igneous Complex of Lundy.

E. M. Durrance, Department of Geology, University of Exeter—Chapter One: Introduction, Chapter Two: The Devonian Rocks, Chapter Four: The Variscan Structures.

R. A. Edwards, Institute of Geological Sciences, Exeter—Chapter Nine: The Tertiary Sedimentary Rocks.

E. C. Freshney, Institute of Geological Sciences, London—Chapter Four: The Variscan Structures, Chapter Nine: The Tertiary Sedimentary Rocks.

M. B. Hart, School of Environmental Sciences, Plymouth Polytechnic—Chapter Eight: The Marine Rocks of the Mesozoic.

J. R. Hawkes, Institute of Geological Sciences, London—Chapter Five: The Dartmoor Granite and Later Volcanic Rocks.

D. J. C. Laming, Herrington Associates, Exeter—Chapter One: Introduction, Chapter Seven: The New Red Sandstone.

C. Nicholas, ECC Quarries Ltd, Exeter—Chapter Twelve: Industrial Minerals.

R. C. Scrivener, Institute of Geological Sciences, Exeter—Chapter Six: Metalliferous Mineralisation.

E. B. Selwood, Department of Geology, University of Exeter—Chapter Two: The Devonian Rocks, Chapter Four: The Variscan Structures.

J. M. Thomas, Department of Geology, University of Exeter—Chapter Three: The Carboniferous Rocks

C. D. N. Tubb, South West Water Authority, Exeter—Chapter Thirteen: Hydrogeology.

A. Vincent, Watts, Blake Bearne and Company PLC, Newton Abbot—Chapter Twelve: Industrial Minerals.

Contents

List of Plates

1. Devonian and Carboniferous Fossils from Devon
2. Slumped bed at junction of Baggy Sandstone and Pilton Shale, Baggy Point (SS 435401)
3. Pillow lava, Chipley (SX 807722)
4. Agglomerate, Brent Tor (SX 470804)
5. Folds in Bude Formation, Warren Gutter (SS 201110)
6. The Granitic Rocks of Dartmoor
7. Hound Tor (SX 743790)
8. Permian volcanic neck, Hannaborough (SS 529029)
9. Permian breccias overlying Meadfoot Slate, Thurlestone Sands (SX 675420)
10. Aeolian sand (Dawlish Sands), Dawlish (SX 967768)
11. Otter Sandstones overlying Budleigh Salterton Pebble Beds and Littleham Mudstones, Budleigh Salterton (SY 061816)
12. Blue Lias with *Metophioceras*, Kilve, Somerset (ST 145445)
13. Chalk overlying Cenomanian Limestone and Upper Greensand, Beer (SY 230890)
14. Cretaceous Fossils from Devon
15. Bovey Formation (Southacre Member) clays and lignites, East Golds Pit (SX 860730)
16. Pleistocene Raised Beach overlying Devonian limestone, Hope's Nose (SX 950637)
17. Pleistocene Raised Beach overlying slate, Pencil Rock, Croyde Bay (SS 423402)
18. Valley of Rocks, Lynton (SS 705495)
19. Head, Middleborough (SS 429398)
20. China clay pit with working high-pressure water jet, Lee Moor (SX 572629)

Acknowledgements

We wish to thank all the authors for their contributions to this book, and for their patience with the editors' quibbles throughout the different stages of its development. Thanks go also to Delphine Jones, Jeffery Jones and Terry Bacon for their skill in producing the text-figures, for which Owen Pook did the typesetting; to John Saunders and Andrew Teed for their photography, and to Gwanwyn Wright, Brenda Hosking and Brenda Spivey for preparing the typescript. Lastly our thanks go to Professors D. L. Dineley and J. W. Murray for their generous helpful advice and comments, and to Barbara Mennell for shepherding the book through all the stages of its production.

Exeter, January 1982

ERIC DURRANCE
DERYCK LAMING

The University of Exeter gratefully acknowledges the support of the following organisations:

Gaffney, Cline and Associates Ltd
Elf UK Ltd
Herrington Associates
Selection Trust Ltd
Amax Hemerdon Ltd
Unionoil Company of Great Britain
Cluff Oil Ltd
Esso Exploration and Production UK
Texaco Ltd
ECC Quarries Ltd
Arco Oil Producing, Inc.
Cities Service Europe—Africa Petroleum Corporation
Horizon Exploration Ltd
Clyde Petroleum Ltd
Phillips Petroleum Company UK Ltd
Sussex Mutual Building Society
Western Geophysical Company of America
The Midland Bank
K. Wardell and Partners
Mr G. N. Rayson

Institute of Geological Sciences photographs reproduced by permission of the Director, Institute of Geological Sciences: Crown copyright/NERC copyright.

Chapter One

Introduction

Devon is one of the largest of the English counties: it measures 118 km east to west and almost the same north to south, and has an area of 6710 sq. km. It also has one of the longest coastlines, with some 100 km bordering the Atlantic Ocean and Bristol Channel in the north, and 200 km along the English Channel in the south. Within its area lies one of the most varied displays of geology to be found anywhere in the British Isles. This is reflected in the great variety of its landscape: the plateau and valleys of the Cretaceous rocks of east Devon, the rich red earth on the Permian and Triassic lowlands, and the 'dunland' pastures on the Carboniferous of mid Devon—these all form a marked contrast to the granite upland of Dartmoor and the Devonian sandstone moorland of Exmoor. Unlike more northerly counties, glaciation has played only a small part in shaping the landscape, so the relationship between bedrock and hill form is more faithful. Moreover, outcrops of the rocks are common, especially along the coastline, making this a favourite place for geological research, student field parties and amateur geological study. The soft climate, easy accessibility of important localities and the pleasant surroundings contribute to the county's appeal, for geologists, tourists and residents alike.

This account of the geology of Devon is intended for all categories of geologist, aimed at presenting an overview of the present state of knowledge in fundamentally factual terms, communicating to specialists and generalists alike but avoiding too great a complexity of terminology. An attempt has been made to update the information to near the date of publication, but the understanding of Devon's geology is developing at a quick pace, with many people from both within and outside the county contributing their research efforts and reporting their results to scientific meetings several times a year. Inevitably, some of the results reported herein will become dated and possibly inaccurate in the light of future work; in particular, the interpretation of the county's history in terms of crustal dynamics—plate tectonic theory—is now only in its infancy and may well see radical changes of attitude in a very few years.

The county is distinguished as being the only one in the British Isles to give its name to a geological system of world-wide recognition. The Devonian System was the name introduced in 1839 by Rev. Adam Sedgwick and Sir Roderick Murchison for those strata that were deposited after the Silurian slates and limestones of Wales but before the coal-bearing Carboniferous rocks of the English Midlands. The Old Red Sandstone, which stratigraph-

1

ically lies between them, lacks marine fossils; so these distinguished geologists travelled to Devon to investigate a sequence of limestones and slates with marine faunas that could be correlated with this interval. It is said that this project had to be undertaken in somewhat of a hurry, to forestall the redoubtable James Hall describing rocks of the same age in New York State and naming the system after Lake Erie. Subsequently, the succession in Devon proved very complex and geologists chose stratigraphic sequences in Belgium and elsewhere to act as world standards of comparison.

PREVIOUS ACCOUNTS OF THE GEOLOGY

Sedgwick and Murchison built upon accounts of geology made before 1839 and published in a variety of places. One of the first observers of Devon geology was W. G. Maton (1797) whose general account of the nature of the soils and rocks showed a remarkable perception of the complex character of the county, but an earlier account was given by J. Milles (1750) and R. Polwhele's major work on the county (1793) included much on the geology. Rev. J. J. Conybeare (1823) gave a more comprehensive description and the first classification of the rocks of Devon, but it was not until the 1830s that detailed geological work started in the area. A major advance was the report on the geology of Cornwall, Devon and west Somerset in 1839 by Sir Henry De la Beche, a gentleman from Trinidad who became Director of the Geological Survey of England and Wales. The results of palaeontological studies which helped Sedgwick and Murchison to propose the Devonian System were published by Professor W. J. Phillips in 1841, followed by R. A. C. Godwin-Austen in 1842 with an important report on the geology of south and east Devon. William Pengelly, whose name occurs frequently in any bibliography of Devon geology, published an account on the stratigraphic value of the New Red Sandstone in 1863, but soon after that another name began to appear in published work: W. A. E. Ussher, a quiet, painstaking and thorough field geologist who in 1870 began the great task of revising the Old Series Geological Survey maps of the South West of England.

It is to Ussher that we owe much of the basis of our present knowledge of the geology of Devon, though the Institute of Geological Sciences (IGS), now with a regional office in Exeter, have in the last two decades systematically resurveyed many of his published map sheets. To honour his name and contribution, and to advance the science of geology in South West England, a society was formed in 1961 bearing his name. With a membership approaching 250, the Ussher Society holds an annual meeting somewhere in the region: its published Proceedings contain a wide variety of short papers on all aspects of the geology of South West England, in many cases giving advance information of work later published in more detail elsewhere.

Comprehensive descriptions of Devon geology are, nevertheless, few. A. W. Clayden's work, published in 1906, is now a classic; Professor S. Simpson, to whom this volume is dedicated, produced a report on the occasion of the 1969 British Association meeting in Exeter which gave a particularly useful review. The special commemorative volume of the Royal Geological Society of Cornwall for 1964 (edited by K. F. G. Hosking and G. J. Shrimpton) also contained a number of valuable articles, but probably the best source, summarising much recent work, is the *Regional Geology of South-West England* published by the IGS (Edmonds *et al.*, 1975). Metalliferous mining is dealt with most comprehensively by H. G. Dines (1956) with a history of working, geology and output for all known sites. The physiography of the area has been reviewed by A. H. Shorter *et al.* (1969). For non-specialist geologists, two books by J. W. Perkins (1971, 1972) give useful information on south and east Devon and the Dartmoor–Tamar Valley areas respectively.

DEVON AS A GEOLOGICAL UNIT

With such diversity of geology in the county, it is hard to visualise any unifying factor; yet although it is 'of a piece' with Cornwall's granites and metamorphic rocks on the one hand and Dorset's Mesozoic rocks on the other, each of which overlap the county boundary a long way, Devon uniquely presents the relationship between the two. The change from Palaeozoic deep marine sedimentation and orogenic-magmatic activity to shelf sea deposition and planation in the Mesozoic and Cenozoic (Tertiary and Quarternary), is superbly represented, with the important terrestrial red-bed episode of the New Red Sandstone in between.

Figure 1.1 shows that very few geological periods and epochs go unrepresented in Devon geology, especially between the lowest Devonian and the Jurassic; indeed, in the Tertiary, Oligocene beds present in lacustrine basins in Devon have few counterparts anywhere else in the British Isles.

In geological terms, southern England may be viewed as consisting of two geological 'kingdoms': the kingdom of Wessex, contiguous with the Weald and the Cotswold country, formed of Permian, Mesozoic and Cenozoic rocks with generally soft landscapes, the realm for many centuries of the Saxon tribes; and the kingdom of Cornubia, of Palaeozoic metamorphic and igneous rocks, with rugged, deeply dissected scenery that provided a refuge for the Celtic tribes. The boundary between the two runs through Devon from south coast to north, and passes within a stone's throw of the ramparts of the Roman fortress of Exeter (*Isca Dumnoniorum*)—one can speculate on how much the geology of the county has influenced its history and the development of its people.

Whosoever drew the county boundaries of Devon, however, had little regard for topography and even less for geology, so the task of describing the geology is complicated by most formations passing unconcernedly into neighbouring counties. Accordingly, for some aspects of the geology, it has been decided unilaterally to annex certain strategic parts of Cornwall, Dorset and Somerset to the area under consideration so that no artificial discontinuities shall arise; we trust that neighbouring counties will feel more honour than hurt that these parts should be considered worthy of inclusion in Devon's ambit.

In the west, the steep-sided valley of the River Tamar forms a clear topographic boundary with Cornwall for most of its length, from near the Atlantic coast southwards to Plymouth Sound: but across the valley the same scenery of open hills and wooded valleys continues, with the same tendency for metallic mineral occurrence. The Devonian and Carboniferous rocks, striking predominantly east-west, continue unchanged in character as far west as Bodmin Moor, where a fault trending northwest–southeast brings up rocks of different

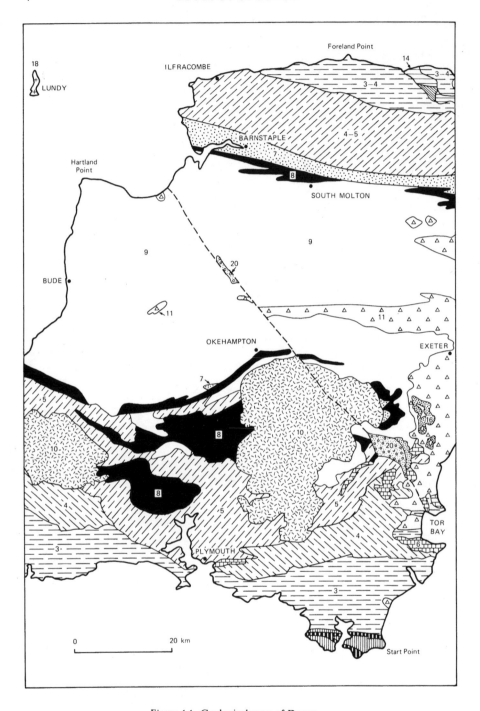

Figure 1.1. Geological map of Devon

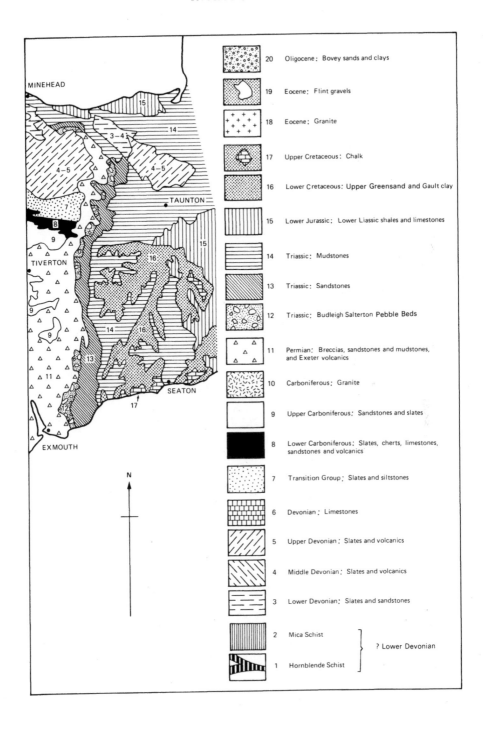

20 Oligocene: Bovey sands and clays

19 Eocene: Flint gravels

18 Eocene: Granite

17 Upper Cretaceous: Chalk

16 Lower Cretaceous: Upper Greensand and Gault clay

15 Lower Jurassic: Lower Liassic shales and limestones

14 Triassic: Mudstones

13 Triassic: Sandstones

12 Triassic: Budleigh Salterton Pebble Beds

11 Permian: Breccias, sandstones and mudstones, and Exeter volcanics

10 Carboniferous: Granite

9 Upper Carboniferous: Sandstones and slates

8 Lower Carboniferous: Slates, cherts, limestones, sandstones and volcanics

7 Transition Group: Slates and siltstones

6 Devonian: Limestones

5 Upper Devonian: Slates and volcanics

4 Middle Devonian: Slates and volcanics

3 Lower Devonian: Slates and sandstones

2 Mica Schist   } ? Lower Devonian

1 Hornblende Schist

facies to form a logical boundary to the area. West Somerset has the Cothelstone Fault, which serves the same purpose for the Devonian of north Devon, which continues eastwards to the Quantock Hills. The eastern margin of the exposed Permian and Triassic south of Taunton is also a natural boundary, but the Cretaceous hilltop cappings which extend as far west as the Haldon Hills before dipping out of sight beneath the Bovey Basin, and the Jurassic rocks which begin to appear at Axmouth, are gradational with the important stratigraphic sequences found further east in Dorset and the Hampshire Basin.

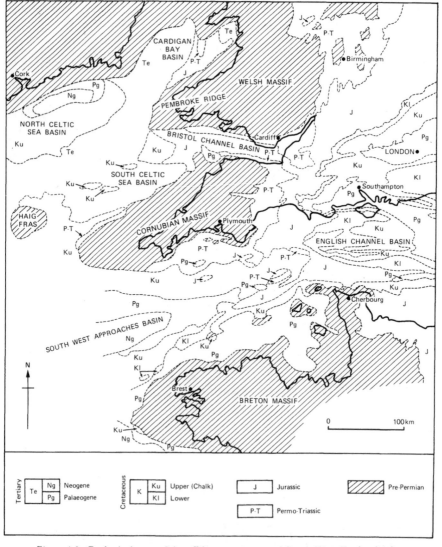

Figure 1.2. Geological map of the offshore areas around South West England (after Warrington and Owens, 1977)

Offshore geology has come to be known very much better in recent years as a consequence of an increasing interest in possible oil and gas exploration. The Cretaceous rocks of east Devon, with their relatively limited outcrop, extend to cover a much larger area of the adjacent English Channel in a wide arc to beyond Cornwall, disappearing southwards under Tertiary strata some 50 km from the coast (*Figure 1.2*). Mesozoic rocks also occur off the coast of north Devon, where a thick Jurassic clay sequence is preserved in part of a major basin that continues into Somerset and Dorset. While these areas may be regarded as appertaining to Devon for the purposes of this volume, it is the land areas which will mainly concern most readers; and it may be said that, disregarding the overstepping Cretaceous and younger strata of east Devon, the county was largely formed as we now see it by the end of the Triassic. It is likely that Cretaceous, Jurassic and maybe Triassic strata once covered all or part of the county; but erosion has removed much of these to reveal a land surface of former times, much modified from its original mountainous desert, but presenting geological and scenic diversity of unparalleled quality that geologists, perhaps more than any others, can appreciate most.

The *deep structure* of South West England has been investigated over recent decades with increasingly detailed work using refined geophysical instrumentation. An important contribution was by Bott *et al.* (1970) who, using seismic techniques, showed that the crust in this region extends to a depth of about 27 km, where it has a sharp horizontal boundary with the mantle. Two layers were found in the crust, the upper consisting essentially of granite, with sedimentary and metamorphic rocks: this was shown to be about 12 km thick. Below this, with a very gradational contact, lies the lower layer; this has never been sampled, as it lies beyond the reach of drilling, but on the basis of seismic velocities and density estimates it is thought to consist of rock of dioritic or granodioritic composition. These layers probably represent the result of magmatic and metamorphic processes operative during the Variscan Orogeny and possibly earlier—granites mobilised during the orogenic process contributing to the upper layer, the depleted lower layer consisting of more mafic material. There is no indication of oceanic crust beneath South West England, but in the Lizard area of Cornwall there are periodotites and other rocks of oceanic crustal origin (Bromley, 1979).

SUMMARY OF GEOLOGICAL HISTORY

In this section, the contents of the later chapters are introduced in a non-technical and generalised way, especially for the reader with only a brief familiarity with geological concepts. It is also hoped that this outline will provide the framework on which links may be established between the separate subjects covered in detail by the later chapters. *Table 1.1* summarises the geological history of Devon.

Devonian

The oldest rocks in Devon of which certain dates are known (some may be older) are the Lower Devonian, deposited as sediments in a marginal marine basin between 395 Ma (million years) and 370 Ma ago. Fragments of older formations are included in rocks of the same age in Cornwall, so earlier crustal elements, land masses or volcanic island arcs, must have been present.

It appears that sea-floor subsidence and sedimentation continued, with minor breaks, until the end of the Carboniferous, when a major episode of folding, faulting, metamorphism and granitic intrusion, comprising the Variscan Orogeny or mountain-building episode, created the Palaeozoic massif of Devon and Cornwall, known to geologists as 'Cornubia'.

Table 1.1 SUMMARY OF THE GEOLOGICAL HISTORY OF DEVON

	Age (Ma)	
QUATERNARY		
Recent		erosion; local deposition of alluvium, beach deposits, peat
----------	0.01	
Pleistocene		erosion; local deposition of boulder clay, head, river terrace gravel, raised beach deposits
----------	1.7	
TERTIARY		
(hiatus)		erosion, minor folding; movement on tear faults
----------	22.5	
Oligocene		fluvial and lacustrine clay, sand and lignite deposition in basins along tear faults; erosion
----------	39	
Eocene		residual and fluvial gravel formation; intrusion of Lundy granite; lateritic weathering in tropical climate; erosion
----------	55	
(hiatus)		minor folding and faulting
----------	65	
MESOZOIC		
Cretaceous		glauconitic sand and chalk deposition in shallow tropical sea and shorelines
----------	110	
(hiatus)		minor folding and faulting; erosion
----------	180	
Jurassic		clay and limestone deposition in shallow shelf sea
----------	195	
Triassic		fluvial conglomerate, sandstone and mudstone deposition in hot desert
----------	230	
PALAEOZOIC		
Permian		breccia, sandstone and mudstone deposition in hot desert; sand dune formation
----------	280	
Carboniferous:		
Stephanian		red-bed deposition commences in hot desert; elvan dykes, rhyolitic lava; basaltic lava; last stages of orogenic and intrusive activity
Westphalian and Namurian		turbidite sandstone and shale deposition in deep sea basin; followed by Variscan Orogeny and granite intrusion
Dinantian		shale, chert sandstone and limestone deposition in marine basin; submarine volcanism
----------	345	
Devonian:		
Upper and Middle		shale, sandstone and reefal limestone deposition in marine basin; submarine volcanism
Lower		fluvial shale and sandstone deposition on continental margin
----------	395	(base of succession not seen)

Details of this most important era of development of Devon's structure are being unravelled by the efforts of many geologists, most of whom work in a small area of outcrop or take one segment of the stratigraphic sequence, but whose studies contribute to the unfolding picture. It is evident that there was an elongate subsiding marine trough during the Devonian Period, on the margins of a large continental massif which extended northwards to Wales, Scotland and beyond. This is known as the Old Red Sandstone continent, and was the result of earlier orogenies which created the mountain areas of Wales, Cumbria, Scotland and Norway. Under a tropical desert climate, this continent was largely devoid of land plants; weathering and erosion were unchecked by a vegetative cover, and large expanses of river deposits were laid down on the flanks of the mountains, deposits which spilled over into the adjacent seas as deltas to form extensive sandstone and mudstone layers. In the tropical conditions all iron compounds oxidised to haematite, and thus the sandstones became red in colour. The famous 'Old Red Sandstone' is of Devonian age and is, therefore, the lateral equivalent of the marine strata preserved in Devon: indeed, in north Devon there are red sandstone layers which were formed as extensions to the Old Red Sandstone in response to increased erosion and sedimentation in central and south Wales, with consequent out-building of coastal plains and deltas. At other times when the coastline was further north—the Bristol Channel had not yet been formed—shales and some limestones with marine fossils were deposited in north Devon.

The Devonian succession in south Devon shows that deposition of non-marine sandstones occurred only in the very early part of the Period, with shales and limestones forming during the remainder of the time in seas that were at times deep, but at other times shallow enough for stromatoporoid-coral reefs to grow. The accumulation of sedimentary strata would have built up to sea level except for the counterbalancing effect of subsidence, and when subsidence was more rapid the waters became deeper: sedimentation was unable to keep pace. Shales were formed in these times, but when seas became shallower, calcareous marine life abounded and their shells accumulated to form limestones. The reefs developed in clear water areas, forming the massive limestones around Torquay and Plymouth.

Volcanic activity was a periodically recurring feature of the seas of south Devon, with pillow lavas being extruded from fractures in the sea floor over a large area. Dolerite sills and dykes were emplaced at moderate depths below. An interesting relationship existed between volcanism and the growth of stromatoporoid-coral reefs, as the latter appear to have died out when lavas and tuffs were deposited in the sea, due no doubt to the effect of the poisonous volcanic gases on the marine life of the reefs.

Carboniferous

Both the sedimentary and volcanic relationships of the Devonian continued into the Carboniferous Period, with deposits forming in a marine trough to the south of a large area of coastal lowlands upon which, in late Carboniferous times, swamp forests grew—now forming the South Wales Coalfield—and continuing submarine volcanic activity. During the early Carboniferous (345-325 Ma) a sequence of shales, thin deep-water limestones and radiolarian cherts were formed, evidently in a very deep trough, accompanied by volcanic and minor intrusive activity seen today as large dolerite masses interlayered with the sedimentary strata, and as volcanic tuff and agglomerate horizons. The famous Carboniferous Limestone of south Wales and the Bristol district is not represented in Devon, though near Westleigh, where the trough margin used to lie in north Devon, a series of limestones occur that appear to have been deposited by turbidity currents carrying soft carbonate sediment swept off the nearby shallow-water shelf areas. The nearest limestones of shallow water type are found at Cannington Park near Bridgwater in Somerset.

Turbidity current deposits of the more conventional kind (which formed sandstones) are common in the later Carboniferous strata in the mid Devon area—very well seen in the western coast sections. Some sandstones of more deltaic affinity on the northern margin of the Carboniferous outcrop, how-ever, testify to great delta swamps which were forming to the north, in south Wales, where peat layers subsequently became coal seams within the Coal Measures. No coal was formed in Devon—a factor that caused the industrial revolution largely to pass the county by—though small pockets of sooty 'culm' were exploited in the past, giving the name to the Carboniferous strata in Devon. These Culm Measures extended into late Carboniferous time (325-280 Ma), but by then volcanic activity had ceased in Devon and the marine trough had been infilled with a thick sequence of turbidite sandstones and shales.

Palaeogeography and the Variscan Orogeny

Palaeogeographic reconstructions of the distribution of land and sea in Devonian and Carboniferous times show that the Atlantic Ocean, as we know it, did not then exist. North America lay to the west of Europe, and the Iberian peninsula occupied the present Bay of Biscay: the Atlantic, as an ocean, did not begin to appear anywhere to the west of Britain until the Cretaceous Period, some 100 Ma ago.

An oceanic basin, however, existed between a northern land area, consisting of Britain north of the Thames-Avon line, the Baltic Shield area and much of northern Europe, and a southern area consisting of a number of micro-plates—islands and connecting shallow seas—to the north of the African continent. Like the main land areas, the micro-plates were composed of conti-nental crust. This southern area was the site of subduction, where oceanic crust was plunging slowly but inexorably beneath continental crust, while the land areas to the north and south were being eroded to form the Devonian and Carboniferous sedimentary formations. Slow but steady closing of the oceanic basin as subduction proceeded, probably accompanied by large scale transcurrent (lateral) movements, caused compression that produced the fold and thrust structures seen today.

These tectonic events are collectively known as the *Variscan Orogeny* (known also as the Armorican or Hercynian), which culminated in the late Carboniferous but began as several lesser earlier tectonic episodes which left their mark as unconformities or minor episodes of folding. During the time before the culmination of the orogeny, the crust of the oceanic basin had generally been consumed by subduction, but in the Lizard area of Cornwall a portion of oceanic crust has apparently been preserved by being thrust (obducted) over a thickness of continental crust rather than under it (sub-ducted) at an early stage. The detail of plate movements and their underlying causes are still, of course, a matter of debate, but a deeper understanding of the causes and effects of the events taking place in Devonian and Carboniferous times comes in the light of an understanding of plate tectonic processes as they operate today in other parts of the world.

Frictional heating between plates, probably aided by tectonic thickening of the crust, would have caused partial melting of the lower continental crust at plate margins; this formed granitic magmas at depth. On rising through the upper levels of the continental crust, these magmas finally reached cooler zones, where they solidified to form the granite batholith of South West England. Even today, after millions of years of erosion, only the top domal parts of this vast body can be seen at the surface, forming separate granite bodies such as Dartmoor and Bodmin Moor. Evidence from gravity surveying shows that these granites are linked at depth as part of a mass stretching from Dartmoor to Land's End, and even beyond to the Isles of Scilly and the submerged reefs of Haig Fras. They are all part of the same huge granite batholith, about 300 km long and 35 km wide.

The Variscan Orogeny produced low-grade metamorphism in the Devonian and Carboniferous rocks, with shales being converted to slates in some areas, but the grade becomes higher when traced southwards to the Start Point area. Here a major tectonic line separates a southern zone of quartz-mica and hornblende-chlorite schists (derived from sedimentary and volcanic rocks respectively) from slate of lower metamorphic grade lying to the north. These schists are resistant rocks forming the spectacular headlands of Start Point and Bolt Tail, the southern rampart of the county both geographically and geologically. Folding spread much further north, however, with remarkable chevron, sinusoidal and near-isoclinal folds having been produced in many areas and seen to good advantage between Bude and Hartland Point. Elsewhere the observer mostly sees inclined strata, but distinct folds, commonly cut by steeply inclined faults, can be examined at many points along both north and south coasts and in inland localities such as the banks of the River Exe in Exeter.

Metamorphism was also produced by the intrusion of the granite, which is dated at about 290 Ma, after the main pulse of the orogeny. Rocks already folded were cut by the granite, and a metamorphic aureole over 1 km wide was formed in which hornfelses, skarns and spotted rocks testify to the heat that flowed out of the granite as it cooled. Even today heat still flows out of the granite, in anomalously high amounts, and represents a potential geothermal energy resource; this is attributed mostly to the decay of radioactive minerals present in the rocks.

Similar events to those of the Variscan Orogeny can be traced in central Europe and in eastern North America; in the latter region, the Appalachian Mountains were formed at this time and give their name to the orogeny on that side of the Atlantic.

Permian and Triassic

Probably the last event of the Variscan Orogeny was uplift. The intrusion of a large body of relatively low-density granite resulted in an isostatic response, and the seizing up of the plate margin meant that subduction ceased, so there was no longer a downward drag on the fold-belt. The mountains that rose must have been at least 3,000 m high, probably much more in places, but beside them and among them were deep valleys created by erosion. The new mountain range emerged into a sub-tropical arid climate: infrequent but violent rain storms caused large amounts of sediment to be swept down into the valleys and out onto the fringing desert plains. From the end of the Carboniferous and through the Permian Period, great alluvial fans built up on the mountain fringes, gradually infilling the valleys and swamping the lower slopes; thorough post-depositional weathering turned all iron compounds into red oxides, and the rocks became red throughout. Thus the *New Red Sandstone* began, with coarse breccias and sandstones, some of them wind-blown and some laid by impermanent rivers; at the same time, mafic and intermediate lavas erupted in the Exeter area and westward, intermingling with the sedimentary deposits. Quartz porphyry, now seen only as boulders in the red-beds, appears once to have been abundant across the top of the Dartmoor dome, but erosion was vigorous and cut down even as far as the granite itself. Infilling of the valleys led to wider expanses of desert surface, across which sand dunes were blown from the southeast. As the desert basins developed into wide flood plains, farther away from the mountainous source areas, so finer muds and silts were laid down, and coarser beds became rare. The scenery of Devon in those days must have been quite similar to the southern fringes of the Sahara Desert today, though an important difference is that, during the Permian, many land plants of the desert fringe had not then evolved and the desert would then have been much more lifeless.

The beginning of the *Triassic Period*, from 230 Ma, saw the deposition of conglomerates in the bed of a great river, characterised by quartzite pebbles whose nearest source now appears to be western France. This implies that the English Channel was then a single desert basin across which a large river could flow. When the river ceased flowing, wind-blown sand covered the channel, and a flood-plain regime fed by small impermanent streams was established. In areas north and east of Devon, large salt pans gave rise to thick halite deposits, though it appears the environment was not suitable in most of Devon for them to form here.

Jurassic

The end of the Triassic, at about 195 Ma, was marked by a marine incursion which laid down the 'White Lias', followed in Dorset and parts of Devon by the long shelf-sea deposition of the Jurassic Period. Most of Devon continued as a land mass, as did Cornwall, but all other adjacent areas began to experience rift-valley and trough development. One such trough was the Somerset Basin, which extended into the Bristol Channel, where a 450 m thickness of marine Lower Jurassic deposits accumulated. Another, formed probably at the same time, cut across central Devon from north of Exeter to Crediton and Hatherleigh, preserving within it the long tongue of New Red Sandstone that forms such a striking feature of the present-day geological map. However, it is possible that this zone of faulting may have had its origin in Permian times, as distributional trends in the red-beds show the presence of a valley there, and volcanic rocks were also associated with its margins. It is probable that this rifting and subsidence was a side-effect of an early attempt at the opening of the Atlantic Ocean.

Cretaceous

The sea was probably absent from all but a small area of southeast Devon during most of the Jurassic Period, but re-entered in the Cretaceous Period. About 105 Ma ago the Upper Greensand was deposited in a shallow sea which extended to cover large areas but which probably left Dartmoor and Exmoor as islands. The sea continued to advance, with occasional minor regressions, but Dartmoor was certainly still an island during the deposition of the Lower Chalk and was only covered by the sea when Middle and Upper Chalk were forming elsewhere.

Minor phases of folding and faulting affected parts of Devon during the Cretaceous Period. One phase, well marked in Dorset, produced an easterly tilt to the Mesozoic strata of east Devon immediately before the deposition of the Upper Greensand, and in the Torquay district faulting and tilting of the New Red Sandstone reached more vigorous proportions. Deformations of a minor nature affected the Upper Greensand around Beer and in the Haldon Hills before the Chalk was deposited, and gave rise to a series of gentle periclinal folds. Also in the Cretaceous, volcanic activity recommenced in South West England; Epson Shoal, off Land's End, is composed of phonolitic lava dated at about 130 Ma.

Tertiary

The sea finally retreated from Devon in the late Cretaceous or early Tertiary, but the geological evolution of the county did not then cease. Lundy was created as a granite intrusion in Eocene times (dated at 54 Ma), and may have been the centre of a Tertiary dyke complex, as happened in the Scottish islands at the same time. Uplift of the land area, under a sub-tropical climatic regime, led to deep weathering and erosion of the Chalk overlayer, leaving behind large expanses of residual flint gravel which were partly redistributed by rivers to form Eocene and later deposits. During Oligocene times, around 30 Ma, localised basin subsidence took place along the line of the Sticklepath Fault, a major right-lateral wrench fault from Tor Bay to Bideford Bay, and thick deposits of clay, sand and lignite accumulated under river and lake conditions in the Bovey and Petrockstowe Basins.

The wrench faulting that affected South West England in the Tertiary was a distant product of the Alpine Orogeny that affected the Mediterranean region and reached a climax in Miocene times. Devon experienced little else from these earth movements, only some minor faulting and stratal warping such as affected the Chalk at Beer.

Quaternary

Substantially, the county achieved its final form by the end of the Pliocene, though almost certainly the land extended well beyond the present coastline for much of the remaining time, nearly two million years, eroding back to the present cliffs under the action of the sea. Erosion of the land surface continued also; and during the Quaternary some significant changes occurred. Most of the British Isles, in common with many other parts of the globe, were affected by the repeated advance and retreat of vast ice sheets, which scoured the land surface deeply and covered large areas with a thick mantle of glacial drift. The ice sheets from the north reached Devon only once during this time, however, having crossed the Bristol Channel and lodged against the cliffs of Bideford Bay. Glacial deposits in the Taw Estuary are exploited today as pottery clays. The higher parts of Dartmoor and Exmoor, possibly the Haldon and Blackdown Hills too, were affected by a variety of cold-climate weathering and erosion processes which altered slope and valley forms and played a significant role in fashioning the scenery of today. Periglacial solifluction deposits within the valleys and mantling the hillsides (the 'head' deposits) are widespread throughout the county.

Sea-level lowering during the glacial epochs resulted in coastal fringes which grew forests and marshes: these are seen today as submerged forests in places like Tor Bay and Westward Ho!, while river courses are marked by deeper-water channels offshore. During the interglacial periods, sea level was sometimes higher than at present and beaches developed at those times are now preserved as raised beaches along many parts of the coast. The recovery of the world-wide sea level from its last glacial minimum was largely completed about 6,000 years ago, with a rise which drowned the forested fringes and the estuaries of many Devon rivers. In some cases, such as the estuaries of the Rivers Exe and the Taw-Torridge, the channels infilled with sediment to a large extent, with accumulations of wind-blown sand as at Dawlish Warren and Braunton Burrows providing areas of botanical interest; other estuaries, such as those at Dartmouth and Salcombe, remained reasonably clear.

Modern Times

Has the process finished? On a human life-span it would seem that Devon, as a county, has been completed. Geological processes may appear so extremely slow that no change can be detected, but nevertheless changes can be seen and measured almost everywhere. At the seashore, sedimentation goes on—comparison of early and more recent charts shows that large amounts of sand are moved each year, silting up harbour entrances. Beaches grow, or are eroded: usually both processes are accompanied by complaints from those living nearby. The rivers bring down mud, sand and gravel fed by intermittent rockfalls and slides on hill slopes, and mud is spread across the flood plains or discharged to the sea. Floods occasionally sweep aside a few caravans or wreck houses—exceptionally, with loss of life. And, as part of a world-wide phenomenon due to the melting of glaciers, the sea level is rising by about 1.3 mm a year—five inches in a century. Those who maintain the sea defences and tidal banks, or have care of the rivers, have to be aware of all these processes and seek to understand them so that their adverse effects can be guarded against. Geologists have a significant part to play in this work.

Chapter Two

The Devonian Rocks

Devon earns a worthy place in the annals of geology, for it was here that the Rev. Adam Sedgwick and Sir Roderick Murchison first recognised the Devonian System. The preliminary account of the creation of the new system was made to the British Association for the Advancement of Science in 1839 and the case more fully documented in the *Transactions of the Geological Society of London* (Sedgwick and Murchison, 1840). Palaeontological arguments did much to persuade them that the new system was needed. Foremost amongst these was William Lonsdale's view that the fossil corals of Devon were intermediate in type between those of the Silurian and the Carboniferous. He suggested that the highly fossiliferous rocks of Devon and Cornwall might be the marine equivalent of the continental Old Red Sandstone (which was then becoming increasingly well known in Wales and the Welsh Borderland). This perceptive observation was later to be physically confirmed in north Devon and south Wales where the Old Red Sandstone interfingers with marine Devonian rocks.

Despite initial enthusiasm, Devonian studies in South West England were to progress but slowly. The pioneers were principally frustrated by difficulties in interpretation of the complex Variscan orogenic structures of the area; difficulties which were compounded by indifferent inland exposure and by the metamorphism of many of the sequences. Devon thus turned out to be singularly ill-equipped to serve as the type area for the Devonian System, and it is scarcely surprising that the formal biostratigraphic subdivision of the marine Devonian did not originate in Britain. These standards were to be established later, in Germany, Belgium and France, where the rocks are both highly fossiliferous and little disturbed. Significant progress in unravelling the complexity of the occurrences in Devon had to await these developments. At present the Devonian strata are disposed in two east-west tracts lying north and south of the so-called Culm Synclinorium of South West England. The separation of these two areas is of the order of 40 km, but it can be safely assumed that the separation was substantially greater before tectonic deformation took place, though by how much is a difficult question. Since the two regions show quite distinct geological characteristics it is appropriate as well as convenient to deal with each individually.

In the following account details of faunas and their localities are not included. The reader is referred to the relevant Geological Survey Memoirs and to a review by House and Selwood (1966). More recent references and a summary account of the Devonian geology of South West England are to be found in House *et al.* (1977). Those interested in the broader picture of the marine Devonian, against which this summary is set, are referred to Professor House's Presidential Addresses to the Yorkshire Geological Society (1975a and b).

SOUTH DEVON

There can be little doubt that the official geological survey of south Devon, completed in the early part of this century, constitutes the most significant phase of geological investigation in the south Devon area. The maps and memoirs provide a unique record of data which is of inestimable value for modern research. Apart from the geological map for Newton Abbot (IGS 1:50,000 scale, Sheet 339), which has been produced recently as a result of a co-operative effort between the University of Exeter and the IGS, all current maps date from this time. These reveal no orderly sequence of Devonian strata across this area, though in a general way the older rocks occur in the south and the younger in the north. Geological successions are not easily established from direct mapping because of complex facies changes and very considerable tectonic disruption of strata. The early surveyors were aware of these difficulties, but, within the limits of geological science at the time, could not resolve them. Nevertheless, apart from the ground between Bodmin Moor and Dartmoor (Sheets 337 and 338) where the colouring of Devonian on the map bears little relation to reality, the maps do give a fair representation of the geology as far as the map legends permit. *Figure 2.1* summarises some of the more important successions in the area.

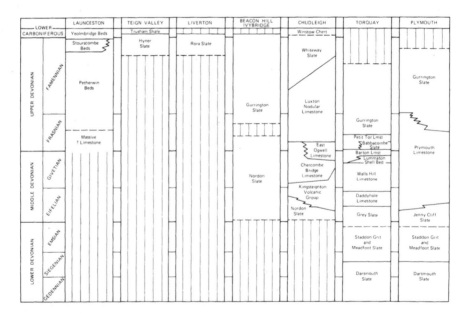

Figure 2.1. Devonian succession in south Devon (after House *et al.*, 1977)

Lower Devonian

The Dartmouth Slate is the oldest formation (or mappable stratigraphic unit) recognised in south Devon, where it crops out in the core of the Dartmouth Antiform in the South Hams (*Figure 4.4*). A substantial thickness of beds appears to be represented; a figure of 3,100 m was quoted by Hobson (1976a) south east of Plymouth but this could be considerably exaggerated by folding. Slates in various shades of red and green predominate; they are interbedded with medium to fine grained sandstones and intraformational conglomeratic bands, including locally derived clasts and exotic pebbles. Within the succession Dineley (1966) recognised a variety of lithofacies, conspicuous amongst which are fining upwards cycles of conglomerate, sandstone, siltstone and slate (*Figure 2.2*). These compare closely with the fluvial cyclothems identified in the Lower Old Red Sandstone rocks of the Welsh Borderland. Dineley additionally suggested that strongly laminated strata within the succession represent intertidal deposits and that all deposition took place near to sea-level with open marine conditions not too far away. Periodic marine incursions would certainly provide a satisfactory explanation for Ussher's record of bellerophontid gasteropod fossils within the beds (Ussher, 1904, 1907).

Poorly preserved pteraspid fish are found, not only in the fluvial channel deposits represented by the intraformational conglomerates, but also in fine-grained rocks, where in places they are so abundant as to suggest catastrophic mortality. These fish provide the only means of dating the strata in this area. Dineley recognised two forms which indicate an age within the range of Middle Dittonian to Breconian (Lower Devonian). There is thus no evidence to suggest that these beds, as presently exposed, are any older than Siegenian.

During the time when the Dartmouth Slate was being formed, spilitic volcanism was initiated, with sodic basalt lavas being extruded on the sea floor. This was to become an increasingly important feature of later Devonian times. Water laid tuffs and agglomerates constitute 15–20 per cent of the Yealm Formation, a unit recognised by Dineley (1966) within the Dartmouth Slate, where they occur finely interbedded in a slate and sandstone succession. Minor mafic igneous intrusions are widespread.

At the close of Dartmouth Slate time, fairly rapid subsidence led to a spread of fully marine conditions throughout south Devon. The principal outcrops of the succeeding Meadfoot Slate now appear in the South Hams, flanking the Dartmouth Antiform (*Figure 4.4*). At Scabbacombe Sands (SX 517923), south of Brixham, a transitional passage from the Dartmouth Slate to Meadfoot Slate can be observed (Richter, 1967). It is most unlikely that this passage is everywhere synchronous. An interfingering of the Dartmouth Slate and Meadfoot Slate must be expected, but there is no palaeontological evidence to support the assertion by Hendriks (1951) that the beds are of the same age. The major interdigitation shown on the One Inch Scale Geological Survey maps is almost certainly due to folding and faulting.

The Meadfoot Slate takes its name from Meadfoot Bay, Torquay, where in a tectonic setting less complex than most other parts of Devon, the beds are perhaps best displayed. Prior to deformation, the Meadfoot Slate consisted of a variable sequence (*Figure 2.2*) of lenticular mudstones, siltstones and sandstones with sporadic thin shell bands. Sedimentary structures are abundant,

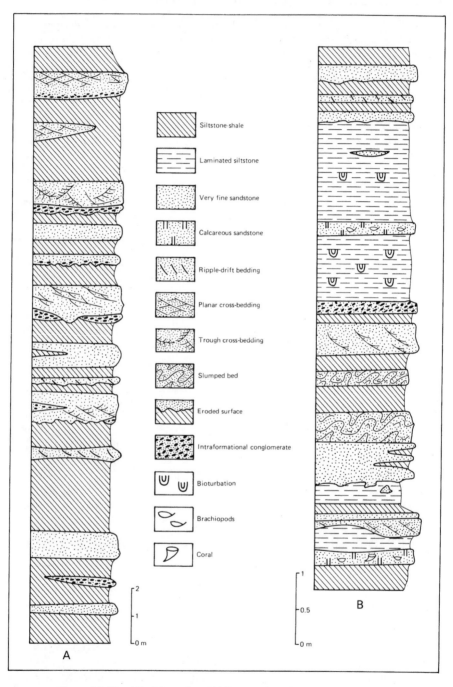

Legend:

- Siltstone-shale
- Laminated siltstone
- Very fine sandstone
- Calcareous sandstone
- Ripple-drift bedding
- Planar cross-bedding
- Trough cross-bedding
- Slumped bed
- Eroded surface
- Intraformational conglomerate
- Bioturbation
- Brachiopods
- Coral

A

B

Figure 2.2. Stratigraphic sections (A) Dartmouth Slate (B) Meadfoot Slate

and these clearly indicate deposition in a high energy environment. Analysis of these structures (Richter, 1967) showed a preferred east–west orientation with erosional features indicating currents flowing from the east. Scour and fill structures, however, reveal that currents flowed both from the east and the west, and strongly suggest tidal influence. This possibility is supported by the occurrence of finely interlaminated slates and siltstones. There is no evidence of actual emergence; probably a tidal lagoon is indicated, as this would be consistent with the high degree of bioturbation also shown by these beds. The orientation of slumped sediment masses supports the idea of the sea floor deepening towards the south.

The shell beds, which are found at many levels in the Meadfoot Slate, represent posthumous concentrations of shell fragments by currents. The brachiopod and trilobite faunas give a Middle Siegenian to Emsian age (Simpson, 1951). Although thicknesses are difficult to estimate, Richter (1967) quoted a figure in excess of 200 m for the formation as a whole.

In the cliffs of Staddon Heights on the east side of Plymouth Sound, the Meadfoot Slate (locally known as Bovisand Beds) is succeeded by the Staddon Grit, a series of massive purplish grey grits and sandstones interbedded with thin purple and grey slates and some conglomerate beds. In this, the diverse sedimentary structures of the underlying Meadfoot Slate are not found, and the overall character of the formation suggests considerable shoaling of conditions of deposition.

The Staddon Grit is present discontinuously eastwards from Staddon Heights to Sharkham Point, south of Brixham, and again about Torquay. Whilst this outcrop pattern has undoubtedly been modified by tectonism, all the evidence suggests that its lenticularity is a primary depositional feature and that the Staddon Grit represents a local facies development within the upper part of the Meadfoot Slate (Dineley, 1961; Simpson, 1951).

Hendriks (1959) interpreted the Staddon Grit as a 'ridge' deposit in a complex basin and rise marine environment, but it is simpler to interpret the Lower Devonian in terms of deposition at the seaward side of a fluvial coastal plain extending south from the Old Red Sandstone continent. Here the land-derived sediment discharged by rivers into the sea would have been redistributed by marine currents to give a linear sandy shoreline with offshore sand bars running parallel to the coast.

Recent shorelines of this type (for example the northwest Gulf of Mexico) show an alluvial coastal plain, passing gradationally seawards through tidal flats, lagoons and offshore bars and barrier islands, to the open sea. The picture in Devonian times is consistent with a sea advancing steadily northwards, so giving an upward sequence of alluvial beds (Dartmouth Slate), tidal flat and lagoon deposits (Meadfoot Slate) and offshore bar sands (Staddon Grit). The boundaries between these formations can be expected to be markedly diachronous.

Developing this model, further northward transgression of the sea introduced muds of the open marine shelf into the south Devon area, deposited well below wave base. Slates of Eifelian age are indeed found overlying the Meadfoot Slate and Staddon Grit throughout the area. This lithological passage is best observed in Jennycliff Bay on the east side of Plymouth Sound where the Staddon Grit passes up into black and grey slates (Jennycliff Slate) with occasional thin limestones. Orchard (1977) noted that the upward frequency of these limestones increases, and that at Rum Bay (SX 490525) they have yielded conodonts of late Emsian age, and conodonts of early Eifelian age at Dunstone Point (SX 488526). The important Lower–Middle Devonian boundary thus lies within the slate succession; this serves to remind us of the hazards of equating lithostratigraphic boundaries with biostratigraphic boundaries.

A conformable passage from Staddon Grit to slates of Middle Devonian age has also been reported by Smythe (1973) from South Bay (SX 930544), Sharkham Point, but here only 15–20 m of red slates with thin limestone bands separates the Staddon Grit from overlying limestones which yield Lower Eifelian conodonts. The attenuation of these slates, and perhaps the colour change too, suggest a significant reduction in the rate of sedimentation in this area. Smythe attributed this to shoaling associated with the development of a submarine rise. The position of the Lower–Middle Devonian boundary at South Bay has not been determined.

The Rocks of Start Point and Bolt Tail

At Start Point and Bolt Tail there are schists, rocks which are unlike any elsewhere in Devon, representing the product of a higher grade of regional metamorphism. In these rocks no fossil evidence is found, and the Start Boundary, a line separating them from strata of known Lower Devonian age to the north, is a fault. Thus the age of the original rocks from which these schists have been formed can only be assessed by indirect methods.

On land two main groups of rocks are recognised: mica schists and hornblende-chlorite schists. The outcrop of these two groups is shown in *Figure 1.1*. To the east of the Salcombe estuary the mica schists occur as a band trending east–west, centred upon Start Point and bounded both to the north and south by the hornblende-chlorite schists. To the west of the estuary both bands of hornblende-chlorite schist converge to a single outcrop, flanked by mica schist, which extends west to Bolt Tail. Offshore the schists are known to extend westwards to the West Rutts Reef and southwards for about 10 km to where they are overlain by New Red Sandstone rocks. Also offshore an additional rock type, a dolomitised limestone, forms the East Rutts Reef (Edmonds *et al.*, 1975).

The mica schists have a simple mineralogy, consisting essentially of quartz and muscovite, in places with chlorite and albite. Accessory sphene, tourmaline, iron oxides, epidote, zircon, rutile, apatite and ilmenite may also be present. Quartz is developed in areas separating bands of muscovite and chlorite; the grains are commonly elongated and show mutual intergrowths. The chlorite appears to be a primary metamorphic mineral and not secondarily derived from metamorphic biotite (Tilley, 1923). These schists are clearly derived from an initial sedimentary sequence consisting of shale, siltstone and sandstone. In most areas the mica schists have been cut by quartz and quartz-albite veins.

The hornblende-chlorite schists are characterised by an almost universal green coloration. Though usually showing a distinct schistose foliation, commonly with the invasion of quartz veins along the foliation planes, more massive types are not uncommon. Two subdivisions can be recognised, depending upon whether hornblende or chlorite is the dominant mineral; the rocks also contain epidote and albite, with accessory sphene and amphibole.

An increasing proportion of hornblende in the rock indicates an increasing grade of dynamic metamorphism. The schists have been formed by the metamorphism of mafic lavas or sills.

Transitional rock types showing all stages of gradation between mica schist and hornblende-chlorite schist occur at the junction of the two main lithological units. These rocks are thought to have originally been mafic volcanic ash. They form a natural link between the schists derived from sedimentary rocks and those derived from igneous rocks. Mineralogically these transitional types are relatively complex, containing chlorite, epidote, albite, calcite, sphene, quartz, muscovite, rutile, iron oxides, garnet and hornblende. Subdivision into two groups on the basis of the presence or absence of garnet is possible (Tilley, 1923).

The major structure in the area to the east of the Salcombe estuary was interpreted by Tilley (1923) as a large anticline with its axis trending east–west: the mica schists form the core and the hornblende-chlorite schists the flanks. West of the estuary, the convergence of the two bands of hornblende-chlorite schists therefore indicates a westerly plunge for the anticline, so that the mica schists in this area must occupy a higher stratigraphic position than those to the east. The stratigraphic sequence established by Tilley is thus:

> Upper mica schists
> Hornblende-chlorite schists
> Lower mica schists

However, Marshall (1965), from a study of the minor structures in the schists, has described the presence of east–west trending recumbent folds from an early tectonic phase, which are overturned to the north. Refolding by more open upright folds, which also trend east–west, probably gave rise to the distribution of the schists around a major antiformal fold. On this basis the Upper and Lower mica schists may be equivalent, their repetition being due to the early phase of folding. A third interpretation was given by Hobson (1977) who suggested that shearing occurred along the contact between the hornblende-chlorite and mica schists which resulted in a reversal of the initial sequence, although two separate mica schist formations were still considered to be present.

The fold styles exhibited by the schists were considered by Marshall (1965) to be very similar to those found in the Lower Devonian slates which lie to the north of the Start Boundary, the only exception being that deformation started earlier in the schists. But this view was contradicted by Hobson (1977) who showed that the early phases of folding are common to both areas, with deformation continuing for a longer period in the schists.

The Start Boundary was described by Marshall (1965) as a sub-vertical fault formed contemporaneously with the upright folds. The concept that the boundary is a major thrust, which was originally introduced by Sedgwick (1852) and later supported by the work of Hendriks (1937, 1939), is not now accepted (Sadler, 1975; Robinson, 1981).

Radiometric dating of the schists from Hallsands and West Cliff by Dodson and Rex (1971) using the potassium-argon method gave ages of 299 and 298 Ma for the samples collected at Hallsands, and an average of 305 Ma from three determinations on those collected at West Cliff. These ages are very similar to a value of 299 Ma also obtained by Dodson and Rex from samples of Lower Devonian slates taken from Hallsands. Both the schists and the slates thus appear to have undergone a metamorphism at 300 Ma (late Carboniferous) which probably accompanied the folding of these rocks. The structural and metamorphic continuity of the schists with the Lower Devonian slates led Marshall (1965) and Hobson (1977) to consider that the original sedimentary/igneous sequence which was the parent of the schists was essentially formed in Lower and Middle Devonian times. The differences between the two areas arose because the precursors of the schists were deposited in a rapidly subsiding zone where tectonism occurred over a longer period of time, and, due to the greater depth of burial, metamorphism reached a higher grade. Sadler (1975) also amplified this view by suggesting that the Start Boundary marks the position of an important facies change within the Devonian of south Devon.

Middle Devonian

An apparently uniform supply of fine sediment to the whole of the south Devon area during early mid Devonian times is demonstrated by the extensive outcrop of Middle Devonian slates. Nevertheless, evidence from the Newton Abbot area (*Figure 4.3*), where the beds are known as the Nordon Slate, indicates that the depositional environment was not everywhere the same. For example, the robust thick-shelled brachiopod and bivalve fauna of some sections contrasts with a delicate thin-shelled tentaculid fauna in others. Ecologically the former might be expected to inhabit turbulent, relatively shallow water, whilst the latter are believed to be pelagic: their occurrence, in the general absence of the thick-shelled fauna, suggests deposition in deeper water. This relatively quiescent depositional phase was soon to be interrupted by intense volcanic activity, and by the initiation of a major phase of carbonate rock deposition.

The carbonate rocks—limestones and dolomites—are today distributed in a series of isolated sheets around Plymouth and between Newton Abbot and Torbay. Although it is now recognised that this distribution is the result of considerable tectonic disruption, involving the development of recumbent folds and thrusting, it is evident that carbonate deposition was never continuous between the two areas. Tectonic instability, linked with increasing volcanic activity, led to the development of shallow shelf areas that acted as the sites of carbonate deposition; these were separated by basinal areas where argillaceous deposition of Nordon Slate continued, interrupted only by the deposition of vast quantities of volcanic material—mainly pyroclastic tuffs and agglomerates with some lavas.

This change in the morphology of the sea floor was already signalled, for, where shallow water faunas are found in the Nordon Slate, carbonate deposition generally followed. A notable exception is the limestone succession north of the Teign estuary where a thick pile of volcanic rocks (the Kingsteignton Volcanic Group) caused shoaling from Nordon Slate of deeper water facies. Spilitic crystal tuffs, agglomerates and lavas are included in the group, and lenticular biogenic limestones within it testify to several unsuccessful attempts by reefal organisms to colonise the mound.

Within the carbonate areas a bewildering abundance of local rock-unit names are witness to the variable nature of the limestone sequences. The elucidation of facies relationships has been hampered by difficulties of corre-lating tectonically isolated exposures in poorly exposed ground, and by the fact that many of the faunas, although locally abundant, are in urgent need of re-examination and revision. Additionally, recognition that many of the macrofaunas appear to change their characters more dramatically laterally than they do vertically has not eased matters.

A vital recent development has been the application to the area of the conodont biostratigraphy established elsewhere in northwest Europe. These microfossils (of unknown affinities) appear less affected by facies consider-ations than other fossils and have enabled the determination of a number of datum horizons throughout the area. Within this framework, brachiopod, trilobite and particularly rugose-coral sequences can be used with more confi-dence. Much remains to be done, but an intriguing picture is already begin-ning to emerge.

Within the areas now recognised as constituting the shallow marine carbo-nate shelf, carbonate deposition did not begin everywhere at the same time. Between Torbay and Newton Abbot it first appeared in the south near Brixham in early Eifelian times, but it was not until late Eifelian and Givetian times that carbonates spread farther north. Around Plymouth, the main car-bonate deposition began in mid to late Eifelian times.

Currently, the area between Torbay and Newton Abbot is more fully understood than that around Plymouth, and it will usefully serve as a model for the interpretation of the Middle Devonian limestones. The outline presented here is based largely on a summary account by Scrutton (1977) which integrated the results of much research (*Figure 2.3*).

In Torquay, the first limestones appeared in mid Eifelian times, consisting of dark grey well-bedded micrites, interbedded with dark coloured slates par-ticularly in the lower horizons. The Daddyhole Limestone sequence shows that carbonate deposition was initiated under quiet conditions, with only weak currents able to winnow the substrate and disturb locally abundant corals from their growing positions. Two episodes of shoaling follow, demon-strated by the appearance of abundant small stromatoporoids and by the reworking of fossils and matrix within the sediment. This feature is most strikingly preserved at Hope's Nose quarry (SX 949637) where an unconfor-mity shows where erosion terminated the first phase of shoaling. Scrutton (1977) envisaged that 'in shallow conditions, a bank of lime-mud was swept into the area and largely killed off a flourishing coral-stromatoporoid fauna. Erosive current activity helped to shape the surface of the mud bank, follow-ing which a relatively sharp subsidence resulted in the deposition of shaly limestones and finally shales.'

The subsequent shoaling event led into the deposition of the Walls Hill Limestone, a massive pale grey limestone with a micritic or bioclastic matrix and a fauna dominated by stromatoporoids. The sequence of stromatoporoid forms and the associated faunas within the formation reflects the development of a stromatoporoid reef in early to late Givetian times. This limestone is capped by a bioclastic coralline limestone (the Barton Limestone) which con-tains the famous Lummaton Shell Bed at its base. This limestone represents shell banks which continued to accumulate under relatively high-energy en-vironments—wave and current activity—during the final submergence of the reef.

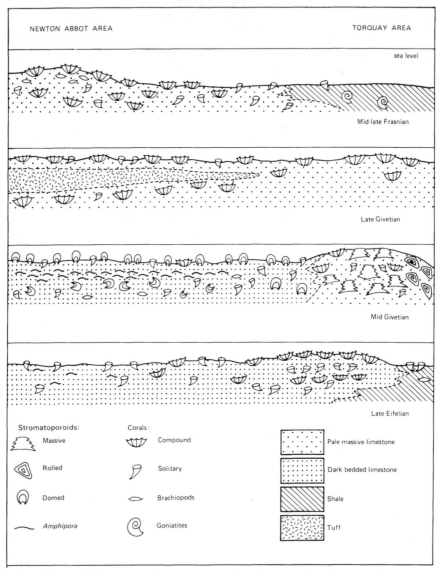

Figure 2.3. Environmental reconstruction of the Torbay reef complex of the Middle and Upper Devonian (after Scrutton, 1977)

North of Torquay, the Newton Abbot Limestone Group generally starts with the Denbury Crinoidal Limestone, a grey limestone of highly variable thickness composed almost entirely of crinoid debris. This formation seems to have filled hollows in the Nordon Slate platform which lay behind the carbonate banks already established in the Torquay area. Possibly the shelter provided by these banks allowed the luxuriant development of crinoids on their landward flanks, and derived debris was swept into adjoining hollows, eventually to provide the foundation for the developing reefal complex.

By Givetian times the stromatoporoid reef of Torquay progressively isolated the interior shelf, causing severe restriction of sea-water circulation. The lateral equivalent of the reefal beds in the Newton Abbot area is the Chercombe Bridge Limestone, which is well-bedded, fine-grained and dark grey in colour. Coral and stromatoporoid horizons are common in the lower levels, but these are less common higher up and are replaced by bands of *Amphipora* set in dark, well-bedded micrite. This change is correlated with the increasing isolation of the interior platform and the development of typical back-reef conditions.

In parts of the area, now represented in the Lemon Valley, back-reef conditions were brought to an abrupt end by the deposition of a thick sequence of tuffs (the Foxley Tuff) in mid to late Givetian times. This event more or less coincided with the final drowning of the reefal complex. At the southern margin of the complex, the submergence of the reef initiated freer circulation of waters in the back-reef areas and beds of Barton Limestone facies gradually spread northwards across the former back-reef area. These beds, the East Ogwell Limestone, are pale, essentially bioclastic limestones including some reefal masses, particularly in the higher horizons. The carbonate platform finally subsided in mid to late Frasnian times. Scrutton (1977) considered that the Brixham Limestone represents a buildup of the reef facies separate from that at Torquay, possibly associated with a sea-floor topographic high created by the accumulation of the Ashprington Volcanic Group.

The back-reef Chercombe Bridge Limestone, which rests on the Bickington Thrust at Chudleigh, can be traced in a similar structural position west of the Bovey Basin through Bickington to Buckfastleigh, where it passes laterally into a thick sequence of volcanic deposits which continue to the granite margin. This Buckfastleigh Volcanic Group, which contains carbonate bioherms in places, overlaps the Chercombe Bridge Limestone eastwards at much the same stratigraphic level as the Foxley Tuff appears in the Lemon Valley sequence. Here, however, it is succeeded by the deep-water Gurrington Slate. Undoubtedly, the carbonate complex had a very considerable lateral extent.

Between the carbonate areas, sediments now represented by the Nordon Slate continued to accumulate. Locally, limestones are developed; for the most part these are highly crinoidal and sporadically graded limestones which include rolled corals and stromatoporoids. They seem to have been derived from the flanks of the carbonate platforms and were possibly introduced into the area by turbidity flow. Throughout the Middle Devonian, the basinal area intermittently received considerable thicknesses of spilitic lava, tuff and agglomerate. The Ashprington Volcanic Group represents the activity of a most important volcanic centre throughout much of early and mid Upper Devonian times. Lenticular biogenic limestones included within the volcanic deposits probably represent patch reefs formed during quiescent periods in the general build up of the volcanic pile.

The important faunal break marking the Middle–Upper Devonian boundary does not coincide with any lithological change in south Devon, for the widespread volcanicity and the deposition of massive limestones persist from the Middle Devonian into the Frasnian, the lowest stage of the Upper Devonian. In the latter part of this stage the progressive deepening of the sea spread more uniform conditions throughout the area. Goldring (1962) termed this important episode in the Devonian story the 'Bathyal Lull', a period of deep-water sedimentation uninterrupted by tectonic activity. Although all would not agree with the precise bathymetric interpretation, a major quiescent phase was certainly initiated at this time, that in many areas was to persist to the close of the Lower Carboniferous. The Upper Devonian succession described in the following section is essentially the story of this phase.

Upper Devonian

The transgression of the Devonian sea left the sites of the reef complexes far removed from the Upper Devonian shoreline, so that only the finest-grained of land-derived sediment was carried into the area. The submarine morphology inherited from Middle Devonian times, however, still had a profound effect on sedimentation. In the basinal areas, the muddy sediments characteristic of the Middle Devonian continued to accumulate, though there is a marked decrease in the frequency of volcanic intercalations. Upwards through the slate succession, the dark grey colours of the Nordon Slate give way to purple and green: the significance of this colour change is not understood, but faunal evidence reveals that it is not stratigraphically uniform, appearing at different times in different areas. Apart from a few thin siltstones which apparently represent distal turbidites (or contourites) the slates are remarkably uniform in their lithological character. The faunas are moderately abundant and characteristically pelagic, with ostracods, ammonoids and, in the Frasnian, styliolinid gasteropods predominating. Apart from rare trace fossils, benthic elements are generally absent.

The Upper Devonian basinal slates are known as the Gurrington Slate in the Newton Abbot area (*Figure 4.3*) where they are seen to pass stratigraphically upwards into the Lower Carboniferous with no change of lithology. Traced laterally towards the former carbonate shelves, the Gurrington Slate shows a facies change; nodules of fine-grained limestone occur which increase in size and abundance towards the reefal complexes. At the same time the fauna seems to increase in abundance and some benthic elements are present. Around Newton Abbot, this facies is known as the Luxton Nodular Limestone, where it appears to flank and cover the former carbonate sea-floor rise. Passing up the flanks of the rise, the formation thins dramatically, mainly by a progressive reduction in the proportion of slaty material. The carbonate nodules thus come to lie closer and closer together and, where seen on the crest of the rise, as at Chudleigh, they may become continuous and form limestone bands some of which show flaser structure. These limestones are highly fossiliferous and the faunas affirm that the spectacular reduction in thickness is essentially a function of a reduction in the rate of sedimentation rather than of penecontemporaneous erosion.

An identical situation to that described above has been recognised for many years in the Upper Devonian of the Rhenish Schiefergebirge in Germany. Schmidt (1926) introduced the concept of *Schwellen* (rises) and *Becken* (basins) controlling sedimentation. Fine material is believed to have been winnowed from the surfaces of the rises by gentle agitation of the substrate, and transported to the adjoining basinal areas (*Figure 2.4*). The winnowing action effectively concentrated the relatively dense shells which collected over the rises so that the resulting beds (*Cephalopodenkalk*) are highly fossiliferous.

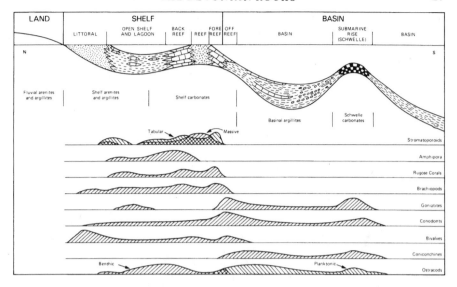

Figure 2.4. Facies and faunal variations in the Devonian (after House, 1975b)

Pelagic faunas presumably fell to the sea floor of the basinal areas at a similar rate but these beds (*Cypridinenschiefer*) appear less fossiliferous because individual horizons are separated by greater thicknesses of sediment.

Much of the carbonate of the cephalopod limestones underwent very early lithification, with cementation taking place preferentially within and around fossils. During periods of non-deposition, extensive local submarine solution took place with occasional erosion, so that derived faunas occur in a few places with fossils derived from earlier strata (Orchard, 1975). Subsequent compaction and pressure solution resulted in the movement and squeezing of the argillaceous sediment between the nodules. Tucker (1974), who noted the transitional nature of limestone nodules into limestone bands, and nodules into argillaceous matrix, suggested that it was the clay content and its later effects which determined the texture of the sediment. In a detailed account of the sedimentology of Devonian pelagic limestones he concluded that a depth of formation of several tens to hundreds of metres is likely; this is considerably less than many other authors have conceived. Orchard (1975) indicated that the Plymouth rise might even have been emergent in Famennian times.

As a result of progressive subsidence in the Upper Devonian, the ostracod slates were deposited over the rises in south Devon, effectively terminating significant carbonate deposition. These slates (known as the Whiteway Slate in the Newton Abbot area) are of reduced thickness when compared to their basinal equivalents and demonstrate the continuing effect of the rise on sedimentation. The Whiteway Slate is greenish-grey in colour and bears calcareous nodules and some siliceous bands; the fauna is abundant and contains important benthic elements. The Devonian–Carboniferous boundary lies within the formation.

The northern boundary of the Gurrington Slate outcrop is delimited by the Bickington Thrust, but the Rora Slate, identified in the Liverton tectonic unit

(Waters, 1970) east of Dartmoor, indicates that purple and green slates were also developed at the northern margin of the basinal area in late Famennian to lowest Carboniferous times. The abundance and diversity of the benthic fauna in this area strongly suggest northward shoaling.

The Rora Slate is separated from the main outcrop of the purple and green slate by the Kate Brook Slate. This formation, which constitutes a thick monotonous sequence of hard greenish-grey slates with sporadic thin sandstones, is now allochthonous (found tectonically removed from its original location) and is everywhere fault-bounded. On the east side of Dartmoor it lies between the Holne and Bickington Thrusts; here the local occurrence of purple slates within the formation suggests a southward lateral passage into this basinal facies. West of Dartmoor, slates identical to the Kate Brook Slate occur in an extensive east–west tract of country intruded by the Kit Hill and Gunnislake granites. Work in progress indicates that these slates pass gradationally upwards into a series of black, locally siliceous slates with occasional sandstones and volcanic horizons which encompass the Devonian-Carboniferous boundary. In this district too, ostracod slates occur at the northern margin of the basin, between the southward extension of the relatively shallow water Teign Valley facies and the Kate Brook Slate.

Through a considerable thickness of the succession, the Kate Brook Slate is quite barren of fossils, but locally horizons a few centimetres thick contain numbers of thick-shelled spiriferid and rhynchonellid brachiopods. The lack of species diversity in these faunas suggests that they were opportunistic forms colonising a generally hostile environment as and when conditions temporarily ameliorated. Unlike other Upper Devonian basinal slates, no entomozoid ostracods are recorded.

The remarkable lithological uniformity of the Kate Brook Slate, even allowing for repetition of beds by folding, appears to have persisted through a very considerable thickness of strata. Basinal deposition is suggested by the lithology, in which the thin sandstones were introduced as distal turbidites. The evidence indicates that accumulation took place in the axial part of a basin that was receiving purple and green sediments on its north and south flanks.

The faunal character of the Kate Brook Slate calls for special explanation. Rare fossiliferous horizons show that the substrate was not inimical to life, and since there is no evidence of very rapid sedimentation which could have hindered the development of the benthos, it seems that the depth of water was a limiting factor. The development of anoxic conditions must also rank as a possibility. The absence of pelagic ostracods is more difficult to explain; conceivably, their restriction to the areas bordering rises was a function of the nutritive content of the waters.

On the east side of Dartmoor (*Figure 4.2*), the Hyner Slate represents the oldest formation exposed in the autochthonous (tectonically unmoved) Teign Valley succession. These beds, which span the Devonian-Carboniferous boundary, consist of hard bluish-green to dark grey slates with thin calcareous and siliceous bands and lenses. They indicate the further northward extension of the ostracod facies, though the presence of a locally rich benthic fauna suggests considerably shallower water conditions than the Gurrington Slate. This Teign Valley facies of the topmost Upper Devonian is extensively developed, and can be identified north of Dartmoor in the core of the Meldon anticline (Meldon Slate-with-Lenticles Formation), and irregularly on the west of Dartmoor where it forms part of an autochthon or parautochthon (tectonically unmoved or little moved), overthrust by Devonian and Carboniferous strata.

In Lydford Gorge, Isaac (1981) has demonstrated that the River Lyd Slate-with-Lenticles Group (Dearman and Butcher, 1959) is exposed through a tectonic window and is not, as hitherto supposed, in conformable contact with the enclosing Carboniferous rocks. He divided the group into three formations showing varying grades of contact and dynamic metamorphism, each representing different sedimentary environments and separated from one another by thrusts. The slates and sandstones with calc-silicate lenses, exposed at the lowest tectonic levels, are interpreted as a mixed carbonate-clastic rise sequence (the Whitelady Slate), whereas the overlying formations which are nodular phyllitic slates with various sandy and volcaniclastic lenticular beds are essentially basinal in character. One of these formations is interpreted as a slump, in which soft-sediment deformation structures have subsequently been highly tectonised. All these formations yield late Famennian dates.

At higher levels in the allochthon, seen in the Launceston area, Stewart (1981a) employed conodont correlations to reveal complex facies changes involving the Petherwin Beds and Stourscombe Beds (Selwood, 1971b). In broad terms, the Stourscombe Beds are now seen to represent various slope and basinal environments which existed contemporaneously with the considerably shallower water environments recognised within the Petherwin Beds. As presently exposed, the Petherwin Beds consist of a series of greenish-grey slates and thin bioclastic limestones yielding a rich brachiopod-bivalve fauna; conodonts indicate a maximum range within the Platyclymenia to Wocklumeria Stufe (Devonian biozones established in Germany), and considerable condensation of thicknesses. These beds pass laterally into green slates which resemble the basinal Kate Brook Slate. The character of the fauna must indicate temporary shoaling, possibly associated with a general mid Famennian marine regression or a submarine rise. Although no longer exposed, the presence of cephalopod limestones within the Petherwin Beds favours the latter. Museum collections of fossils indicate that the cephalopod limestones of the Petherwin Beds may range down into the Manticoceras Stufe (Selwood, 1971b). A comparison with the Chudleigh rise is tempting.

The Stourscombe and Petherwin Beds extend no farther westward than the Cambeak Fault Zone which forms the geological limit of this volume.

NORTH DEVON

Unlike south Devon, where stratigraphic sequences have had to be painstakingly put together, the succession of beds in north Devon has been known for many years (*Figure 2.5*). Coastal sections provide almost complete exposure in a dip section extending from Lynton to Barnstaple. Inland, the beds may be traced along the east-northeast strike into west Somerset where they disappear beneath the New Red Sandstone. Folding and faulting do little to disrupt this essentially simple picture. To the northeast, the Quantock Hills form an inlier set apart from the main north Devon section, but here poor exposure has for a long time inhibited geological interpretation.

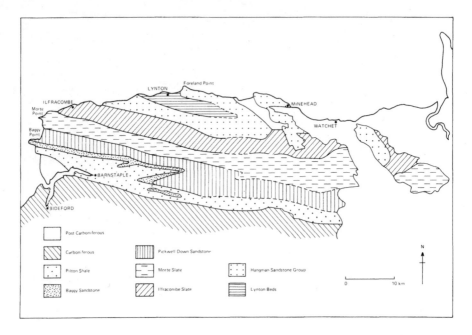

Figure 2.5. Geological map of north Devon and west Somerset (after Edmonds *et al.*, 1975)

Although different, the geological problems presented by north Devon geology have proved no less intractable than those of the south. Deformation is a good deal more intense than geological maps suggest; the alternation of major competent and incompetent stratigraphic units has led to much internal deformation within formations and to the development of slaty cleavage in fine-grained rocks. Fossils are consequently almost always deformed and often recrystallised.

Transitions between marine and non-marine conditions dominate the stratigraphic history of north Devon. The resulting facies changes and the accompanying restriction and specialisation of faunas means that correlation presents major problems, not only for local geological events but also with the standard marine chronology, but a view is presented in *Figure 2.6.* Only the Devonian-Carboniferous boundary, which lies within the Pilton Shale has been satisfactorily defined.

The geological problems of north Devon figure less prominently in the geological literature than those of the south of the county. In part this reflects the inaccessibility and daunting challenge of the cliffs, but much more important has been the absence of an official geological survey map. Apart from their own contributions, these maps invariably stimulate and serve as the basis for much individual research. Currently, north Devon is being actively investigated by the IGS and primary 1:10,000 scale mapping is well advanced. Some of the first 1:50,000 scale maps have been recently published; the remaining maps and memoirs are awaited with great interest.

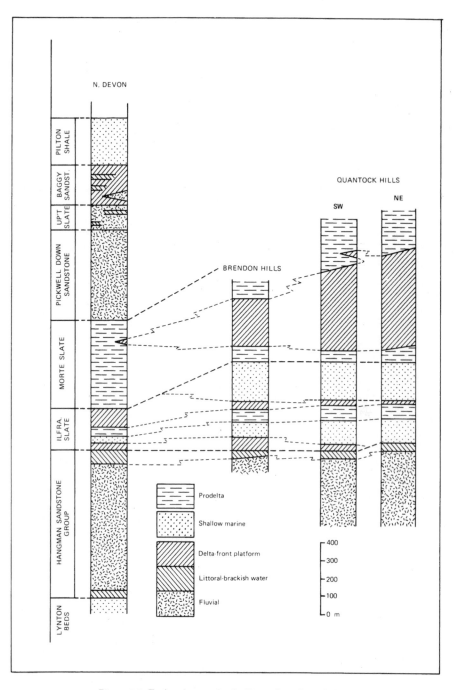

Figure 2.6. Facies changes in the Devonian of north Devon

Lower and Middle Devonian

The oldest strata exposed in north Devon are the marine Lynton Beds; some 400 m thickness of thinly bedded fine-grained sandstones, siltstones and grey mudstones. The lower parts of the sequence are sandier than the upper, with thick parallel-bedded sandstones and delicately interlaminated fine sands and silts. This lamination, which characterises much of the formation, is frequently disturbed by burrowing: *Chondrites* (Simpson, 1957) is particularly abundant. Some thin bioclastic limestones occur within the argillaceous upper parts of the Lynton Beds, together with the bands crowded with fenestellid bryozoa and brachiopods. At the top of the formation the beds are less clearly laminated and become increasingly sandy so that there is a gradual passage into the overlying Hangman Sandstone Group.

The Lynton Beds yield locally abundant shelly fossils, with bivalves predominating. However, species diversity is low and preservation poor. Simpson (1964) referred the beds to the late Emsian or early Eifelian stages.

Although muddy shelf sea conditions are indicated, precise interpretation of the depositional environment is difficult. The lithology is remarkably uniform and the occurrence of sedimentary structures is very restricted; ripple marks are found only in the thicker sandy horizons.

The Lynton Beds are almost certainly associated with the initial marine transgression flooding the Old Red Sandstone continent in what is now the Bristol Channel area. One might reasonably forecast that unexposed lower members of the Lynton Beds or earlier formations, not presently accessible to observation, would reveal in downward succession a progressive shoaling of depositional environment, perhaps through offshore and intertidal sediments, into terrestial beds. Considering the age of the Lynton Beds, it appears that here are the results of the same transgressive event as that associated with the Meadfoot Slate in south Devon. This rapid northward spread of the sea in Emsian times is consistent with the concept of a wide alluvial plain extending south from the Old Red Sandstone uplands.

Simpson's (1964) correlation of the Foreland Grit with the Hangman Beds (now Hangman Sandstone Group), confirming the original view of Sedgwick and Murchison (1840), is now generally accepted. These beds mark the entry into the area of sedimentary material derived from the continent to the north.

Tunbridge and Whittaker (1978) formally divided the Hangman Sandstone Group into five formations totalling approximately 1,600 m in thickness, which record a major regressive and transgressive event. The basal Hollowbrook Formation is made up of interbedded parallel-laminated and crossbedded quartzitic grey sandstones. Sedimentary structures suggest a well developed beach facies with subtidal deposits filling channels which cut the beach sands. The overlying Trentishoe Formation consists mostly of parallel-laminated red sandstones occurring in thin but laterally persistent beds which are interpreted as sheet-flood deposits. Interbedded red siltstones and very fine sands with desiccation cracks are believed to have accumulated in ephemeral lakes. The succeeding Rawns Formation, again of probable fluvial origin, consists of coarse sandstones with exotic clasts.

The two highest formations record an overall transgression, but within the lower, the Sherrycombe Formation, repeated coarsening-upward sandstone cycles resting on scoured surfaces attest to minor transgressive and regressive events. Tunbridge and Whittaker interpreted the sequence in terms of estuarine or possible fan-delta deposits. The Hangman Sandstone Group is completed by the Little Hangman Formation, a sequence of cross-bedded sandstones and mudstones deposited under near-open marine, but tidally influenced, conditions. Webby (1965a) noted that the transgressive beds at the top of the Hangman Sandstone Group in the Brendon Hills, which he suggested accumulated in an intertidal environment, are thinner than their equivalents in north Devon and that this northeasterly attenuation is continued into the Quantock Hills. He correlated this with a gradual east-northeasterly transgression of the strand line. Inland from the north Devon coast, and in west Somerset, lack of adequate exposure does not allow the divisions recognised on the coast to be mapped.

Above the Hangman Sandstone Group comes the Ilfracombe Slate, which figures prominently in the geological literature of north Devon. Whittaker (1978a) included a brief historical review which acknowledged the fundamental contribution of Holwill (1962), who first detected important repetition of beds by folding. With the geological survey of the Ilfracombe Slate by the IGS presently nearing completion, the outline of results by Whittaker is particularly important.

On forthcoming IGS maps the Ilfracombe Slate is to be divided into four formations. The grey slates with thin sandstones, siltstones and limestones of the basal Wild Pear Slate record the return to fully marine conditions initiated in the upper part of the Hangman Sandstone Group. An identical horizon, the Mansley Beds, is recognised in the Brendon Hills (Webby, 1965a) and in the southern part of the Quantock Hills (Webby, 1966a); in that area they thin out and pass laterally northwards into a group of maroon siltstones which there succeed the upper Hangman Sandstone Group. Webby (1965b) suggested that the Wild Pear Slate was deposited in a delta-front environment and that its equivalents in the north Quantocks are intertidal.

The succeeding Lester Slates and Sandstones consist of about 40 m of cross-bedded sandstones, siltstones and mudstones with bioclastic limestones in places. The faunal content of these limestones indicates that shallow, offshore conditions had become established. Webby (1965b) suggested that the sandstones may represent the shoreward spreading of reworked earlier sands over the delta-front deposits of the Wild Pear Slate and Mansley Beds. Equivalent, but thinner, beds are recognised in the Brendon Hills and throughout the Quantocks.

The overlying Combe Martin Slate contains four main limestone horizons and some subordinate sandstones. Coral faunas give somewhat equivocal ages: Holwill (1962) considered that the fauna of the Combe Martin Beach Limestone was of Frasnian age and that the two underlying limestone beds were Givetian, but more recent work has placed all the coral faunas of the Combe Martin Slate in the Givetian. Webby (1966b) interpreted the changing character of the limestones through the succession in terms of varying bathymetry. In the Quantock Hills the limestones, bearing abundant stromatoporoids and tabulate corals, are thought to have accumulated under shallow-water conditions. To the southwest, in the Brendon Hills, the limestones yield only a few stromatoporoids and tabulate corals but abundant solitary and compound rugose corals suggest moderately shallow conditions. This character is maintained farther southwest into the two lower limestone horizons of

Combe Martin Slate in north Devon. The Combe Martin Beach Limestone, which is characterised by a fauna of abundant small solitary rugose corals, but few tabulate corals and no stromatoporoids, is taken to indicate a relatively deep-water environment. It probably coincides with the maximum deepening of the sea. The picture obtained from the limestones of progressive shallowing towards the north-northeast is consistent with the overall view of transgression given by Webby.

The sandstone and siltstone layers introduced in the higher beds of the Ilfracombe Slate (Kentisbury Slate) indicate the shallowing and subsequent regression of the sea that was to culminate, albeit after minor fluctuations, in the establishment of a second continental phase of deposition.

Upper Devonian

The Morte Slate succeeds the Ilfracombe Slate conformably, and includes, 1,500 m of grey and purple slate with scattered calcareous nodules and thin sandstone beds of moderately shallow water (possibly pro-delta) marine origin. Traced eastwards into west Somerset the beds are non-calcareous and become much more sandy, with laminar and cross-laminar siltstones and fine sandstones predominating, particularly in the middle part of the formation. Webby (1965a) interpreted this development as characterising a delta front area within the regressive pro-delta/delta platform complex initiated in late Ilfracombe Slate times. The beds have yielded poorly preserved brachiopods and bivalves including *Cyrtospirifer verneuili*; a Frasnian and possible lower Famennian age is generally accepted. The impoverished fauna may correlate with rapid sedimentation inhibiting colonisation of the sea floor; reduced salinity in the delta mouth area may also be a contributory factor.

At the close of Morte Slate times the regressive stage culminated in the deposition of the Pickwell Down Sandstone. This formation is made up of 1,200 m of red, purple and green cross-bedded and ripple-marked sandstones interbedded with greenish-grey siltstones and slates. No sedimentological account of these beds is available, but they are generally accepted to be of fluvial and lacustrine origin and comparable with the Old Red Sandstone. Generally the beds are unfossiliferous, but fish remains have been obtained from the Bittadon 'Felsite', a keratophyric tuff band that marks the base of the formation; these indicate an Upper Devonian, presumably early to mid Famennian, age.

The Pickwell Down Sandstone grades upwards into the Upcott Slate, some 250 m of greenish-grey slates and graded siltstones with occasional fine cross-bedded sandstones. Apart from highly comminuted shell debris the beds yield no fossils. Goldring (1971), who noted that the beds maintain their character for some 64 km along the strike without change of facies, suggested that they had been deposited in a back-swamp alluvial environment or in shallow freshwater lakes.

The Baggy Sandstone is made up of 440 m of massive cross-bedded sandstones, thin-bedded sandstones and siltstones associated with intraformational conglomerates, slumped masses and sporadic thin crinoidal and gastropodal limestones. These beds mark the beginning of a major transgressive phase that was to be maintained well into Carboniferous times. The junction with the underlying Upcott Slate is sharp and not marked by any evidence of erosion; submergence of the continental area must have been particularly rapid, for the basal Baggy Sandstone is not transitional.

The Baggy Sandstone is completely exposed in the cliffs of Baggy Point, where it has been the subject of a detailed sedimentological study by Goldring (1971). He recognised both nearshore marine and non marine delta-like facies, with the latter accounting for 16 per cent of the total succession.

A recurrent association of dark grey shales, siltstones and thin sandstone forms the background to this diverse sedimentary sequence. The sandy sediments are intensely bioturbated and studies of the trace fossil *Diplocraterion yoyo* reveal evidence of repeated phases of erosion and deposition. Cross-stratification indicates on-shore sediment transport with a southerly provenance. The group is believed to have accumulated on a delta-front platform. Periodic shoaling is indicated by interbedded sets of graded fine sandstones with sharp undulose soles and by sub-beach and intertidal channel-fills. A contrasting environment is revealed in the lower Baggy Sandstone, where tabular units of bioturbated and penecontemporaneously eroded graded sandstones and shales with patches of disoriented *Lingula* and thin bellerophontid limestones are interpreted as lagoonal or restricted bay deposits.

Amongst the non-marine facies recognised within the Baggy Sandstone, channel-fill deposits predominate. Most common is a flat stratified and cross-bedded sandy fill with abundant plant debris, which is attributed to fluvial origin. *Dolabra* which is associated with these beds could have been a fresh-water bivalve. Striking support for the delta model of the Baggy Sandstone is represented in a fine-grained siltstone and shale sequence which appears to have been deposited in fresh-water lakes. Directional evidence in the non-marine facies suggests a northerly provenance. In his palaeogeographic model, Goldring envisaged distributaries discharging into a marine area associated with a barrier island chain, broken at intervals by channels, which extended laterally from the distributary mouths. This barrier formed a partially enclosed bay environment. The seaward gradient was low. Generally all the land surface, and a broad strip subparallel to and seawards of the high water mark, has not been preserved. There is, however, some evidence to suggest the presence of tidal pools near the mouth of each distributary and a fresh-water lake.

Fossils, other than trace fossils, are not common in the Baggy Sandstone, but brachiopods and spores from the upper part of the formation do not suggest a different age from the overlying Lower Pilton Shale, which is referred to the Wocklumeria Stufe. The age of the base of the formation is unknown.

There is a gradual upward passage from the Baggy Sandstone into the fossiliferous neritic sandstones and slates of the Lower Pilton Shale, coinciding with a major northward transgression of the shoreline which terminated all traces of deltaic sedimentation in the area. Goldring (1970) documented in detail the lithological and faunal changes associated with the continued transgression; overall, fine sandstone and siltstone deposition decreased and thin cherty seams characteristic of the Upper Pilton Shale appeared in the succession. This lithological change was more or less coincident with the major faunal changes marking the passage into the Carboniferous System. The shallow water marine faunas of the Pilton Shale are extremely abundant and varied with brachiopods. bivalves, echinoderms and trilobites predominating.

The turbulent conditions of deposition represented by small scour structures means that virtually no fossils are found in growth positions.

DEPOSITIONAL FRAMEWORK

From the foregoing, a cautious interpretation can now be placed on the overall sedimentary setting in the Devon area during the Devonian Period (*Figure 2.7*).

In Lower Devonian times, continental deposition spread beyond the southern limit of the present south Devon area to give a wedge of sediments that thickened dramatically southwards. In the Roseland succession of south Cornwall, Sadler (1973) recorded the accumulation at the same time of a relatively thin sequence of fully marine rocks, which must have formed off the southern shore in the Old Red Sandstone continent.

During late Lower Devonian times, a marine transgression resulted in the flooding of the lowland coastal plain. Differential subsidence commenced that led to the development of a series of persistent sea-floor basins and rises, that were to play a fundamental part in controlling the nature of Devonian and Carboniferous sedimentation and subsequent deformation. It was at this time that the South Devon Basin, lying between the Dartmouth Rise to the south and the Teign Valley Rise to the north, was initiated.

For Cornwall, Matthews (1977a) emphasised the similarities between the depositional framework in Germany and South West England and proposed a model for south Cornwall which is of great relevance to the interpretation of south Devon geology. Matthews argued that basins were developed suc-

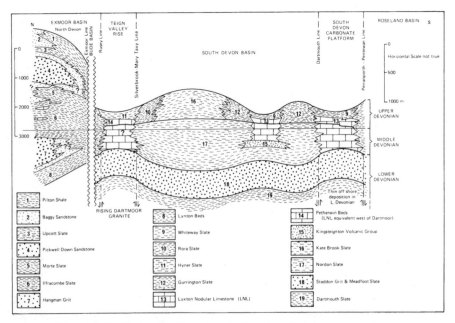

Figure 2.7. Profile showing Devonian stratigraphic relationships across Devon

cessively from south to north during Upper Palaeozoic times, forming between fundamental east–west fractures on which vertical movements were to take place during deposition. In south Devon, Matthews drew attention to the major east–west fault present near the northern margin of the Dartmouth Antiform to which Hobson (1976a) attributed a northerly downthrow of no less than 3,700 m. He argued that the fault is one of a series, all with the same sense of movement, the first of which was initiated in late Lower Devonian times approximately at the northern limit of thick sediments which had been deposited seaward of the Old Red Sandstone coastal plain, that is, close to the then shoreline. This fault series left the area at present represented by the Dartmouth antiform, and an undefined area farther south, as a rise trending east–west between the pre-existing basinal area to the south and the South Devon Basin developing north of it. The resulting positive feature and its flanks could well have provided the site for the Middle and Upper Devonian carbonate reef complex, whilst in the basinal areas, the argillite equivalents accumulated both to the north and south. Such a situation would most satis-factorily reconcile Scrutton's contention (1977) that the carbonate complex around Torbay accumulated at or near a shelf edge facing deep water to the south, and Braithwaite's (1966) view (mainly developed in the Plymouth dis-trict) that the complex was fronted by deep water to the north. Both appear to have grown on the one axial ridge.

The depositional basins, though contrasting bathymetrically with the rises, need not have been of great depth; the substantial thicknesses of sediment represented in the South Devon Basin are best explained by subsidence keep-ing pace with deposition, both being fairly rapid.

The northern limit of the South Devon Basin is now to be found along faults marking the southern limit of the Teign Valley facies. These faults appear to have complex histories, and throws are probably substantial. A case can be argued that they were in operation during deposition, for there is strong circumstantial evidence that the beds constituting the Teign Valley facies accumulated in relatively shallow water, in contrast to the basinal slates immediately to the south. This is seen in the relative abundance of the benthos and emphasised by the occurrence of cephalopod limestones in the Whitelady Slate. The latter are so like those on the carbonate rise at Chud-leigh that it must rank as a possibility that unexposed Middle Devonian rocks developed on this rise may include massive limestones.

The South Devon Basin appears to be the direct lateral equivalent of the Trevone Basin (Matthews, 1977a). In north Cornwall the northern limit of the basin is not clear; this may well be linked with the fact that the Teign Valley facies is not exposed west of the Cambeak Fault. The situation west of the fault would suggest that these beds have been covered by overthrust units.

It can be argued that the Rusey Line, which limits the Teign Valley facies to the north of the Launceston area, is a complex deep-seated structure. Perhaps it, too, was in existence during deposition, and formed a boundary to the Bude Basin (Matthews, 1977a) to the north.

In north Devon, the interfingering of marine and continental deposits places the region at the seaward margin of the Old Red Sandstone continent for much of Middle and Upper Devonian times. During this period more than 5,000 m of sediment accumulated. Because of the overlying Carboni-ferous cover, the relations with rocks lying farther south are unknown; but

Matthews (1975, 1977a), by interpreting geophysical evidence and making comparisons with Upper Palaeozoic successions in southern Ireland, argued strongly that sediment thickness increased progressively northwards from the southern margin of Exmoor. Thus denying the existence of the Exmoor Thrust, he postulated a major east–west fracture downthrowing to the north, separating the thick sedimentary rocks of the 'Exmoor sink' from thinner, more basinal, successions to the south. This interpretation is particularly persuasive in the light of the sedimentary framework proposed by Matthews for Cornwall, and is here considered extended to south Devon.

LOCALITIES

South Devon

2.1. *Outer Hope* (SX 675402): Start Boundary. From the village of Malborough on the A381 take the minor road to Outer Hope where parking is possible near the ramp leading to the beach. On the northern side of the beach dark grey Meadfoot Slate occurs, dipping south with vertical cleavage. Numerous quartz veins are present, some of which are folded. On the south side of the beach the grey slates are in contact with light-coloured mylonitic rocks of the Start Boundary and, in the cliff which marks the southern limit of the beach, bands of grey schist can be seen.

2.2. *North Sands Bay, Salcombe* (SX 731383): Hornblende-chlorite and mica schists. From the A381 at Salcombe, descend the narrow winding road (Sandhills Road) to the car park at North Sands Bay. On the west side of the beach green-coloured hornblende-chlorite schists are exposed, showing excellent examples of multiphase folding and quartz rods. Around the headland, separating North Sands Bay and South Sands Bay, grey-coloured mica schists are interlayered with the green schists.

2.3. *Meadfoot Bay, Torquay* (SX 936633): Meadfoot Slate (Lower Devonian, Emsian). From the car park (SX 936633) traverse the cliff section eastwards towards Hope's Nose. The cliffs provide excellent exposure of the varied sedimentary structures represented in the Meadfoot Slate. Cross-lamination, ripple-drift lamination, erosion channels and scour-and-fill structures are present. Brachiopods, bivalves, crinoids, corals and rare trilobites occur in thin calcareous sandstones. The trace fossil *Chondrites* is common. The section is considerably folded and faulted.

2.4. *Triangle Point, Torquay* (SX 928628): Daddyhole Limestone (Middle Devonian, Eifelian). Leave Meadfoot Sea Road at (SX 930630) and walk to end of promenade (SX 929628). The limestones at the point are downfaulted to the south by important east–west faults. The steep southwest dipping beds here form the inverted limb of a major fold overturned to the east-northeast. A number of highly fossiliferous horizons yielding abundant stromatoporoids and rugose and tabulate corals are present in the quarry. The rich fauna exposed on the large bedding surfaces at the northeast end of the quarry appear to be mainly *in situ*. Bun-shaped stromatoporoids, frequently intergrown with *Syringopora*, can be seen surrounded by branched tabulate corals. The latter appear to have acted as sediment traps for much shell debris. The Daddyhole

Limestone, in inverted succession, is also exposed in Knowles Quarry above Triangle Point; it is succeeded by the shales represented beside the promenade leading to the point.

2.5. *Hope's Nose* (SX 947635): Daddyhole Limestone, Eifelian slates. (Middle Devonian, Eifelian). Access by footpath leaving Ilsham Marine Drive at (SX 945638). The 25-foot raised beach at the southern end of Hope's Nose, rests on well cleaved, thinly-bedded limestones and calcareous slates. These beds may yield deformed fossils including corals, crinoids, trilobites, bryozoa and brachiopods. The last named may show crystalline infills. The beds are cut by important thrust planes; the structure hereabouts is picked out by thin, deeply eroded tuff bands. The best vantage point lies on the 'reefs' exposed at low tide seawards of the raised beach. The tuff bands also act as useful markers in interpreting the deformed sections which can be examined northwards towards the sewage outfall.

In Shennell Cove (SX 947634), the Daddyhole Limestone can be seen to be faulted against Eifelian slates. To the west of the cove, a prominent fault introduces the slates and sandstones of the Meadfoot Slate against the Eifelian slates. The former dip gently northwards and show many sedimentary structures.

The quarry (SX 949637) at the north end of Hope's Nose shows thin-bedded Eifelian limestone and slate resting on massive stromatoporoidal limestone.

2.6. *Lummaton Quarry* (SX 913665): Walls Hill Limestone and Barton Limestone (Middle–Upper Devonian, Upper Givetian–Lower Frasnian). Access is via Happaway Road (SX 913667), Barton, Torquay. Permission must be sought in advance from the quarry manager. The present working quarry is in a massive grey limestone, locally highly dolomitised, at the top of the Walls Hill Limestone. Stromatoporoids and rugose and tabulate corals are locally abundant. The quarry is famous for the Lummaton Shell Beds which yield a rich fauna including brachiopods, bryozoa, trilobites, ostracods, crinoids, gastropods, bivalves and algae. The shell bed, which is variably exposed at levels near the top of the quarry, forms the basal member of the bioclastic Barton Limestone which is now rather poorly exposed in the northern part of the quarried area.

2.7. *Ransley Quarry* (SX 844702): East Ogwell Limestone (Upper Devonian, Frasnian). North side of Ogwell Road, 300 m east of the village green, East Ogwell. The massive pinkish-grey limestone represents part of a bioherm, the upper surface of which is irregular and infilled by red shales and shaley limestones. The fauna includes rugose and tabulate corals, brachiopods, goniatites, gastropods and ostracods.

2.8. *Palace Quarry, Chudleigh* (SX 868787): Chercombe Bridge Limestone (Middle–Upper Devonian, Givetian–Lower Frasnian). Access is from Rock Road, Chudleigh (SX 867789). Permission should be sought at the adjoining garage. The fine-grained, thick-bedded, dark grey limestone exposed in the quarry is characteristic of the back-reef facies in the Newton Abbot Limestone Group. Scattered rugose corals and brachiopods occur and *Amphipora* is locally abundant, particularly in horizons near the top of the quarry. Blocks of fossiliferous Upper Devonian slate, which overlies the limestone, are commonly found on the quarry floor. They appear to have been derived from Quaternary cavity infill at the top of the quarry. The Bickington Thrust, which carries the Chercombe Bridge Limestone over the Kate Brook Slate, can be examined beneath the waterfall in the Kate Brook, west of the quarry.

2.9. *Chipley Quarry* (SX 807721): Spilite in Gurrington Slate (Upper Devonian, Famennian). Leave A383 at Chipley Mill (SX 811719), proceed to Chipley Farm (SX 81037210) where permission must be sought and a small charge paid. A series of quarries north of the road from Bickington to Chipley reveals spilite with well defined pillow structures. Greyish-green slates at the top of a 15 m face (at SX 807722) yield ostracods of the hemisphaerica-dichotoma Zone in fallen blocks. At SX 809722 cherts are represented between the pillows.

2.10. *Ashburton*, road sections (SX 753692): Gurrington Slate (Upper Devonian, Famennian), Kate Brook Slate (Famennian) and Chercombe Bridge Limestone (Middle Devonian, Givetian). Leave the A38 at junction for Ashburton (west), follow the route signed to Landscove, and park in the widened section of road immediately above the westbound carriageway.

The adjoining cutting reveals grey and black Gurrington Slate mainly overlying, but also interbedded with lithic tuffs. The latter include blocks of stromatoporoidal limestone which may be *in situ*. In the cuttings along the approach and exit road immediately below this section these slates and tuffs are highly disturbed immediately above the Bickington Thrust. The overridden Kate Brook Slate, although similarly disturbed at the thrust junction, does yield brachiopods and crinoid debris. These thrust relations can be further examined in cuttings adjacent to the eastbound carriageway.

Highly dolomitised Chercombe Bridge Limestone, lying above the thrust, can be examined in a cutting at SX 757695. The limestone shows penecontemporaneous brecciation and passes up into calcareous tuffs. There is a significant development of umber at the base of the tuffs which appears to occupy hollows in the limestone.

2.11. *Holne Bridge* (SX 730706): Kate Brook Slate (Upper Devonian, Famennian). Roadside and adjoining river sections west of Holne Bridge. This locality is of particular structural interest; it shows Kate Brook Slate thrust over Upper Carboniferous slates and sandstones on the Holne Thrust. The thrust is recognised in a topographic hollow immediately below a roadside bluff of Kate Brook Slate (SX 729705) and can be examined in continuous exposure in the river section. The thrust passes under the arch of the road bridge. Considerable disturbance of the slates is associated with the thrusting. Thin calcareous siltstones within the Kate Brook Slate yield specimens of *Cyrtospirifer* in places.

North Devon

2.12. *Lynmouth* (SS 728496–SS 735495): Lynton Beds (Lower Devonian, Emsian). The lithologies of the Lynton Beds are characteristically exposed in sections lying eastwards along the beach from Lynmouth. The slates are strongly cleaved and many of the interbedded mudstones and siltstones are strongly bioturbated. Some brachiopods and bivalves occur, but these are much deformed. At SS 735495 the Lynton Beds are faulted against the Hangman Sandstone Group.

2.13. *Combe Martin*: Ilfracombe Slate (Middle–Upper Devonian, Givetian–Frasnian), and Hangman Sandstone Group (Middle Devonian, possibly Givetian). A variety of localities can be reached within easy walking distance of the car park at Combe Martin (SS 578473). The section east of Combe Martin Beach exposes the Lester Slate and Sandstones within the Ilfracombe Slate. The beds include prominent cross-bedded sandstones and siltstones together with some shelly limestones. A limestone, marked by prominent solution hollows, yields corals and brachiopods.

Within the Combe Martin Slate exposed to the west of the beach the Combe Martin Beach Limestone is conspicuous. This yields tabulate and rugose corals, brachiopods and gastropods. Other limestones within the sequence can be examined at Sandy Bay (SS 569474) which can be reached by a footpath leaving the A399 500 m west of the village (SS 572472).

The Wild Pear Slate, which forms the lowest unit of the Ilfracombe Slate, can be examined in Wild Pear Beach; this can be reached from the cliff path east of the village. The grey slates are strongly deformed and include thin sandstones and limestones. At the eastern end of the beach (SS 582478) these slates are in faulted contact with the cross-bedded sandstones and mudstones of the Little Hangman Formation, which is the youngest formation in the Hangman Sandstone Group. A bed crowded with *Myalina* marks the base of this formation. At low water the upper part of the underlying Sherrycombe Formation can be examined; this shows repeated upward-coarsening sandstone cycles above erosion surfaces.

2.14. *Baggy Point to Croyde Bay*: Baggy Sandstone and Pilton Shale (Upper Devonian, Upper Famennian). Leave cars at SS 433397 where the metalled road to Baggy Point ends and follow the cliff path to SS 422409 where, from the cliff top, the sharp base of the Baggy Sandstone can be seen resting on Upcott Slate. The rich variety of sedimentary structures described by Goldring (1971) from the Baggy Sandstone can be examined in the spectacular cliffs of Baggy Point. Although the broad features of the formation can be appreciated from exposures adjacent to the footpath, detailed examination necessitates precarious descent to sea level. This is *dangerous* and should not be undertaken lightly. Details of the section and suitable points for descent are given by Goldring (1978).

The top of the Baggy Sandstone is marked by a major slump (SS 435401). The characteristic development of the Pilton Shale can be examined from here southwestwards into Croyde Bay. A notable locality is Laticosta Cave (SS 427404) where a variety of brachiopods are displayed on bedding surfaces. In the section extending into Croyde Bay the Lower Pilton Shale fauna can be obtained from decalcified sandstone bands. A thin tuff band occurs at SS 433396.

Fossiliferous Pilton Shale can also be conveniently examined on the foreshore at Downhead (SS 4338), the southern headland limiting Croyde Bay. Here two tuff bands form useful markers in the interpretation of the strongly folded and faulted sequence.

Chapter Three

The Carboniferous Rocks

The sedimentary rocks which represent most of the Carboniferous Period in South West England were defined as *Culm Measures* by Sedgwick and Murchison (1840). This lithostratigraphic term, of supergroup status, has remained appropriate to this day, especially during the time when its bio-stratigraphic age was in considerable doubt (De la Beche, 1839): throughout South West England, no clear lithological changes coincide with the boundaries of the Carboniferous Period as defined by the use of fossils, and litho-stratigraphic names are therefore important. The area of outcrop of the Culm Measures in Devon is shown in *Figure 3.1*.

The overall picture of Carboniferous sedimentation in South West England is of a deep water marine trough which accumulated a thick succession of shales and sandstones. At the end of Devonian times, the sea bed had shown considerable topographic variation, but more rapid subsidence and less sediment supply at the commencement of the Carboniferous resulted in an overall deepening and consequently more uniform deep-water conditions. The trough persisted throughout most of the period, shallowing in many places before Variscan deformation began.

In the early part of the Carboniferous this newly formed trough was largely starved of sediment (the 'Bathyal Lull' of Goldring, 1962), slow sedimentation of suspended clay being interrupted only in the area to the north and west of Dartmoor, where more shallow-water sands and silts were deposited. In areas to the east and north of Devon (the Mendip Hills and south Wales) limestones were being deposited in shallow water in the early Carboniferous on a subsiding marine shelf, forming a thick sedimentary succession; some of this carbonate debris was swept into the deep water trough by turbidity currents, and formed thin beds of limestone interbedded with shale which can be seen today in mid and north Devon. Much of Devon received very little sediment and thin sequences of radiolarian chert form most of the Lower Carboniferous succession here. Just before the end of early Carboniferous time the supply of carbonate sediment ceased; slow deposition of clay typified the early Middle Carboniferous, followed by rapid deposition of widespread sand sheets. These were carried into the basin by turbidity currents during late-Middle and Late Carboniferous times, and built up to a considerable thickness of strata. The last stages of infill appear to have taken place in shallower water, probably due to a combination of sediment build-up and crustal movements which reduced the rate of subsidence or even caused uplift.

Mucrospirifer sp. × 1½ (L. Dev.)

Dechenella setosa × ¾ (M. Dev.)

Parawocklumeria laevigata × 2 (U. Dev.)

Trimerocephalus mastophthalmus
× 1¾ (U. Dev.)

Dechenella setosa × ¾ (M. Dev.)

Gastrioceras subcrenatum × 1 (U. Carb.)

Platyorthis circularis × 1 (L. Dev.)

Prolecanites hesteri × 1¾ (L. Carb.)

Posidonia becheri × ¾ (L. Carb.)

Plate 1. Devonian and Carboniferous fossils from Devon

Plate 2. Slumped bed at junction of Baggy Sandstone and Pilton Shale. Baggy Point (SS 435401)

Plate 3. Pillow lava, Chipley (SX 807722)

Plate 4. Agglomerate, Brent Tor (SX 470804)

Plate 5. Folds in Bude Formation, Warren Gutter (SS 201110)

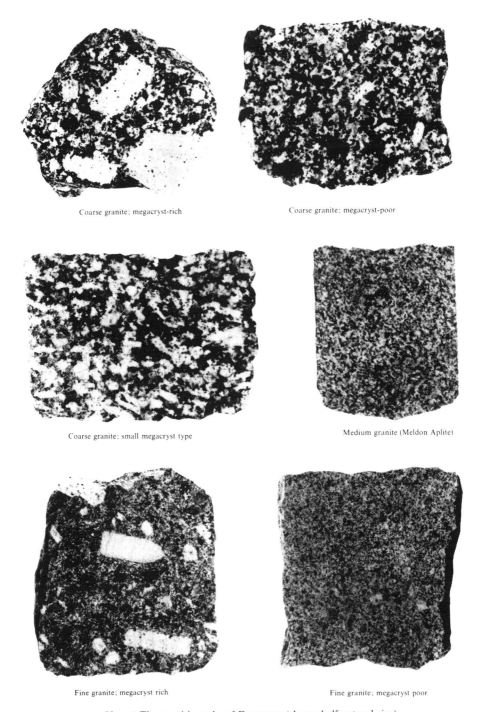

Coarse granite; megacryst-rich

Coarse granite; megacryst-poor

Coarse granite; small megacryst type

Medium granite (Meldon Aplite)

Fine granite; megacryst rich

Fine granite; megacryst poor

Plate 6. The granitic rocks of Dartmoor (shown half natural size)

Plate 7. Hound Tor (SX 743790)

Plate 8. Permian volcanic neck, Hannaborough (SS 529029)

Plate 9. Permian breccias overlying Meadfoot Slate, Thurlestone Sands (SX 675420)

Plate 10. Aeolian sand (Dawlish Sands), Dawlish (SX 967768)

Plate 11. Otter Sandstones overlying Budleigh Salterton Pebble Beds and Littleham Mudstones, Budleigh Salterton (SY 061816)

Plate 12. Blue Lias with *Metophioceras*, Kilve, Somerset (ST 145445)

Plate 13. Chalk overlying Cenomanian Limestone and Upper Greensand, Beer (SY 230890)

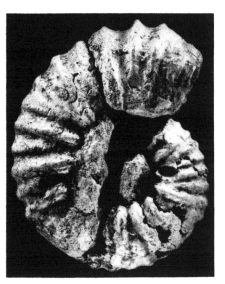

Calycoceras sp. cf. C. naviculare ×½

Mantelliceras sp. cf. M. dixoni ×⅔

Holaster laevis ×1

Tiaromma benettiae ×1½

Turrilites costatus ×1

Orbirhynchia wiestii ×1

Chlamys (Aequipecten) aspera ×¾

Plate 14. Cretaceous fossils from Devon

Plate 15. Bovey Formation (Southacre Member) clays and lignites, East Golds Pit (SX 860730)

Plate 16. Pleistocene Raised Beach overlying Devonian limestone. Hope's Nose (SX 950637)

Plate 17. Pleistocene Raised Beach overlying slate, Pencil Rock. Croyde Bay (SS 423402)

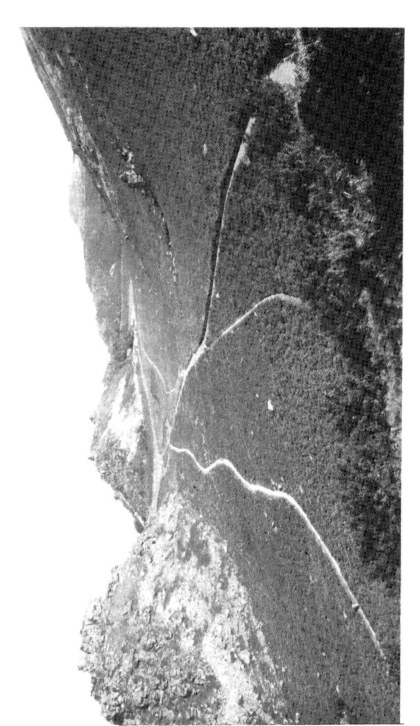

Plate 18. Valley of Rocks, Lynton (SS 705495)

Plate 19. Head, Middleborough (SS 429398)

Plate 20. China Clay pit with working high-pressure water jet, Lee Moor (SX 572629)

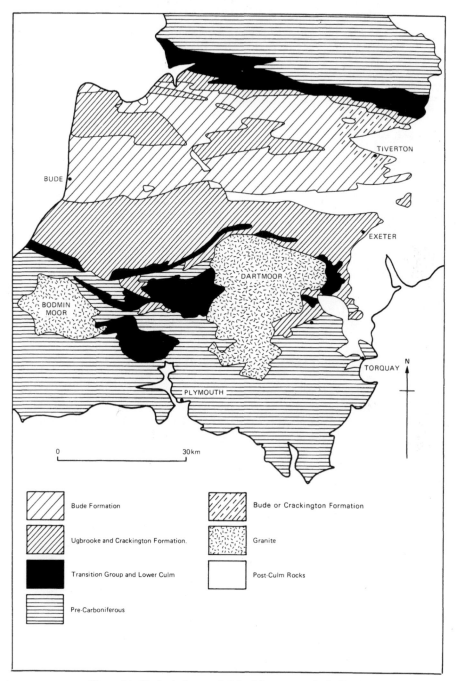

Figure 3.1. Geological map of Culm Measures rocks in Devon

The basin was part of a chain of similar ones which extended from central Europe along the southern edge of a major continental mass which occupied northern Europe, Greenland and the Canadian Shield; these had comparable stratigraphic histories. The south Wales shelf area fringed a part of this land mass, and deltaic deposition that took place there had its effects in north Devon where some sandstone beds are obviously of deltaic origin. To the south, the situation is less clear, but there may have been another land mass which contributed sediment to the trough.

STRATIGRAPHIC TERMINOLOGY

In this description of the Carboniferous rocks of Devon, it is more convenient to use lithostratigraphic terminology. The rocks here are very different to the better known Carboniferous Limestone, Millstone Grit and Coal Measures of Wales, the Midlands and the Pennines, but fossils have shown that they are of comparable age. Correlation has been attempted, based principally upon goniatites and conodonts (*Table 3.1*) (House and Selwood, 1966), but these faunas and the depositional situation lend themselves more to a comparison with the Kulm of the Rhineland (Paproth, 1969), rather than the carbonate shelf, coal-swamp and deltaic sedimentation of the rest of Britain. The cycles of sedimentation described from the more northerly areas of Britain cannot be identified here, so the chronostratigraphy for the early Carboniferous presented by George *et al.* (1976) is not appropriate to South West England.

The Carboniferous succession is conformable on Upper Devonian strata and is continuous up to beds equivalent to the Lower Coal Measures (Westphalian). There are three unequal lithostratigraphic sequences: the Transition Group (so called because it continued Devonian sedimentation into Carboniferous times), the Lower Culm Group and the Upper Culm Group; the two latter groups form the Culm Measures. The first sequence is the thinnest, while the Upper Culm is thickest and may once have been thicker, before post-orogenic erosion. The base of the Carboniferous is difficult to position within the Transition Group, and the top of the Upper Culm is defined lithostratigraphically as the highest recognisable horizon involved in the Variscan deformation. Laming (1965), on the basis of radiometric ages, suggested that the earliest part of the unconformable cover of New Red Sandstone deposited after the Variscan Orogeny was of latest Carboniferous (Stephanian) age, but this is continuous with the Permian succession, and therefore dealt with in Chapter Seven. Thus neither upper nor lower limit of the Carboniferous System can be identified as a clear boundary within the stratigraphic succession in South West England. Lithostratigraphic successions and faunal marker horizons used throughout Devon are summarised in *Figure 3.2.*

THE TRANSITION GROUP

In north Devon, the shelf-sea clastic sedimentation of the late Devonian Pilton Shale continued unchanged into the Carboniferous. These soft slates with a lustrous cleavage are found in an east–west belt from the Devon–Somerset border near Ashbrittle to the Taw–Torridge estuary. Local lenses of sandstone, many of which are medium to coarse-grained quartzitic sandstone

Table 3.1 CARBONIFEROUS ZONAL SCHEMES

Stage	British Goniatite Zones		Conodont Zones	German Goniatite Zones	
Westphalian	Gastrioceras listeri	G2b			
	G. subcrenatum	G2a			
Namurian	G. cumbriense	G1b			
	G. cancellatum	G1a			
	Reticuloceras superbilingue	R2c			
	R. bilingue	R2b			
	R. gracile	R2a			
	R. reticulatum	R1c			
	R. nodosum	R1b			
	R. circumplicatile	R1a			
	Homoceratoides prereticulatus	H2c			
	Homoceras undulatum	H2b			
	Hudsonoceras proetus	H2a			
	Homoceras beyrichianum	H1b			
	H. subglobosum	H1a			
	Nuculoceras nuculum	E2c			
	Cravenoceratoides nitidus	E2b			
	Eumorphoceras bisulcatum	E2a			
	Cravenoceras malhamense	E1c			
	Eumorphoceras pseudobilingue	E1b			
	Cravenoceras leion	E1a			
Visean	Sudeticeras costatum	P2c			
	Goniatites subcircularis	P2b			
	G. granosus	P2a	Paragnathodus nodosus	G. granosus	CuIIIγ
	G. koboldi, Neoglyphioceras spirale	P1d		N. spirale	
	G. falcatus	P1c		G. striatus	CuIIIβ
	G. striatus	P1b			
	G. crenistria	P1a	Gnathodus bilineatus	G. crenistria	CuIIIα
	G. maximus	B2		Entogonites grimmeri	
	G. hudsoni			E. nasutus	CuIIδ
	Beyrichoceras hodderense	B1	Gn. texanus	Pericyclus	CuII β/γ
	Merocanites applanatus		Scaliognathus anchoralis-Doliognathus latus		CuIIα
			Gnathodus typicus		
			Siphonodella isostichia-upper S. crenulata		
Tournaisian			Lower S. crenulata		CuIIα
			Upper S. duplicata	Gattendorfia crassa	
			Lower S. duplicata	Ga. subinvoluta	CuI
			S. sulcata		
Upper Devonian			S. praesulcata	Wocklumeria	

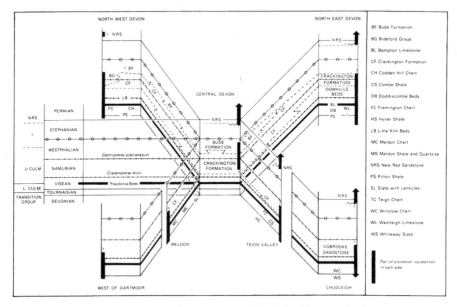

Figure 3.2. Stratigraphic successions of the Culm Measures in Devon

with some bedding surfaces rich in white mica flakes, occur irregularly along the outcrop belt. Limestone composed of shell debris occurs locally, some beds yielding lowermost Carboniferous faunas (Edmonds, 1974; Matthews and Thomas, 1974; Edmonds *et al.*, 1979).

In southwest Devon and east Cornwall, Selwood (1971b) showed that, in the structurally complex area around Launceston, the Devonian–Carboniferous boundary may lie within the Yeolmbridge Beds. These green and black slates contain limestone, and some sandstone bands, and have yielded fossils of earliest Carboniferous age representing the *Gattendorfia* zone. In the Teign Valley, green and grey slates with some siltstone bands, the Hyner Slate, occupy a similar structural position and, from their fauna, appear to span the Devonian–Carboniferous boundary.

Between the Teign Valley and the Launceston area the Transition Group is represented by more sand-bearing units, probably all referable to the Meldon Slate-with-Lenticles Formation, which has been recognised in the Belstone, Okehampton, Lifton and Lewdown areas (Dearman, 1959; Dearman and Butcher, 1959). Although in all these areas the formation underlies Lower Culm units, no diagnostic fauna has been found to prove their basal Carboniferous age, and its stratigraphic relationship is also problematical on the grounds that it shows more complex folding than the adjacent rocks. The dark grey slates contain more bands of sandstone and siltstone than their lateral equivalents to the east and west (Hyner Slate and Yeolmbridge Beds), and the lenticular form of the hard bands is partially sedimentary and in some places tectonic in origin.

Transition Group units also occur within some of the superimposed overthrust (allochthonous) units in south Devon, associated with the southern facies of the Devonian (*Figure 3.3*). For example, the Gurrington Slate, a thick

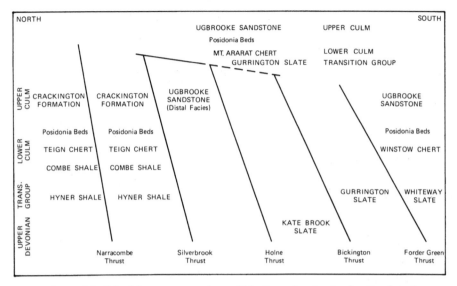

Figure 3.3. Culm Measures successions within the major structural units of south Devon

sequence of purple, green and grey slates with tuffs and lava flows, although yielding Upper Devonian faunas in most sections, do contain lowermost Carboniferous ostracods at a few exposures. Similarly the very thin, condensed sequence of the Whiteway Slate around Chudleigh yields lowermost Carboniferous faunas from the highest parts. Almost all these Transition Group formations continue the sedimentation pattern already established in the Upper Devonian and are only included in this chapter on their proven or predicted earliest Carboniferous age. The main lithological changes in the succession took place within the Carboniferous Period.

LOWER CULM GROUP

There appears to be a marked lithological change in the successions in many parts of South West England, with the widespread development of black or dark grey shales above the varied Transition Group lithologies. The first appearance of these dark slates is taken as the base of the Lower Culm Group, but in almost all areas this part of the succession is devoid of stratigraphically useful fossils: so this lithological boundary may not be a chronological one. The main part of the group consists of either chert or interbedded limestone and shale, and the top is defined as the last bed of chert or limestone in the succession. The topmost Lower Culm in most of Devon consists of a distinctive thin sequence of grey shales, usually partly calcareous and commonly slightly siliceous, and highly fossiliferous in many places; it is informally named the 'Posidonia Beds' or the 'Neoglyphioceras spirale Beds', on its fossil content. However, although this informal member is not usually thick enough to be mapped as a separate unit, it forms the best marker horizon in the Culm Measures of Devon, and the overlying dark shales are regarded as the basal formation of the Upper Culm. The faunas of the Posidonia Beds are high Visean zones P1b and P2a (Cu IIIα to IIIγ of the Kulm nomenclature) (*Table 3.1*). Detailed summaries of the state of knowledge of the biostratigraphy of the Culm Measures were given by Butcher and Hodson (1960), House and Selwood (1966), Edmonds (1974) and Edmonds *et al.* (1975).

North Devon

In northwest Devon, Prentice (1959) recognised two distinct facies in the Lower Culm Group southwest of Barnstaple; he regarded these as separate facies of a single formation known as the Chert Beds, which extend from the base of the lowest recognisable chert horizon to the top of the highest chert. The northern facies, the Fremington Chert, consists of dark shales with numerous, laterally continuous dark chert beds and some impure limestones, totalling about 46 m in thickness. The type section is on the west bank of Fremington Pill near its mouth on the Taw Estuary (SS 514333). Prentice recognised a sequence of unfossiliferous dark grey silty slates beneath these cherts, which he regarded as the topmost part of the Pilton Shale. He named these slates Group D, and assumed that they overlie the uppermost Pilton Shale, earlier defined as Group C, which are of lower Tournaisian (Lower Carboniferous) age. These dark slates should thus perhaps be regarded as a member within the Lower Culm Group, but whether or not these dark slates form a mappable unit remains to be seen.

In the area to the south, the Lower Culm Group is represented by the Codden Hill Chert, the central facies of the Chert Beds; it consists predominantly of bedded chert with little interbedded shale and no interbedded limestone, although the pale coloured open-textured chert seen in all weathered outcrops might have had a dispersed carbonate content before weathering. The thickness probably totalled more than 90 m. A reference section may be seen in Templeton Quarry, near Tawstock (SS 543297); no type section was designated when the unit was described, and this quarry is not on Codden Hill, but it is representative of the lithology and stratigraphy. Biostratigraphic horizons lower than those in the northern facies are present in these cherts, and Prentice (1959) suggested that the lowest parts of the unit might be the lateral equivalents of the Group D Pilton Shale in the north. He recognised a *Goniatites spiralis* bed forming the top of both chert units, which is overlain by the black shales of the Limekiln Beds of the Upper Culm Group, and reported that the two facies were brought together by thrusting, but Edmonds (1974) stated that, of the two Lower Culm chert facies, 'mapping shows the one to pass into the other'.

The Lower Culm Group may be traced eastwards from Barnstaple to within 3 km of South Molton, although little information has been published on the north Devon succession. The Lower Culm appears to thicken eastwards, with thin limestone beds well developed in the upper part of the succession in the South Molton area, giving an interbedded limestone and dark shale sequence comparable with the Kulm Plattenkalk facies of the German succession. For about 22 km eastwards from South Molton, the Lower Culm is cut out by the Brushford Fault, but it reappears west of

Dulverton, from whence there is a continuous zig-zag outcrop across a series of plunging fold axes eastwards to beyond Ashbrittle where it disappears beneath the New Red Sandstone.

In this eastern area the underlying Pilton Shale yields *Siphonodella* zone conodonts (*Table 3.1*), but the basal black laminated slates of the Lower Culm Group, the Doddiscombe Beds, are unfossiliferous. However, the overlying Bampton Limestone, which completes the Lower Culm Group, yields some stratigraphically useful fossils except in the lowest shaly Hayne Beech Member. The chert, shale and thin limestone beds of the Kersdown Chert Member above those shales have yielded Upper Tournaisian–Lower Visean fossils, whilst the Bailey's Beds higher up, also of member status, include the rest of the Visean stage up to horizons rich in *Posidonia becheri* together with the goniatite *Neoglyphioceras spirale* (Matthews and Thomas, 1974). The Bailey's Beds, which have been extensively quarried around Bampton for limestone, represent a late Visean phase of limestone deposition which extended throughout much of Devon.

The succession can be traced eastwards from Bampton in a strip of ground about 1 km wide into west Somerset, with limestone less in evidence and shale with chert predominant, until the succession disappears beneath unconformable New Red Sandstone near Tracebridge (ST 072210). However, to the south of this strip, in the area around Holcombe Rogus and Westleigh, the Lower Culm is represented by the Westleigh Limestones (Thomas, 1963a and b). Their relationship to the Pilton Shale and Doddiscombe Beds is uncertain as the age of their base is unknown, but they appear to be laterally equivalent to the whole of the Bampton Limestone. The zone of contact between the strips shows blocks of Westleigh limestone, up to 800 m across, apparently set within the Bampton Limestone outcrop and overlain by Upper Culm Group.

The lower Westleigh Limestones are seen only around Westleigh and are sparsely fossiliferous. They consist of thin bedded, fine-to-medium-grained detrital limestone in laterally continuous units with few interbedded, slightly calcareous shales, and sporadic thick medium-to-coarse grained detrital limestone units. Many of the thin beds show bioturbation, evidence of sea-floor burrowing organisms, but, where this has not destroyed original textures, graded bedding is visible. The upper Westleigh Limestones occur more widely and consist of laterally continuous thick detrital limestones and, in places, limestone conglomerate units interbedded with shales. The limestones yield a sparse, transported fauna of rolled corals, brachiopods, bryozoans, crinoid ossicles, etc, characteristic of the Carboniferous Limestone further north in Wales; the shales, however, contain some indigenous benthonic faunas of brachiopods and crinoids showing little evidence of having been transported far, and faunas of free-moving goniatites, orthocone cephalopods and *Posidonia*.

Some of the limestone beds show clear graded bedding; almost all have sharp bases and gradational tops and are interpreted as limestone turbidites. Fossils belonging to zones ranging from B2 to P2a (*Table 3.1*) have been identified, so evidently limestone sedimentation continued later here than elsewhere in South West England. Shales yielding *Neoglyphioceras spirale* and even *Goniatites granosus* have thick limestones interbedded, whereas everywhere else in Devon these occur in shales of the Posidonia Beds overlying the limestones.

South Devon

The Lower Culm sequences in the south of the county differ from those of the north in that they contain appreciable amounts of igneous material; interbedded tuffs and agglomerates or penecontemporaneous dolerite sills are commonly found. Many of the south Devon successions also appear to be thinner than those in the north, but because the structures are much more complex there is uncertainty about some of their details.

In the area of the IGS Newton Abbot Sheet (339), the Lower Culm Group appears in various tectonic units (*Figure 3.3*). In the lowest unit, in the middle and upper Teign Valley, the Devonian–Carboniferous boundary lies in the Transition Group, upon which the Lower Culm succession begins with the black shale of the Combe Shale—again, unfortunately, poorly fossiliferous and containing widespread dolerite intrusions which can be seen in roadside sections north of Ashton Cross (SX 840847).

Above this, the Teign Chert is a clearly distinguished sequence of thin-bedded radiolarian cherts with dark grey interbedded shales and containing interbedded units of tuff ranging from rhyolitic to basaltic in composition (seen in roadside sections north of Spara Bridge, Ashton, from SX 843841 to SX 841844). Both of these Lower Culm formations are invaded by thick, laterally continuous sills of alkali dolerite, some of which appear to have been intruded into wet sediment as there are vesicular zones in the dolerite and vesicular adinoles in the underlying and overlying shales; intrusion must have taken place close to the sea floor with no great confining pressure to inhibit steam formation. The shales, cherts and igneous rocks can be followed from the Bridford Thrust (Selwood and McCourt, 1973) southwestwards as far as Haytor Down (SX 767779) where they are cut by the Dartmoor granite.

The chert beds have yielded few stratigraphically useful fossils, but they pass upward into the calcareous shales with impure siliceous limestone and chert of the fossiliferous Posidonia Beds, seen in the Teign Valley just north of Spara Bridge in the roadside section (SX 843841). The Posidonia Beds yield abundant *Posidonia becheri* and spirally striate goniatites, suggesting a high Visean age (P1b–P1d) equating with the Posidonia Beds elsewhere; it also indicates that the underlying chert beds span the usual age range for South West England, from high Tournaisian to mid Visean. The overlying black shales of the Ashton Shale Formation equate with the Dowhills Beds and Limekiln Beds in northeast and north Devon respectively, as the basal black shale unit of the Upper Culm Group.

In the area around Ilsington and Liverton, south of the Narracombe Thrust, Lower Culm lithological units similar to the Teign Valley succession are recognised in the Liverton structural unit, but here the structures are more complex. In topographically low ground, near Ilsington, Transition Group lithologies occur associated with Combe Shale and Teign Chert. However, on some hill tops the Mount Ararat Chert appears to overlie both the Posidonia Beds and the Upper Culm, giving an inverted succession (such as north of Bickington, SX 7973). Waters suggested that these relationships were due to large scale recumbent folds, but thrusting is more likely: this zone can be traced westwards into the area north of Ashburton, where similar cherts and Upper Culm rocks are seen to be overthrust northwards across other Culm sequences (on Ashburton Down, SX 7672 and 7772, and Woodencliff Wood SX 7571).

In the areas to the south of these Teign Valley sequences, large scale overthrusting from the south has taken place, with different sequences present in each thrust slice. Units of the Lower Culm Group occur only in the area south and east of Chudleigh, where the condensed deposits of the Whiteway Slate of the Transition Group pass upwards into a very thin condensed sequence of the Winstow Chert. These cherts appear to represent most of the Tournaisian and the Lower Visean, with Posidonia Beds at the top, overlain by the Ugbrooke Sandstone.

Lower Culm Group units occur discontinuously westwards from these areas both around the north of Dartmoor to beyond Okehampton, and south of Dartmoor to Ivybridge. However, most of the original lithologies are difficult to identify as the rocks lie within the metamorphic aureole of the granite. Around Okehampton, Dearman (1959) recognised a Lower Culm succession, best exposed in Meldon quarries (SX 5692), which extends round the west side of the granite to Lydford (Dearman and Butcher, 1959). A modified version of their stratigraphic terminology was used by Edmonds et al. (1968), and this is used below.

The lowest stratigraphic unit in most sections is the Meldon Slate-with-Lenticles Formation of the Transition Group (*Figure 3.2*) and, although it shows more complex folding than the rest of the succession, the contact with the Lower Culm has been interpreted as a simple upward change of lithology. The lowest Culm Measure unit is the Meldon Shale and Quartzite Formation, in which is included an igneous member, the Meldon Volcanic Beds (Edmonds et al., 1968). The basal 26 m of this formation appear to have been dark grey slate with thin siliceous bands, similar to the Combe Shale of the Teign Valley, but has since been metamorphosed by the granite to cordierite hornfels. There are more interbedded quartzite units in the higher parts of the succession, some of which originally may have been siltstone and sandstone. These detrital rocks are unusual in the Lower Culm of South West England, but have been recorded at this level also at Lifton, which lies west of the main Dartmoor granite aureole. Tuff and locally even agglomerate beds, varying in thickness up to 160 m, occur within the Meldon Shale and Quartzite Formation at what appear to be different stratigraphic levels, some apparently overlain by more strata of the Meldon Shale and Quartzite Formation whilst others are in contact with the overlying Meldon Chert Formation (Edmonds et al., 1968).

The Meldon Chert Formation, which is the equivalent of Dearman's (1959) Calcareous Group, consists of chert and dark shale, with some limestone beds in the lower part but more especially in the upper part. Although these limestones are more numerous than those in almost all other comparable successions in Devon, unfortunately the only clearly determinable faunas come from near the top of the formation, giving high Visean ages (P1b–P2a). These thicker limestones near the top of the formation have been quarried extensively around South Tawton and Drewsteignton. The Okehampton area is outstanding in Devon for the thickness of the lower parts of the Lower Culm Group, with the development of sandstone beds, as well as for the presence of the Meldon Volcanic Beds. There are also doubtful records of post-tectonic dolerite dykes, following fault planes, which are unusual, as well as the more usual dolerite sills intruded penecontemporaneously into the Lower Culm strata.

Lower Culm successions, similar to those described above from around Okehampton, where described by Dearman and Butcher (1959) from the northwest side of Dartmoor, but later work has shown the geology here to be complicated by Lower Culm rocks occurring at both high and low topographic levels. The structure of this area appears to be an eastward continuation of that around Launceston where overthrust tectonics have created structural units comparable with the southern Teign Valley (Selwood, 1971a). In topographically low positions, in the northern part of the Launceston area, the Yeolmbridge Beds of the Transition Group appear to be associated with a chert sequence which is presumed to represent most of the Lower Culm Group. Small outcrops of calcareous units with *Posidonia* also occur north of Launceston, but these appear to form the lowest part of overthrust units made up of Upper Culm sandstones. Large areas of outcrop of chert with volcanic rocks and intruded by dolerite sills (but with no Posidonia Beds) occur at structurally and topographically high levels across a large part of the northern area south and east of Launceston; these are presumed to be of the usual Upper Tournaisian–Visean age and extend as far south as the Mary Tavy–Trekenner Fault Zone.

Work in progress in the Callington–Tavistock area suggests that quite different Lower Culm sequences occur in some of the overthrust units west of Dartmoor. Coarse-grained quartzose sandstones seem to be associated with dark grey slates and some tuffs in sequences which appear to be of Lower Culm age but whose fossil content is meagre.

Lower Culm Palaeogeography

Compared to the shelf facies of the Carboniferous in the Mendip Hills and south Wales (*Figure 3.4*), the Lower Culm Group is much thinner, less fossiliferous, and consists of shale and chert rather than limestone. This led early authors (such as Vaughan, 1904) to suggest major disconformities within the succession, although this suggestion is no longer accepted. However, this contrast of facies has led later authors to suggest that originally the two facies were separated by a considerable distance which was foreshortened by Variscan thrusting. However, Matthews (1977b) showed, by comparison with similar facies in Ireland and Germany, where there is no large scale thrusting, that there is no palaeogeographic need for major crustal shortening. He also predicted, by comparison with southwest Ireland, that a facies with Waulsortian-type mud banks might be expected in the intermediate areas. Interestingly, these detrital limestones built into mounds by the binding activity of soft bodied organisms have since been identified in a borehole at Cannington Park.

The Carboniferous Limestone of south Wales and the Bristol–Mendip shelf region has been shown to have a deeper water facies when traced towards Devon (George, 1958). The palaeogeographic context normally accepted for the beginning of Lower Culm sedimentation is the establishment of a starved, deep-water basin in early Carboniferous time (the 'Bathyal Lull' of Goldring, 1962), probably with the sea bed near or below the compensation depth for carbonate (below which all carbonate debris is redissolved by seawater) and with some volcanic activity in the southern parts of the basin. Some detrital sand and silt reached the Tavistock–Okehampton area early in Lower Carbo-

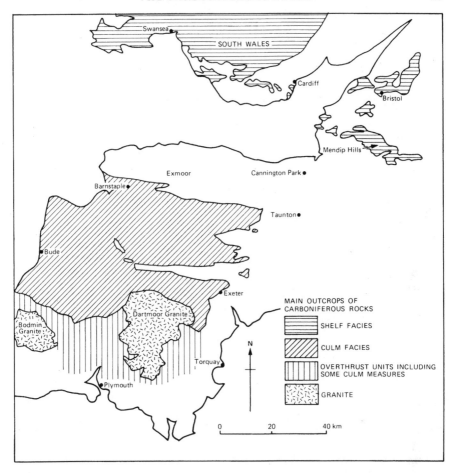

Figure 3.4. Culm facies and shelf facies Carboniferous rocks in South West England and adjacent areas

niferous time, but most of the sediment reaching the floor of the basin was suspended mud. In north Devon carbonate sediment from the shelf seas to the north was redeposited in the basin by turbidity currents, and some of the chert beds may have been carbonate debris silicified on the deep sea floor. In northeast Devon, however, an east–west tongue of deep-water cherty facies extends north of the Westleigh Limestones outcrop, and its presence precludes the possibility of these more proximal turbidities having been derived directly from the exposed shelf-facies areas. Instead, these limestones appear to be part of a westward outbuilding submarine fan from a southward extension of the Mendip Shelf, and thus raise doubts about how far eastward the northern part of the Culm trough extended. Also, as there is no evidence of reefal material in the limestone debris in the basin, it is likely that there was a continuously sloping sea bed from shelf to basin, north and east of Devon, without a marginal reef development.

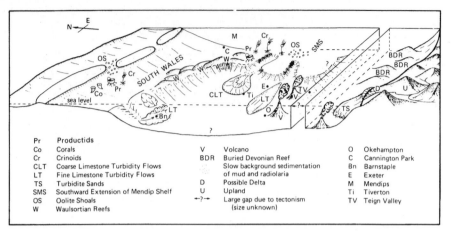

Pr   Productids
Co   Corals
Cr   Crinoids   V   Volcano   O   Okehampton
CLT   Coarse Limestone Turbidity Flows   BDR   Buried Devonian Reef   C   Cannington Park
LT   Fine Limestone Turbidity Flows   Slow background sedimentation   Bn   Barnstaple
TS   Turbidite Sands   of mud and radiolaria   E   Exeter
SMS   Southward Extension of Mendip Shelf   D   Possible Delta   M   Mendips
OS   Oolite Shoals   U   Upland   Ti   Tiverton
W   Waulsortian Reefs   ←-?-→ Large gap due to tectonism   TV   Teign Valley
                                                       (size unknown)

Figure 3.5. Palaeoenvironments reconstructed for the Lower Carboniferous of South West England and south Wales

The northernmost outcrops of the south Devon facies appear to represent deposits in an even more starved basin during Lower Culm times than that of north Devon, with very thin successions largely of radiolarian chert. However, even within this basin there are variations of sediment thickness and local areas of volcanic ash deposition, although no volcanic centres have been recognised. In the southern parts of the area west of Dartmoor (on IGS sheets 337 and 338) recent work (Isaac *et al.*, 1982) has shown that considerable slumping and olistostrome emplacement took place during Lower Culm accumulation, and some southern basins were already receiving sand turbidites characteristic of later beds.

Limestone deposition became much more widespread at the end of Lower Culm times, just before and during the Posidonia Beds deposition. The abundance of shells in this unit suggests slow deposition of detrital sediment compared to organic production, and the preservation of carbonate may be due to some change of sea-bed conditions, possibly better ventilation of the deep-sea basin lowering the carbonate-compensation depth. The incoming of dark clay deposits once more on top of the Posidonia Beds heralded the Upper Culm phase of detrital sedimentation. *Figure 3.5* depicts the Lower Culm palaeogeographic interpretation in a simplified way.

UPPER CULM GROUP

The base of the Upper Culm Group is defined as the beginning of the dark grey and black non-calcareous shale and sandstone sequence, and the characteristic rock type of the group is sandstone. The stratigraphic contact with the underlying Posidonia Beds of the Lower Culm may be gradational or abrupt, depending on locality. The biostratigraphic age spans the whole of the Namurian and apparently may include the highest Visean (as zones P2b and P2c are not identified with certainty in South West England) and the lower part of the Westphalian (Coal Measures), but excludes the igneous rocks and associated strata at the base of the New Red Sandstone rocks, which have given very late Carboniferous radiometric ages (*Figure 3.2*).

In some parts of south Devon, such as Ugbrooke Park southeast of Chudleigh (SX 8678), sandstone deposition appears immediately above the Posidonia Beds, while in others an intermediate unit of black shale with little sandstone is present beneath the sandstone sequence. In the middle Teign Valley this is recognised as the Ashton Shale; in northeast Devon, the Dowhills Beds; and in northwest Devon, the Limekiln Beds. This incoming of sandstones is also diachronous, possibly beginning in uppermost Visean times (directly upon P1 Posidonia Beds) at Ugbrooke, during the mid Namurian *Homoceras* zone in the Exeter area and not until the late Namurian, high *Reticuloceras* zones in north Devon.

The history of the development of the lithostratigraphic nomenclature of the Upper Culm is complex. Many names have been erected for the same rock group in different areas and conflicting views held by various authors of the relative age of unfossiliferous units. However, Freshney and Taylor (1971), Edmonds (1974), and Edmonds *et al.* (1979) have presented simplified lithostratigraphic schemes.

North and Central Devon

In north and central Devon the lowest sandstone sequence usually consists of mature, poorly sorted quartz sandstones, medium to fine-grained with a clay matrix, of the type usually called greywacke. They occur as frequently repeated, laterally continuous sandstone sheets with sharp bases and gradational tops, interbedded with dark grey or black shale. They show most of the characteristic features of a moderately distal turbidite suite, with flute and groove moulds, and graded bedding. They also show the usual 'grouping' of beds, with sequences of coarser and thicker sandstone beds alternating with finer grained sequences. Most of the depositional directional indicators show westward or eastward-moving currents, which are regarded as having flowed along the axis of the depositional trough. This succession, widespread throughout north and central Devon, is named the Crackington Formation; Freshney and Taylor (1971) suggested this name should be used for all these turbidite sequences, which would incorporate the following local stratigraphic terms: the Exeter Grits of the Exeter district; the Kiddens Formation of the middle Teign Valley; the Holmingham Beds of northeast Devon; the Instow Beds of the Fremington area; the Westward Ho! Formation of Westward Ho! and the Welcombe Measures of the west coastal section. The details of this correlation were reviewed by Edwards (1974), Freshney and Taylor (1971), and Edmonds *et al.* (1979). The Crackington Formation can be traced from Holcombe Rôgus at the extreme northeast of its outcrop westwards to the coast at Westward Ho!, and southwards through Exeter to the middle Teign Valley. West of that valley the outcrops extend through the Okehampton area, north of Launceston and on to the west Cornwall coast at Widemouth.

The faunas recognised from the Crackington Formation were listed by Edmonds (1974) and range in age upwards from doubtful *Eumorphoceras* Zone and certain *Homoceras* Zone in the south, through most of the zones of the Namurian, and in some places overlapping the *Gastrioceras subcrenatum* marine band into the lowest Westphalian (Lower Coal Measures), as shown in *Figure 3.2*. Good sections in the Crackington Formation can be seen in the Bonhay Road section beside the River Exe downstream from St David's Station in Exeter (SX 915924), and in the Pinhoe Brickworks on the eastern edge of the city (SX 954947). Also, good exposures are present in the old railway cuttings of the former Teign Valley railway near Dunsford (SX 836894), in roadside quarries in the Exe Valley between Bolham and Bampton (SS 943179) and in the cliff sections on the coast south of Hartland Quay and near Widemouth (SS 195014).

In extreme northwest Devon this basin-fill sequence of turbidites appears to pass upwards gradually into a cyclic sequence of sandstones and shales, 370 m thick, best seen on the coast south of Westward Ho! (SS 420292). These rocks comprise the Bideford Formation (Edmonds, 1974) and include repeated coarsening-upward cyclothems which range from dark shales through silt-stones to sandstones, some sandstone beds occur in downcut channels. The formation has been shown to have been deposited under a very wide range of deltaic environments, from topset to pro-delta, channel to bay, marine to fresh-water swamp, and beach. The thin, regressive phases of the cycles show much bioturbation, possibly including some rootlet disturbance, but no well developed seat earths occur to suggest any long phase of subaerial exposure. However, some seams of high carbonaceous content are present and the best of these horizons, the 'Bideford Black', approaches a poor grade coal which will burn with difficulty. These layers are known locally as 'culm' and appear to be the origin of the name for the whole lithostratigraphic group.

Edmonds (1974) observed that as the Bideford Formation is traced towards South Molton it appears to pass gradually into the Bude Formation laterally, and that the line of separation is difficult to define. Both the Bideford and the Bude Formations appear to be younger than the local Crackington Formation and yield Lower Westphalian faunas, but there is a possibility that both may have some lateral passage (particularly eastwards) into Crackington lithologies (Edmonds *et al.*, 1979).

The Bude Formation is characterised by massive sandstone beds, commonly slightly calcareous, brown weathering, non-graded and interbedded with thin sandstones, siltstones and shales as extensive sheet features. Marine fossils are of limited occurrence and some of the Cornish coastal sections are known to yield abundant xyphosurid trails, left by primitive arthropods. Also, large-scale and small-scale ripple marks occur. The features are suggestive of fairly shallow water deposition, and possibly not truly marine. The thicknesses of the formation suggest deposition in an actively subsiding basin of which the Bude Formation represents a late filling stage, and widespread horizons of soft-sediment deformation which occur in it might be related to earthquake shocks affecting unconsolidated sediment. Palaeontological evidence so far available suggest that the Bude Formation in places can range up to Westphalian C, but the base, defined as the first appearance of thick massive sandstone beds in the succession, may be diachronous. The contact with the Crackington Formation usually appears to be gradational; it is rarely fossiliferous and thus evidence of age is sparse. As the Crackington Formation occupies the southern margin of the Upper Culm outcrop there is, therefore, no direct evidence of the relationship between the Bude Formation and the southern facies of the Upper Culm.

South Devon

The Upper Culm in the overthrust units in south Devon appears to be distinct from that of north and mid Devon, and is defined as the Ugbrooke Sandstone (*Figures 3.2* and *3.3*). It consists of coarse-grained, in places pebbly, usually feldspathic sandstones interbedded with dark grey and black shales. In some parts of the sequence, such as at Ugbrooke Park (SX 869780), massive structureless conglomeratic sandstones from 30 cm to 5 m thick occur, with very little shale, whilst in other sections well graded granule to fine-grained, sandstone units are found, separated by shale units (such as in the Rydon Ball Farm borehole). In other sections the only difference from the moderately distal parts of the Crackington Formation is the slightly coarser grain size of the basal parts of beds and in their slight feldspathic content; separation is consequently difficult. In the Tavistock area McCourt (1975) had to resort to the chemical differences of the shales interbedded with the sandstones to distinguish between the Ugbrooke Sandstone and the Crackington Formation.

In the south, within the highest overthrust structural units (which may have had the most southerly origin) the Ugbrooke Sandstone contains more coarse-grained material, and more thick massive sandstone beds with thinner and fewer shales between them, than the lower, northern overthrust units. This suggests a gradation from proximal to distal turbidites from south to north, and therefore a southern source for the feldspathic sand. Also, Lower Carboniferous fossils have been recorded in derived fragments in Ugbrooke Sandstones and the equivalent rocks in East Cornwall, suggesting that Lower Carboniferous sediment had been lithified and uplifted in Upper Culm times, presumably by earth movements in some area to the south.

The stratigraphic age of the Ugbrooke Sandstone is uncertain. Near Chudleigh, sandstones appear to succeed the Posidonia Beds without any intermediate black shale unit. Thus they may be equivalent to the Crackington Formation, but how high up the Carboniferous zonal scheme they range is unknown. However, their deposition and overthrusting preceded the intrusion of the Dartmoor granite: both sedimentation and deformation must therefore be of Upper Carboniferous age.

Upper Culm Palaeogeography

The overall picture of deposition during Upper Culm times is one of a well established trough or troughs, which, having been rather starved of sediment during Lower Culm times, were receiving it in much greater quantities (*Figure 3.6*). In some places the filling appears to have continued until very shallow-water deltaic conditions were established, not long before tectonic deformation began. This filling was mainly of the typical offshore basin pattern, with turbidity currents derived from the ends and flanks of the basin flowing along its axis and depositing sandstone turbidite beds in the deeper parts of the sea floor. The northern margins, being a tectonically stable land area of relatively low relief, with slow epeirogenic uplift, slow erosion of gentle slopes, and long transport paths, supplied moderately fine-grained quartz sands and clays through such river systems as existed during deposition of the south Wales Namurian and Westphalian deltaic successions. At the same time, somewhere to the south of the Culm Measures trough, large scale earth movements were

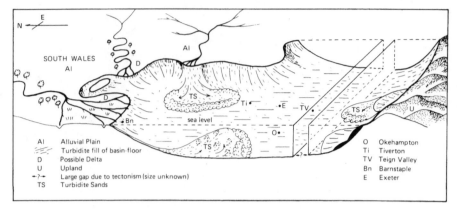

Figure 3.6. Palaeoenvironments reconstructed for the Upper Carboniferous of South West England and south Wales

taking place, producing uplands of youthful relief. Rapid erosion supplied coarse-grained feldspathic sediment to the southern area, again predominantly as sea-bed turbidites. In the extreme northwest of Devon, this infill sequence (with or without basin-floor subsidence) was so complete that cyclic deltaic sedimentation was able to build southwards close to sea level into the former deep-water basinal area by early Westphalian times. The central parts of the trough may have become moderately well filled also; the Bude Formation, the youngest strata now preserved, show features which suggest less than fully marine, and possibly moderately shallow-water, conditions.

To the south there appear to have been much more vigorous tectonic movements with youthful uplands adjacent to the basin yielding greater quantities of feldspar and rock detritus. Therefore it appears that earth movements caused the uplift of Lower Carboniferous rocks during the Upper Carboniferous times, an early phase of the Variscan Orogeny; these early movements preceded the later overthrusting of these southern rocks in the later stages of the orogeny.

DEVONIAN AND LOWER CARBONIFEROUS IGNEOUS ACTIVITY

There is a considerable contribution of penecontemporaneous igneous rocks to the Upper Palaeozoic sequences of south Devon, although they are notably absent in north Devon. Lava flows and tuffs (lithified volcanic ash) predominate, but there are also numerous minor intrusions, sills or dykes. These volcanic and intrusive rocks occur in association with sedimentary rocks from a wide range of depositional environments and are found in the Devonian and Carboniferous sequences at most structural levels.

The most detailed information available on these rocks in Devon concerns those in the Teign Valley. Here the rock types present in the Lower Culm sequences range from felsic quartz keratophyre tuffs to basaltic lava flows, but they can be regarded as parts of one differentiation series with a common origin. The lavas are sodium rich and show good pillow structures in some sections (for example Scannicliff Copse mine, SX 845863) and could be spilites, which were extruded on the sea floor. Most of the tuff bands show welding of the original ash texture, with collapse of glass shards and the formation of a tough interlocking texture; this indicates that the ash fell while still very hot. However, the sedimentary strata interbedded with the tuffs appear to show a continuous marine environment of deposition: some of the less welded tops of the tuff bands contain radiolarian skeletons, suggesting that, improbably, some of this welding took place under submarine conditions!

In the Teign Valley these volcanic rocks are associated with laterally extensive dolerite sills of various thicknesses, many of which have been quarried. In any one section these sills are usually found intruded into the sedimentary-volcanic succession at a level below that of the volcanic suite. Thus their occurrence, like that of the volcanic rocks, is confined to the Lower Culm. The sills commonly show fine-grained chilled upper and lower margins and, although the centres of the thicker sills may appear gabbroic, some have vesicular marginal zones and vesicular textures are shown by the adjacent contact-metamorphosed sedimentary rocks. This suggests that the sills were intruded at no great depth into water saturated sediment. From a study of their chemistry and mineralogy, it appears that the sills are similar in character to the volcanic suite, and could be part of the same differentiation series. The chemistry and differentiation of the whole suite is similar to that of the Hawaiian alkali basalts.

Contemporaneous volcanic and intrusive rock suites also occur in the Lower Culm outcrops around Belstone and Sticklepath (Selwood and McCourt, 1973) where the dolerite sills have been deformed by the late Carboniferous folding in a similar way to those in the Teign Valley. Dearman (1959) described a thick volcanic unit in the Lower Culm at Meldon Quarry. Although metamorphosed by the Dartmoor granite, and of greater thickness, these tuffs appear to be similar to those of the Teign Valley. Large bodies of dolerite also occur in Meldon Quarry, but Dearman (1959) and Edmonds *et al.* (1968) did not believe all these to be sills, and interpreted some as post-tectonic dykes. However, all the dolerite bodies have similar chemistry and mineralogy, and it seems unlikely that there was such a difference in their emplacement.

Selwood (1974) showed that volcanic rocks of various types and ages are found in the ground between Tavistock and Launceston, but that problems of accurate dating exist in this area of complex overthrust tectonics. For instance, spilitic lava flows resting on radiolarian cherts and tuff bands occurring in them east and north of Launceston may be of Lower Carboniferous age, although the thick Meldon Volcanic Beds do not seem to extend into this area. However, most of the volcanic rocks around Launceston, which also extend eastwards to the Tavistock area, appear to be interbedded with Upper Devonian slates. It is possible, therefore, that the lavas at Brent Tor are also of Upper Devonian rather than Lower Carboniferous age.

This problem of age determination is also present in the area south of Dartmoor, although more fossil evidence is forthcoming here. Above the Holne Thrust (*Figure 3.3*) the Kate Brook Slate (Upper Devonian) contains only rare small dolerite bodies, the relationships of which are difficult to determine, but which may be sills (seen, for instance, in the A38 road cutting at SX 748686). However, immediately above the Bickington Thrust, volcanic rocks are more common and in some places dominate parts of the succession. For example, within and immediately north of the towns of Ashburton and Buckfastleigh, the Middle Devonian succession is represented largely by thick calcareous lithic tuff. Areas of tuff are surrounded by Middle Devonian limestone bodies which can be seen to interdigitate with the margins of the tuff masses, as though the latter were originally ash cones on the sea bed. Around these cones abundant organic growth appears to have taken place, perhaps even forming fringing reefs. Eastward from the Ashburton and Buckfastleigh area, the Middle Devonian rocks resting on the Bickington Thrust are largely limestone, although they contain local tuffaceous bands and lenses, and usually the top of the limestone interval is marked by a thick lithic tuff horizon.

The overlying Gurrington Slate (Upper Devonian) contains local volcanic and intrusive rocks in many places. These range from spilitic pillow lava (seen in Chipley Quarry, SX 807721 to SX 808721), to crystal and lithic tuff units and small dolerite sills. Due to the complexity of the Variscan structures in this area, it is uncertain whether the many small patches of volcanic rocks now seen in outcrop were originally parts of more extensive sheets which have been tectonically separated, or were originally lenticular. Similarly, the outcrop of the dolerite sills is of much greater complexity than that seen in the Teign Valley. Nevertheless, an exceptional small quarry exposure near Dipwell Farm (SX 771695), within the Gurrington Slate outcrop, shows what appears to be a volcanic vent; with other tuff beds and dolerite sills nearby. Here, blocks of Middle Devonian limestone from lower in the succession and showing considerable chemical alteration, are found in a matrix of igneous origin, in an area of approximately circular shape; this is apparently the infill of the actual vent. Cross-cutting relationships with other lithologies, however, cannot be seen.

From the southernmost overthrust units in south Devon, Middleton (1960) described a whole range of volcanic and intrusive igneous rocks which range from andesitic to basaltic spilite in composition; their age appears to range from Lower to Upper Devonian. The most important of these, probably originating as a submarine ash cone, replaces the Middle Devonian limestone and slate succession around Ashprington. Richter (1965) suggested that the limestones with local tuff bands of Middle Devonian age in Torbay are probably marginal to this volcanic complex. In the extreme south of Devon, Richter also recorded Lower Devonian crystal tuff bands and small dykes intruding Lower Devonian slates, but as these latter are interpreted as following an early cleavage their intrusion may have been much later than the Lower Devonian. Dolerite sheets in the Ashprington Volcanic Series and in the Torquay area are probably sills which were intruded almost at the same time as the deposition of the sediments in which they were emplaced.

Chemical and mineralogical information on these Devonian and Carboniferous igneous rocks from Devon has been combined with data from similar rocks from Cornwall (where metamorphism has generally converted them to greenstone) and used in discussions about possible plate tectonic regimes in South West England during the Upper Palaeozoic times. Whatever similarities may be accepted between oceanic crust material and the basic igneous rocks of the Lizard and south Cornwall, all authors regarded the igneous rocks of Devon as showing that the crust beneath Devon at the time of their formation was truly continental in character (Floyd, 1972).

PLATE TECTONICS DURING THE UPPER PALAEOZOIC

Studies of palaeomagnetic orientations of rocks from the continents around the present day Atlantic Ocean show that considerable continental movements took place during the Late Palaeozoic, but the various attempts to reconstruct them differ considerably. Some interpretations have depended strongly on the interpretation of the Lizard Complex in Cornwall as an ophiolite sequence, or segment of oceanic crust, overthrust (obducted) onto continental crust. Confusingly, not only is this interpretation used in quite different schemes of crustal movement, but some authors have claimed that the chemistry of the Lizard rocks shows that they cannot be of oceanic origin (Floyd, 1972; Floyd et al., 1976). Indeed, Bromley (1979) showed that the Lizard complex includes sedimentary rocks of continental origin, and that it formed a passive basement on which a Lower Devonian sequence was deposited.

The weight of available evidence, on which plate tectonic interpretations can be made for the Variscan crustal movements affecting South West England, appears to come down against a simple pattern of continental movements and ocean-floor subduction. Throughout the Upper Palaeozoic successions, the rocks of Devon appear to have been formed as part of a continental area. Devon and Cornwall may have been near the edge of this continent, and elsewhere the disappearance of ocean floor crustal material beneath another crustal plate may have allowed considerable relative movements of continents. However, in the Variscan Orogeny there appears to have been a major movement of continents and microcontinents laterally, relative to each other, rather than directly towards each other (Badham, 1982).

Tarling (1979) produced a series of four continental reconstructions for part of the Upper Palaeozoic, based upon the available palaeomagnetic evidence, and these have been superimposed in pairs to draw the successive palaeogeographic maps of *Figures 3.7, 3.8* and *3.9*. During the late Palaeozoic the European, African and American continents were all moving rapidly and progressively northwards, as is shown by the latitude lines for the combined North American–European continent in the diagrams (the Atlantic Ocean as such did not begin to appear until about 100 Ma later). But the other continental masses appear to have moved at slightly different rates and in slightly different directions from this continent: in each diagram, the broken lines show the continental positions of the younger of the two ages, to compare with the solid lines of the earlier age, with the North American–European continent drawn as though it were stationary. The minor continental segments ('microcontinents') of the modern Mediterranean area appear to have been trapped and displaced or broken by the relative movements of the major adjacent continents as they rotated relative to each other. It is uncertain how wide, or how truly oceanic in character, were the seaways between the various masses in Devonian times, but Devon may have been near the continental margin in early Devonian times, becoming a clearly internal part of the continental area by Upper Carboniferous times.

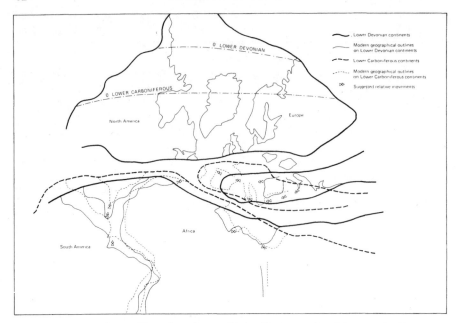

Figure 3.7. Lower Devonian—Lower Carboniferous continental reconstructions and plate movements relative to the North American—European plate

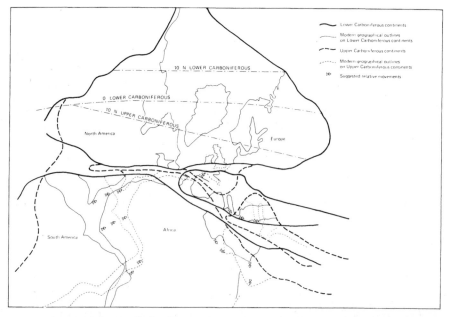

Figure 3.8. Lower Carboniferous—Upper Carboniferous continental reconstructions and plate movements relative to the North American—European plate

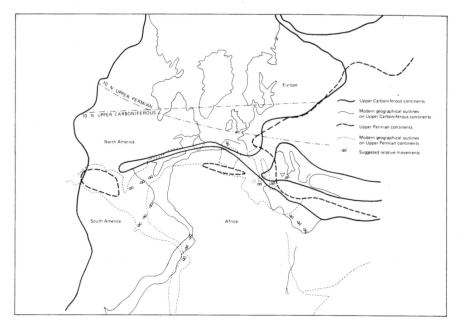

Figure 3.9. Upper Carboniferous—Upper Permian continental reconstructions and plate movements relative to the North American—European plate

LOCALITIES

South Devon

3.1. *Teign Valley road section* (SX 843842): Lower Culm. Northwards from Exmouth Cottages near Spara Bridge, Lower Ashton, the descending succession from Posidonia Beds can be followed in roadside cuttings and quarries through Teign Chert with tuff horizons to Combe Shale with dolerite sills, eventually to Hyner Slate (SX 840848), beside the main B3193 road.

3.2. *Ilsington road section* (SX 788764): Teign Chert. Below Ilsington village, large bedding plane surfaces in the Teign Chert outcrop beside the road.

3.3. *Meldon Quarries, Okehampton* (SX 566923): Lower Culm. A very large complex of working quarries in metamorphosed Lower Culm, worked by British Rail. The whole succession from Meldon Slate-with-Lenticles formation to Meldon Shale and Quartzite Formation can be pieced together from different sections, and large dolerite sills are obvious (advance permission is required to visit these quarries).

3.4. *Bonhay Road section, Exeter* (SX 914927): Crackington Formation. Within the city of Exeter, beside the River Exe north of the Exe bridges and south of St David's Station, river cliffs show Crackington Formation distal sandstone turbidites in a dark shale succession; faulted folds can be seen.

3.5. *River Dart section below Queen of the Dart* (SX 734687): Ugbrooke Sandstone. When the river is moderately low, the eastern river bank shows Ugbrooke Sandstone; rather distal here compared to Ugbrooke Park, but good sole markings occur on graded coarse to fine sandstone beds. Some complex fold structures due to second fold phases refolding earlier structures occur as the Holne Thrust is approached.

North Devon

3.6. *Fremington Quay, River Taw* (SS 514333): Lower Culm. A section along the south bank of the river shows the succession from Pilton Shale to Crackington Formation. The rather faulted outcrops are reached by crossing to the west bank at Fremington Pill on the railway footbridge from the abattoir, and proceeding northwards to the river banks. This is the type section of the northern facies of the Lower Culm cherts and the overlying Limekiln Beds of the basal Upper Culm.

3.7. *Bishop's Tawton Quarry* (SS 568297): Codden Hill Chert. A small disused quarry east of the main Barnstaple–Exeter road (A377), close to the road junction at the southern edge of the village, shows pale-weathered cherts of central or Codden Hill facies.

3.8. *Kersdown Quarry, Bampton* (SS 963222): Bampton Limestone Formation. A working quarry (Scotts of Bampton). The entrance is on the north side of the minor road leading east from the Tiverton Hotel, at the east end of the main street of Bampton. A steeply plunging anticline is present in the Kersdown member, the middle part of the Bampton Limestone Formation, which shows shale, radiolarian chert and thin limestone beds with sole marks and secondary chert. The topmost part of the section grades into limestone-dominated Bailey's Beds.

3.9. *Bailey's Quarry, Bampton* (SS 959218): Bailey's Beds. An overgrown quarry with entrances on the north side of a steep minor road on the east side of the main Tiverton–Bampton road, at the southern edge of Bampton town. The entrance cuttings show an anticline with chert and shale of the Kersdown Beds; in the main quarry the interbedded limestones and shales of the Bailey's Beds, have been worked out, and are now best exposed in the northern parts of the quarry.

3.10. *Westleigh, Whipcott and Pondground Quarries*: Westleigh Limestone. A series of quarries, some actively working, extending northeastwards from Westleigh Quarry, where the main ECC Quarries Ltd office is located (ST 063174), which controls most of these quarries (visit first to obtain permission).

3.11. *Coast section south of Westward Ho!* (SS 423292): Abbotsham Formation. Sandstone turbidites and shale beds are very well exposed and detailed observation of the cyclicity and facies types of this deltaic formation can be made at low tide.

3.12. *Hartland Quay* (SS 224246) to Lighthouse and adjacent coast (SS 229279): Crackington Formation. Folding is spectacular along this coast, and some of the sedimentary features of these turbidites can be clearly seen.

3.13. *Bickleigh Wood Quarry* (SS 945179): Crackington Formation. An abandoned roadside quarry on the east side of the A396 Tiverton–Bampton road near a lay-by on the west side of road. Here the Crackington

Formation turbidite sedimentation is well displayed. The site is an SSSI protected by law; for this reason, do not attempt to remove sedimentary specimens or the usefulness of the site would be destroyed.

3.14. *Bude coastal section* (north Cornish coast): Bude Formation. Cliff and foreshore sections around Bude show good sections in the thick sandstone and shale of the Bude Formation, repetitively folded.

Chapter Four

The Variscan Structures

The Devonian and Carboniferous strata of Devon were deformed during the Variscan Orogeny, which was part of a major period of earth movement spanning late Devonian to late Carboniferous times but which was largely complete before the emplacement of the Dartmoor granite. In Devon this deformation produced a series of major folds and thrusts which trend generally east–west, although most areas show evidence of more than one phase of movement and some structures are present which have an oblique trend. The major aspects of the fold belt have been reviewed by Matthews (1977b, 1981).

A first impression, gained from the geological map in *Figure 1.1* is that the county consists essentially of a large synclinal structure, the so-called Culm Synclinorium. This arises because rocks of Devonian age crop out in north and south Devon, while Carboniferous rocks occupy mid Devon. Such an interpretation is, however, misleading, as will be explained below. Instead of being arranged in a synclinorium, individual folds in the rocks of mid and north Devon display an anticlinorial pattern, as shown by Simpson (1971). Thus in north Devon the major folds are overturned towards the north; in the central part of mid Devon they are upright; and in the south part of mid Devon, in a zone lying near the northern edge of the Dartmoor granite, they are overturned towards the south. The degree of overturning of the folds increases with distance away from the central upright zone; thrusts are found which accompany the overturned folds, the displacements occurring in the same direction as the overturning. The structures of mid and north Devon appear to form a coherent unit, distinct from that in south Devon.

South of Dartmoor, the major folds are overturned towards the north and are relatively flat-lying. The accompanying thrusts also show displacements towards the north (Selwood *et al.*, 1982). Between Dartmoor and Bodmin Moor such structures extend northwards to a major dislocation extending westwards from the coast at Rusey (SX 125940) to the northwest margin of the Dartmoor granite. As this line more or less coincides with the northern limit of the Cornubian batholith (Bott *et al.*, 1958), it is tempting to speculate that the rise of the batholith commenced relatively early in the tectonic evolution of the Variscan orogenic structures, separating the area into two distinct tectonic units characterised by opposing tectonic transport directions. Radiometric ages derived from slates (Dodson and Rex, 1971), and other lines of evidence, indicate that compressional movements started at different times in these two units, with earlier phases of deformation affecting only the southern unit.

Detailed syntheses of the major structural features have been made by Dearman (1970, 1971), Simpson (1969, 1970, 1971), and Sanderson and Dearman (1973); all these authors divided the county into a number of tectonic zones based upon fold style. Individual zones, recognised from detailed structural studies on the well-exposed Atlantic seaboard, have been traced inland, but in the southern tectonic unit, more recent work (Isaac et al., 1982) has shown that the zonal boundaries are not meaningful.

On a wider scale, there have been many recent attempts to explain the development of the Variscan orogenic system in terms of plate tectonic theory. These reflect the equivocal nature of the evidence, much of which does not agree in any simple way with the basic concepts of the theory. Essentially two models for plate movements during Devonian and Carboniferous times have been proposed. In the first model it is considered that during these periods Devon was at the margin of an ocean basin which lay between the Old Red Sandstone continent and a southern European continent, and that the Variscan Orogeny was the result of continental collisions closing this basin. Northward subduction, southward subduction, and both northward and southward subduction acting together, have alternatively been proposed as the cause of collision. In the second plate-tectonic model, it is believed that the deformations of the Variscan Orogeny were associated with the subduction of a Tethyan (proto-Mediterranean) oceanic plate which lay between the southern European continent and an African continent. Again, various suggestions have been made about the direction of subduction.

Modifications of these models have also been made by Badham and Halls (1975) and Badham (1982), who believed that lateral movements were especially important and that a number of small plates were present in the southern European area. This latter interpretation was the basis of the discussion in Chapter Three; however as to which of these various ideas approaches reality is very much open to doubt: indeed, Krebs and Wachendorf (1973) rejected the plate tectonic concept entirely and explained the origin of the German Devonian and Carboniferous basins in terms of intracontinental subsidence. Their explanation of the later deformations of the Variscan Orogeny in general is that they were the result of vertical displacements associated with the rise of granite diapirs, produced by the melting of continental material, due to this subsidence. Such views were followed by Matthews (1977b, 1981) for South West England. However, it is clear from palaeomagnetic data, discussed in Chapter Three, that plate movements were occurring throughout the Upper Palaeozoic and that intra continental subsidence is an idea, like the others, which only explains part of the observed phenomena.

THE STRUCTURE OF SOUTH DEVON

As an aid to discussion of the structure of South West England, it has become customary to divide the region into structural zones. Provided that it is realised that all zones so created are not immutable, this is a useful device. Four zones can be recognised in the south Devon area, and are best demonstrated in the district east and south of Dartmoor (*Figure 4.1*); here they are identified geographically to save confusion with the various numbered schemes of Simpson (1969), Dearman (1970) and Sanderson and Dearman

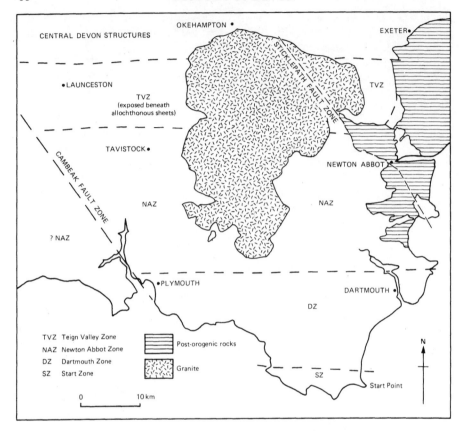

Figure 4.1. Structural zones in south Devon

(1973). Recent work also stresses the importance of distinguishing between allochthonous (tectonically transported), autochthonous (in place) and para-autochthonous (slightly moved) tectonic units. From north to south the zones are:

Teign Valley Zone (TVZ): a sequence of east–west trending, essentially upright folds which locally overturn to the north or south, depending on their position on the major anticlinorial structure. Slaty cleavage is generally not developed. A conformable sequence of beds is represented from the late Famennian into the Namurian (Upper Devonian to Upper Carboniferous).

Newton Abbot Zone (NAZ): a zone of close to tight, gently inclined to recumbent, north-facing folds with well-developed slaty cleavage. Stratigraphic successions, ranging from Lower Devonian to Namurian in age, are severely disrupted by major and minor thrusting. Local refolding of bedding and cleavage beneath thrust planes, indicate a northerly direction of transport. This zone is essentially allochthonous.

Dartmouth Zone (DZ): a region of east–west trending folds, with well developed slaty cleavage, which from south to north become progressively overturned towards the north. At the southern boundary, against the Start Zone, south-facing folds are recognised. Polyphase but coaxial deformation occurred. The beds, which are mainly Lower Devonian, show increasing grade of metamorphism from north to south.

Start Zone (SZ): this zone shows a comparable polyphase deformation history to that of the Dartmouth Zone, but additionally shows high green-schist-facies metamorphism. The beds are probably of Lower Devonian age.

Although the present contact between the Dartmouth Zone and the Newton Abbot Zone is gradational, it is possible that all of the zones are separated by basement fractures involved in complex movements. These may not all have been propagated into the cover rocks.

Teign Valley Zone

The structure of the Teign Valley Zone (*Figure 4.2*) is dominated by an anticlinorial structure, plunging eastwards off the Dartmoor granite and exposing a conformable sequence of topmost Famennian to Namurian rocks. Fold axes strike approximately east–west and axial planes lie in a fan-shaped arrangement, north-facing and south-facing away from a central area of upright folds. As the folds become progressively overturned, they tighten and, in the north, inverted limbs may be cut and replaced by stretch thrusts. The major structure is limited northwards by the Bridford Thrust (Selwood and McCourt, 1973) a south-dipping low angle structure which post dates the folding and stretch thrusting and predates the granite. The overridden Crackington Formation shows south-facing folds which appear to represent the eastward continuation of the fold style observed at Meldon. Selwood and McCourt recognised no fundamental structural or stratigraphic differences across the thrust, arguing that the Bridford Thrust has had the effect of bringing together two anticlinorial areas in a continuous sequence of anticlinorial and synclinorial folds which can be traced southwards from north of Exeter.

The southern boundary of the Teign Valley Zone is a conspicuous tectonic break which introduces north-facing recumbently folded strata with a well developed slaty cleavage. The break is a composite line formed by the Holne and Bickington Thrusts east of the Bovey Basin, and by either the Narracombe Thrust or the Silverbrook Thrust west of the basin: the facing direction of the gently inclined to recumbent folds between the Narracombe and Silverbrook Thrusts has not been determined, so this part of the boundary is uncertain. Since the strata involved are of Teign Valley facies (Selwood, 1971b), they could well represent the south facing continuation of the anticlinorial structure of the Teign Valley Zone; the Silverbrook Thrust would then mark the northern limit of north facing recumbent strata and constitute the southern boundary of the zone.

West of Dartmoor the picture is more complicated and disputed. Dearman and Butcher (1959) held that the progressive southward overturning of folds observed at Meldon (SX 558924) is continued down the west side of Dartmoor, developing into close recumbent folds and eventually merging with an extensive tract of recumbent isoclinal folds. This view was reaffirmed by Sanderson and Dearman (1973) but is at variance with that expressed by Freshney and Taylor (1971) and Selwood (1971a) who believed that a major structural dislocation is represented which corresponds to an easterly continuation of the Rusey Thrust recognised farther to the west. This thrust, which may be traced north of Launceston to Sourton (SX 534903), where it is cut by the Dartmoor Granite, has a long and complicated history and is now represented by a high-angle fault zone. The Rusey line, the thrust and its continuation, marks the boundary between the structures of mid and north Devon and those of south Devon, described in this chapter. Although there is little or no change in the stratigraphic level within the Lower Namurian

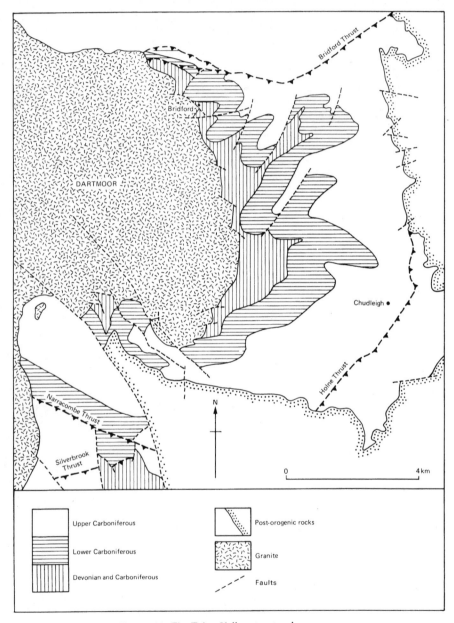

Figure 4.2. The Teign Valley structural zone

succession on either side of the thrust at the coast, there is an abrupt increase in metamorphic grade from a zeolite-facies assemblage, north of and above the thrust, to a chlorite or greenschist-facies assemblage, south of it. There is also a marked increase in the intensity of deformation exhibited by the rocks to the south of the thrust compared with those to the north.

Selwood (1971a, 1975) argued that the disposition of strata between Dartmoor and Bodmin Moor can most satisfactorily be explained in terms of an autochthon, or unmoved block of crust, equated with the Teign Valley Zone on the other side of Dartmoor, overthrust by a series of allochthonous units. This has been confirmed by Isaac (1981) and Isaac *et al.* (1982), although these authors described the ground more in terms of parautochthonous rocks of Teign Valley facies; these can be observed through a series of windows in an area between the Rusey line and a line drawn more or less eastward from Trecarrell Bridge (SX 320770) to south of Mary Tavy (SX 530792). At these lowest tectonic levels, and notably in Lydford Gorge, the simple structures revealed in the Teign Valley are not recognisable; the rocks have been highly disrupted by overthrusting and show the development of mylonitic fault rocks with associated amphibolite-facies metamorphism (Isaac, 1981).

On both sides of Dartmoor, the southern limit of the Teign Valley Zone is overthrust by strata belonging to the Newton Abbot Zone, except where the boundary is replaced by high-angle faults. The latter coincide with the northern boundary of the South Devon Basin and the possibility exists that they represent reactivated elements of fundamental faults which controlled sedimentation.

Newton Abbot Zone

The complex facies changes affecting the Middle and Upper Devonian rocks of south Devon are largely to be found within the Newton Abbot Zone. The contrasts in competence between the limestone of the reef complex and their thick argillaceous equivalents may thus be expected to have played a critical role in the tectonic evolution of the area.

Perhaps the most important feature to appreciate is that in the north and central part of the zone the limestone sequences are allochthonous and highly disturbed by thrusting, so that the lowest structural levels are everywhere of basinal facies. Within the limestones important thrust planes separate major tectonic units that become more disrupted up through the tectonic pile. Continuous sedimentary sequences are seen only in the lowest limestone-bearing unit which was introduced into the area by the Bickington Thrust between Chudleigh (SX 869795) and the Teign Estuary (*Figure 4.3*). This thrust rides over the basinal Kate Brook Slate and reaches locally onto rocks of the Teign Valley Zone. In this unit a complete succession of Middle Devonian to Upper Carboniferous rocks are deformed into three major recumbent folds facing north to northwest, with gentle southeasterly dipping axial planes. No minor folds are exposed, but low-angle thrust faults are common, roughly parallel to the bedding.

Continued west of the Bovey Basin, the limestones overlying the Bickington Thrust occur in normal succession with the basinal Gurrington Slate, though locally there has been sufficient movement at the contact for the limestones to be cut out by northward overriding movements of the slate. The limestones thin and pass westwards into tuffs.

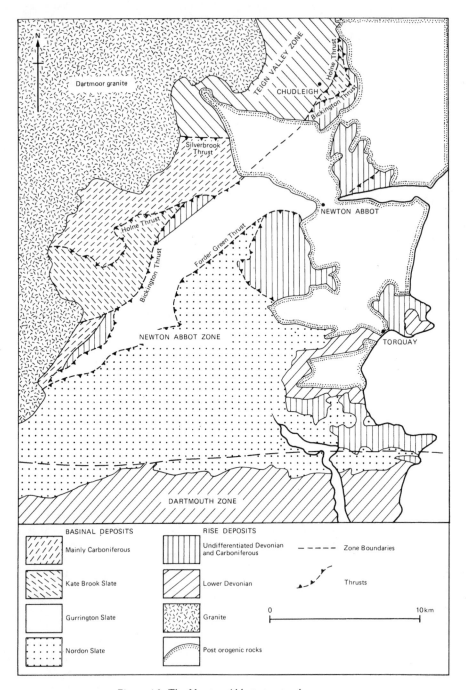

Figure 4.3. The Newton Abbot structural zone

West of Dartmoor, Gooday (1975) indicated a low-angle tectonic break equivalent to the Bickington Thrust, which carries purple and green slate (Gurrington Slate) over Kate Brook Slate.

Southwards, around Newton Abbot, successions are completely dismembered by flat-lying thrust planes. No overall fold pattern can be determined, but beds seem to occur in normal sequence between thrusts. A borehole sunk through the highest tectonic unit exposed in the area at Rydon Ball, Newton Abbot (SX 84366929) revealed that to a depth of 209 m, Middle Devonian, Upper Devonian and Upper Carboniferous rocks were shuffled by major and minor flat-lying dislocations. Where these can be mapped at the surface the thrust slices are completely detached from their source area.

In the Newton Abbot Memoir (Selwood *et al.*, 1982) it was argued that considerable disruption of stratigraphic units took place on low-angle thrust planes prior to their arrival in their present position.

In this context the basal allochthonous unit overlying the Bickington Thrust is distinctive, in that it possesses an undisturbed fold pattern. The implication that it is less far-travelled than the overlying units is supported by its strong basinal affinities. It has been noted in Chapter Two that the limestones contained in this unit became established on a topographic high created in the basinal area by the Kingsteignton and Buckfastleigh Volcanic Groups. Additionally, the unit is at present rooted between basinal slates, that is above Kate Brook Slate and beneath overthrust Whiteway Slate east of the Bovey Basin, and beneath the Nordon Slate to the west. All this suggests that deformation took place more or less in place, within an envelope of argillaceous basinal rocks. The jumble of overlying unrooted slices appears to have slid into the area from a topographic high to the south. Likewise the disordered, unrooted Upper Devonian and Carboniferous thrust sheets represented between Dartmoor and Bodmin Moor and southeast of Dartmoor are thought to have slid into their present position from a southern upland area.

In the absence of recent detailed mapping, the southward extent of the dislocation of the limestone complex is unknown. But it is evident that the limestones at Brixham, and those lying westward along strike at Plymouth, form part of conformable sequences occurring in north-facing fold structures, transitional in style to those of the Dartmouth Zone.

In the Upper and Middle Devonian slates that appear from beneath the allochthonous limestones, bedding and slaty cleavage are generally coincident, dipping at moderate angles to south or southeast. Major folds have not been recognised but minor recumbent to gently inclined folds face north. Within the basinal sequence an overall impression of inversion is obtained for Upper Devonian slates in the north give way through Middle Devonian slates to Lower Devonian slates in the south. This is illusory, for the formational boundaries are moderately low-angle thrusts between which sequences occur in normal succession. These south-dipping structures may not be of great magnitude, but they are remarkably persistent along strike (*Figure 4.3*). The most complete sequence is seen immediately east and south of Dartmoor (Waters, 1970) where from north to south the generally high-angle Silverbrook Thrust carries the Liverton tectonic unit (Upper Devonian to Upper Carboniferous) over the Teign Valley Zone. The much more gently inclined Holne Thrust carries Upper Devonian Kate Brook Slate onto the Liverton

Unit and is followed southwards by the Gurrington Slate and Chercombe Bridge Limestone introduced on the Bickington Thrust, and finally by the Nordon Slate on the Forder Green Thrust. If the Silverbrook Thrust is present east of the Bovey Basin then it is overridden by the Holne Thrust. The thrust sheets within the basinal slates are all rooted and the thrusts would appear to dip more steeply than the flatter lying thrusts which introduced the limestones into the area.

A similar but more complex picture is being revealed west of Dartmoor. Isaac et al. (1982) recognised a complex of overthrust sheets bearing evidence of considerable northward transport. Here, however, the authors pointed to a long deformation history in which this large-scale overthrusting represents the climactic (D_2) event; late Devonian to early Namurian sedimentary strata are involved. Immediately south of the Rusey line the tectonic complexity is such that it is best described as a tectonic mélange; this has been produced by the tectonic disintegration of gravity slides (D_1) generated in late Viséan to early Namurian times. At the higher tectonic levels (Turner, 1981) brittle structures are dominant, accompanied by much silicification, but thrusts propagated within the parautochthon (Isaac, 1981) show the development of mylonitic fault rocks.

The final (D_3) phase of deformation recognised by Isaac et al., includes both normal movements on steeply north-dipping faults and late Namurian and Westphalian underthrusting of the mélange northward beneath the main flysch (Crackington Formation) basin. The latter introduces a major change in the orientation of thrust surfaces south of the Rusey line and is accompanied by large scale back-folding of D_2 thrusts. The subvertical and southerly overturned folds thus generated in allochthonous sheets such as the Lydford 'anticline' (Hobson and Sanderson, 1975) have hitherto been described as primary folds involving conformable sequences.

Dartmouth Zone

The structures of the Dartmouth Zone (*Figure 4.4*) were admirably documented by Hobson (1976a,b) and Richter (1969) who detailed the complex

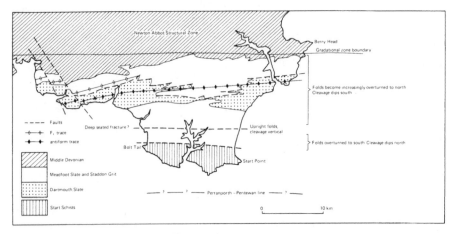

Figure 4.4. The Dartmouth structural zone

antiformal nature of the region. Hobson noted that although the Dartmouth Slate passes conformably beneath the Meadfoot Slate on the southern limb of the antiform, the northern boundary of these formations is mainly represented by a powerful east–west fault downthrowing to the north as much as 3,700 m near Plymouth (Hobson, 1976a). This fault is believed to be an ancient structure influencing not only the deformational history of the area but also exerting some control on sedimentation in post-Lower Devonian times.

It is generally agreed that the form of the present antiform is the result of polyphase deformation. The first (F_1) phase of deformation produced important folding, the major closure of which appears to have been an anticline now located near the northern boundary of the Dartmouth Slate. All F_1 folds trend east-northeast–west-southwest; their axial planes are vertical about a line from Bigbury (SX 668465) to Slapton (SX 822449), but they become progressively less steeply inclined both to north and south to give an anticlinorial structure. There is consequently a fanning of the slaty (S_1) cleavage, but the southerly placement of the crestal part of the anticlinorium means that the cleavage fan is asymmetrical, so that most folds are overturned to and face north. Richter (1969) believed that the cleavage fanning is primary and that it is repeated southwards to the Start Boundary Fault, but Hobson did not detect this in Bigbury Bay; rather he attributed the north dipping cleavage near to the Start Boundary Fault to regional refolding of earlier fabrics by a major monoformal (F_3) fold. This event is thus taken to define the southern limb of the Dartmouth Antiform.

The second (F_2) phase of deformation in the Dartmouth Zone produced no major folds, but small scale folds coaxial with F_1 are associated with the development of a second (S_2) strain-slip cleavage. This cleavage occurs in restricted, steeply dipping zones which are vertical around Stoke Fleming (SX 864484) but fan to the north and south. This fan is unrelated to the S_1 cleavage fan. The 'upthrusting' sense of movement on the south dipping second cleavage planes and 'underthrusting' movement on those dipping north (Richter, 1969) probably serve to emphasise the F_1 anticline near the northern boundary of the antiform. Sadler (1975) associated the S_2 cleavage with the development of the Start Boundary Fault, and argued from geophysical evidence that the Start Schists are represented at no great depth for some distance north of its present fault boundary. He suggested that the zone of north dipping S_2 cleavage between Torcross (SX 822421) and Thurlestone (SX 675430), which also corresponds to a zone of mafic dykes, may be the main displacement.

Both Hobson and Richter recognised south verging F_3 folds giving a monoformal deformation to earlier structures. These southward movements were to continue during the formation of flat-lying kink bands (F_4).

The coaxial nature of F_1 to F_3 folding gives an essential unity to the deformation of the zone, though it was clearly pulsatory. Hobson (1976b) observed that 'the consistent asymmetry to the north of both F_1 and F_2 folds may indicate that the rocks were deformed in a large belt of simple shear, characterised by subhorizontal, northward-directed overturning.' Age relations across the Start Boundary Fault are disputed. Amongst recent workers, Sadler (1975) alone maintained that the schists constitute a 'pre-Variscan basement'. Others argued in favour of a Lower Devonian age (Hobson, 1977) for, when allowances are made for differences in structural level, there is a close similarity in deformation to that of the Dartmouth Antiform. Robinson (1981) recorded merely a pressure difference in the metamorphic grade across the Start Boundary. Hobson equated the Start Boundary Fault with the Perranporth–Pentewan line, an important sedimentary hinge in south Cornwall. This implies that the Start Schists are the metamorphosed equivalents of the basinal Roseland succession. A more likely lithological comparison is with the Lower Devonian north of the boundary fault. Sadler's view that an east–west fault, limiting the Start Schists against the Roseland succession, lies southwards off the Start peninsula is more acceptable.

On the north flank of the Dartmouth Antiform a conformable sequence can be observed from the Meadfoot Slate into massive mid Devonian limestones. Here structures change gradually from south to north, as northward over-turned folds progressively rotate to become recumbent. The recumbency is associated with thrust faulting. This transition, which effects passage into the Newton Abbot Zone, is documented in the Torbay district by Richter (1969) and Smythe (1973) and in the Plymouth district by Hobson (1976a).

Start Zone

The structures which are exhibited by the schists that lie south of the Start Boundary Fault have already been briefly mentioned in Chapter Two. Marshall (1965) recognised the presence of four fold systems. The oldest folds (F_1) are recumbent isoclines which are overturned to the north with axes plunging at $10°$ to $20°$ westward. These folds are cut by an axial plane schisto-sity which dips gently to the south. Later folds (F_2) also trend roughly east–west and have a gentle westerly plunge, but they have near vertical axial planes and are cut by a crenulation cleavage (S_2). Third (F_3) and fourth (F_4) folds are sets of kink bands with steeply and gently inclined axial surfaces respectively.

Hobson (1977) also studied the structure of these rocks and noted that F_1 folds are only infrequently preserved, especially in the mica schists where isolated hinge zones are all that are found. Even in the hornblende-chlorite schists it is only locally (such as around the Salcombe estuary) that these early structures are well developed. In detail, he found the attitude of the F_1 folds to be determined by their position on the major F_2 structures; most F_1 folds east of Portlemouth are recumbent because they lie on a gently inclined limb of an F_2 structure. He also stated that the schistosity (S_1) is only easily seen in the mica schists. With regard to the F_2 folds, Hobson also extended Marshall's observations, and noted that the attitude of the axial planes of small F_2 folds can vary from steeply inclined to almost recumbent. The folds with steeply inclined axial planes are usually present on the more gently inclined limbs of large F_2 folds, while those with recumbent axial planes are on the steeper limbs. In a few localities these recumbent folds are cut by a flat-lying crenulation cleavage (S_3) and Hobson proposed that these folds were produced by a post-F_2 deformation. The affinities of this deformation, if any, with Marshall's F_3 and F_4 events are not clear.

Marshall (1965) considered that the F_1 folds in the schists were formed before the main deformation of the Lower Devonian slates which lie immedi-ately to the north of the Start Boundary, equating the first phase of deforma-tion in the slates with F_2 in the schists. On this basis he used the results of the radiometric age determinations of Dodson and Rex (1971) to suggest that F_1

occurred between about 340 and 360 Ma ago, F_2 between 320 and 380 Ma ago, and F_3 and F_4 between 280 and 300 Ma ago. Hobson (1977), however, considered that there was no objection to a direct correlation of the early phases of deformation in both schists and slates and so implied that F_1 occurred between 320 and 340 Ma ago and F_2 between 280 and 300 Ma ago, with F_3 and F_4 probably occurring soon after the F_2 event.

The Start Boundary is a steep northerly-dipping normal fault, interpreted by Marshall (1965) as forming contemporaneously with and as a small-scale expression of S_2, and accepted as such by Sadler (1975). A number of parallel shear-zones occur north of the Start Boundary, giving the effect of a series of step faults of which the Start Boundary Fault is merely the highest. The abrupt commencement of the positive gravity anomaly associated with the schists several kilometres north of the Start Boundary Fault (Bott and Scott, 1966) probably indicates that the main vertical displacements occurred on the more northerly faults (Sadler, 1975).

THE STRUCTURE OF MID AND NORTH DEVON

The rocks directly affected by the Variscan Orogeny in mid and north Devon range from Lower Devonian to Upper Carboniferous in age. However, faults formed during this orogeny continued to be subject to reactivation, and affect rocks as young as Oligocene. Indeed, some faults have been reactivated in a minor way in recent times, with earth tremors being recorded from time to time.

On a megascopic scale the basic structure of mid and north Devon is one of a large synclinorium. This consists of a central area of Carboniferous rocks flanked to both north and south by Devonian rocks. Its axis, which crosses the north Cornish coast near Duckpool (SS 200114), a few kilometres north of Bude, trends approximately eastwards, passing south of Shebbear (SS 440092) and on through the Winkleigh area to disappear beneath the New Red Sandstone unconformity near Bickleigh (SS 940070). Within this structure the intensity of deformation, assessed by the degree of tightness of folding and the intensity of cleavage, is at its lowest in the axial region, but increases in magnitude to reach a maximum along both the southern (Rusey line) and northern (Bristol Channel) flanks. However, along the southern margin the situation is complicated by the presence of at least one later phase of deformation.

Figure 4.5 is a reconstruction showing the arrangement of the various fold types northwards from the Rusey Line to the north Devon coast near Ilfracombe (SS 520480). This pattern suggests a fold pile with a considerable degree of north–south shortening, effected by both folding and thrusting. The lowest tectonic levels, with attendant high deformation and metamorphism to greenschist-facies, occur along the southern and northern margins and probably at depth below the axial region of the synclinorium. The highest tectonic levels, showing lesser deformation, gravity tectonics and zeolite facies metamorphism, occur in the central area of the synclinorium.

The Main Deformation

The small scale folds which belong to the early phase of the main deformation of the Variscan Orogeny vary considerably in both scale and form, a

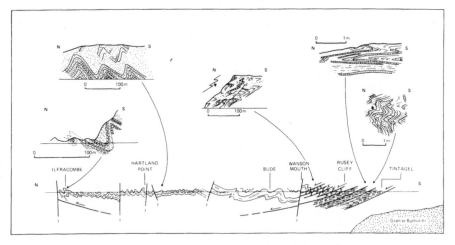

Figure 4.5. Structural section across north Devon and north Cornwall showing the
styles of the early folds. Just north of Tintagel the early folds are refolded

variation which appears to be controlled by the nature of the lithology affected. For instance, the smallest folds, which are only in the order of a few
metres wavelength, occur in the shaley lithologies belonging to such formations as the Crackington Formation, the Pilton Shale, and the Morte Slate.
But the more competent ones such as the Bude Formation and the Pickwell
Down Sandstone, show folds which range up to wavelengths of several
hundred metres.

Intermediate in size between these highly variable small-scale folds and the
major synclinorium there is a series of folds with wavelengths from about
2 km to 10 km. Again the scale of these folds seems to depend to some extent
on lithology. As it is difficult to recognise any definite sequence of development within the range of fold wavelengths, it is possible that the formation of
the synclinorium, the intermediate scale folds and the small-scale folds,
occurred at the same time.

The structural zones of Simpson (1970) and Sanderson and Dearman (1973)
trend roughly east–west and are defined mainly on the basis of fold-facing
directions and fold morphology. In most cases the zones have no sharp junctions, but grade from one to the other; where boundaries are sharply defined
it is commonly due to the presence of faulting.

Within this system, the change of fold style and facing direction between
the zones can be ascribed to variation in both depth and geographic position
within the fold pile. *Figure 4.5* shows a possible arrangement of the major
structures in this pile, with the type of folds present in north Devon occurring
at different positions within it. In the lowest parts of the pile, represented by
structures which outcrop along the southern margin of the synclinorium, the
minor folds are tight to isoclinal with a strong slaty cleavage. This is also true,
but to a lesser extent, of the structures found at the northern margin of the
synclinorium where, near Combe Martin and Ilfracombe, the folds are
strongly overturned to the north and almost recumbent in places, with a
pervasive slaty cleavage in the more argillaceous lithologies.

Passing southwards across Exmoor from its northern coast, the fold recumbency rapidly disappears, although folds are still strongly overturned to the north. This scale also increases because of the massive nature of formations such as are included in the Hangman Sandstone Group which outcrops in this part of Exmoor.

In the northern part of the synclinorium, in Exmoor, there is little evidence of the intense low angle deformation and thrusting that is present near its southern margin. The only thrust which has been postulated to occur is the Exmoor Thrust, which has been suggested as outcropping just off the coast along the northern side of Exmoor. Interestingly, the existence of this thrust was proposed to account for the northerly decrease in Bouger gravity anomaly recorded across Exmoor (*Figure 4.6*), which could be explained by the presence of Coal Measures rocks beneath the Devonian of north Devon (Bott *et al.*, 1958). More recent geophysical work by Brooks and Thompson (1973) is, however, less suggestive of the presence of Coal Measures and hence of the thrust, and Matthews (1975) discounted the presence of the thrust mainly on stratigraphic grounds. Brooks *et al.* (1977) also cast further doubt on the thrust hypothesis, interpreting the gravity and new seismic data to suggest a thick sequence of Lower Palaeozoic or late Precambrian rocks lies beneath the northern coastal area. Nevertheless, the presence of near recumbent folding along the northern edge of Exmoor does indicate that strong northerly movements acted in this area, and there may indeed be some thrusting between the recumbent strata and any less recumbent Devonian and Carboniferous strata lying under the Bristol Channel. Any such thrusting would, as in the south, probably show quite small movements, and is certainly unlikely to be responsible for any major stratigraphic rearrangement.

The presence of a more-or-less east–west trending reverse fault, which dips steeply to the south along the southern edge of Exmoor, was proposed by Whittaker (1978b). It is possible that this structure forms the southern margin of a Variscan-front zone. Whittaker also suggested that this fault may form part of a zone of major strike-slip movement. Various other authors (Badham, 1982; Tarling, 1979) have stressed the importance of transcurrent movements during the Variscan Orogeny.

In contrast to the structures in northern Exmoor, the southern part shows less intense deformation. In general, the northerly overturning of folds in this area is less marked and the pervasive slaty cleavage, while strong in the Morte Slate, is relatively weak in the Pilton Shale. South of Exmoor, in the Clovelly–Bideford area, the shales within the Crackington Formation show cleavage only in the hinge areas of the folds. Clearly, a gradual increase in intensity of deformation occurs northward of the axis of the synclinorium.

On the southern side of the synclinorium in mid Devon, the northward progression from the open, but still more-or-less recumbent folds lying north of the Rusey Line to more open folds is more abrupt than in north Devon. With this transition the pervasive slaty cleavage declines to a fracture cleavage which is best developed only in the hinges of folds. Farther north the change from recumbent or near-recumbent folds to more upright folds also occurs rather abruptly as seen along the coast north of Bude between Millook (SS 185000) and Wanson Mouth (SS 195012). This change is due to a combination of steep east–west normal faulting and northwest–southeast dextral wrench faulting. Inland, in the area around and to the southwest of Okehampton, however, the change from recumbent to upright folds is alleged to be gradual, and has been represented as a transition over the crest of a large overturned fold. Isaac *et al.* (1982) deny this transition; they identified underthrusting of allochthonous sheets beneath the Crackington Formation and interpreted the southerly overturned folds as back-folds lying above, but generated by the underthrusting.

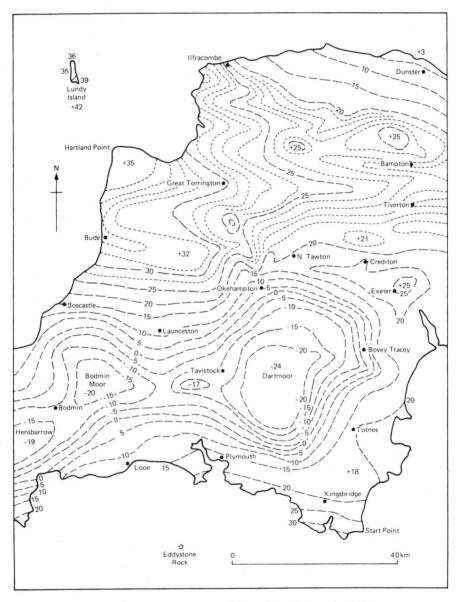

Figure 4.6. Bouguer anomaly map of Devon (after Bott *et al.*, 1958)

The zone of upright to slightly recumbent overturned folds extends inland from the coast between Hartland Point (SS 230269) and Wanson Mouth (SS 195012) to pass beneath the New Red Sandstone between Bampton (SS 960220) and Exeter, and this straddles the synclinorial axis. As the folds to the north of this belt and to the south show a tendency to northerly and southerly overturning respectively, the classic Culm synclinorium could also be referred to as an 'anticlinorium' of fold-facing directions (Simpson, 1971).

In this axial zone there are many features characteristic of folding at a high level in the crust. The most common structures are recumbent folds developed on the limbs of more-or-less upright folds. These recumbent folds have rather variable facing directions but, in general, face down the slope of the major structures and may well have been generated as gravity folds in conditions of low superincumbent load during the development of the major structures. Another modification of the upright folds which is indicative of high level tectonics is the presence of box folds, which are fairly common in the area to the north of Bude.

LATE FAULTING

All Variscan structures of Devon are cut by major and minor northwest–southeast dextral wrench faults. As Dearman (1963) pointed out, these may cause substantial displacements of stratigraphic boundaries and it is clear that their significance on early geological maps had been seriously underestimated. Undoubtedly displacements on some of these faults occurred during the Tertiary, but from their orientation is seems that they represent reactivated Variscan structures. Evidence for this was given by Isaac et al. (1982) who drew attention to the interaction of thrusting and wrench faulting, observing that the wrench fault zones acted as boundaries separating different thrust sheets with contrasting successions at roughly equivalent tectonic levels. Such zones are developed along the River Tamar and more spectacularly between Plymouth and Cambeak (SX 128967); the latter zone (Selwood, 1971b) serves as the natural western limit of the Variscan geology of the Devon area. Matthews (1977b), interpreting changes across this fault, suggested that the wrench movements had the effect of bringing into juxtaposition the normal 'Devonshire' limb, and the inverted 'Cornwall' limb of a major north-facing recumbent structure. When the effects of Tertiary wrench movements (Dearman, 1963) are removed from the fault zone, the facies changes noted in the Devonian across the fault persist: the implication is that fault movements were operating in late Variscan times.

Later tension, possibly associated with the start of Atlantic rifting, reactivated many of the older approximately east–west trending steep faults, and a series of horsts and graben developed; some of the grabens were filled with coarse Stephanian to Permian sedimentary and volcanic rocks (Whittaker, 1975).

STRUCTURAL EVOLUTION

The publication by Dodson and Rex (1971) of dates of metamorphic events in the Upper Palaeozoic rocks of South West England did much to reinforce the view that had developed over many years that the Variscan Orogeny affecting the Cornubian peninsula was initiated in the south and subsequently migrated northwards, ceasing in the Bude Basin in late Carboniferous times. Deformation of the Devonian and Dinantian rocks could also have been proceeding at depth while deposition of rocks of Namurian and Westphalian age occurred at the surface. The metamorphic ages, probably best interpreted as dates of uplift, were incorporated into and form a critical element in the structural zonal scheme for South West England proposed by Sanderson and Dearman (1973). Throughout much of the Newton Abbot Zone and in areas to the north, in mid Devon, Namurian rocks are involved in the primary deformation so that the main evidence for the 320–340 Ma deformation event lay in south Cornwall, although tectonised Upper Devonian pebbles are known to occur in continuous Devonian–Carboniferous successions in the Launceston area. However, Isaac et al. (1982) have recently provided direct evidence of an early deformation in the ground west of Dartmoor. They noted intense synsedimentary (D_0) deformation particularly in the Viséan, which was the prelude to major gravity sliding (D_1) in late Viséan to early Namurian times. This transported northwards, into a region of basinal sedimentation, a thick sequence of essentially Lower Carboniferous sedimentary and volcanic rocks which had accumulated on and adjacent to a major sea-floor rise. The parautochthonous nappe so generated appears to have developed in front of an advancing deformation front, the climactic event of which (D_2) is recognised throughout the Newton Abbot Zone, and involved large scale Namurian overthrusting.

In the same region the contact of the Namurian flysch (Crackington Formation) with the allocthonous sheets at the northern limit of the Newton Abbot Zone dips northwards at 20–50°. Isaac et al. reported that the contact is marked by mylonitic rocks, and indicated that northward underthrusting (D_3) of the allochthon has taken place, producing southerly overturned folds at higher structural levels. Normal movements on steeply dipping faults were also produced at this time.

Although lying beyond the county, it is worthwhile commenting on the geology of the coastal sections south of the Rusey Thrust, for these lie along strike from the areas discussed and, being well exposed, have been the subject of intensive investigation. Similarities are immediately obvious: in particular, north-dipping slides are prominent (Freshney et al., 1972); these are quite comparable to the north-dipping D_3 faults mentioned above. Their outcrop pattern roughly concides with the axial trends of a series of late folds to which they are genetically related. The axial planes of these folds dip fairly steeply southwards and show a transport direction broadly from the south. The folding intensifies towards these slides and indicates down-dip movement away from the area of the granite batholith (Figure 4.7). Such structures persist inland eastwards to the northeast margin of the Bodmin Moor granite, but thereafter their importance decreases. In this area Stewart (1981b) described large scale north-facing D_2 structures (thrust sheets introduced from the south), contained within slices defined by these late north-dipping faults. Small scale folds predating D_3 are generally absent inland but strong mineral lineations are found on foliation planes with a near north–south orientation. These persist through to the coastal sections where they have been ascribed by Dearman and Freshney (1966) to intensive shearing within the foliation planes in a north–south direction; this would be consistent with the northern transport indicated above.

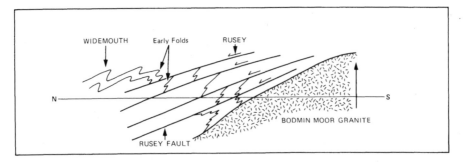

Figure 4.7. Low-angle faults north of the Bodmin Moor granite (after Freshney, 1965)

However, studies of small scale structures on the coast repeatedly report south-facing structures. This contradicts the evidence of Stewart, and is at variance with the model proposed by Isaac *et al.*; an accommodation has yet to be reached. Likewise, there is some difference of opinion on the nature of the north-dipping D_3 faults. Whereas Freshney (1965) interpreted this deformation phase as gravity sliding in response to the upward movement of the Bodmin Moor granite in an essentially tensional regime, Iaaac *et al.* favoured association with active underthrusting during the compressive phase of the orogeny: this they linked to the tightening of the flysch trough adjacent to the rising granite batholith.

Matthews (1977b) proposed an alternative model in which the varying attitudes of folds developed in a superstructure were to a very considerable extent controlled by an already fractured and irregular basement. As has been noted in Chapter Two, such deep fractures appear to have controlled sedimentation; there should thus be a close relationship between the pattern of sedimentation and of folding, so that persistent sedimentary rises could be expected to correlate with tectonic culminations. Deposition possibly took place on a basement of Armorican-type Precambrian (present in Brittany), overlain by a relatively thin sequence of stable-shelf Lower Palaeozoic strata of a kind still preserved in Brittany and Normandy. This Precambrian basement, which may have extended over the whole of the Cornubian peninsula, probably contains within it deep-level granites, some migmatitic, which may have been the source of the high-level granite occurring within the Upper Palaeozoic rocks of Devon and Cornwall. The rise of the high-level granites seems to have occurred in geanticlinal zones; thus the Dartmoor and Bodmin Moor granites are associated with a ridge dividing the Bude Basin from the Trevone and South Devon Basins. Possibly the slow rise of the granite within this zone was responsible for the generation of the persistent sedimentary rise noted in Chapter Two. Subsequently, this ridge appears to have played a vital role in deformation; as the compressional folding commenced earlier in the Trevone Basin than in the Bude Basin, northerly facing folds would have developed against this ridge and these could later have been overriden by south-facing folds developing along the southern margin of the Bude Basin. However, movement on the relatively steep, early fractures postulated by Matthews (1977b) would not have been able to accommodate the crustal shortening incurred during the main compressive phase of the deformation, and it is highly probable that low-angle fracturing occurred in the basement to allow for this. Major faults such as the Rusey Thrust may be one such structure which has propagated into the cover rocks.

In south Devon a basically simple sedimentary framework has been recognised, with a basinal area trending east–west separating two persistent rises, possibly of basement origin. In structural terms, these are now seen to form the Newton Abbot Zone, lying between the Teign Valley Zone in the north and the Dartmouth and Start Zones in the south. Deformation on the rises generated high-level structures of upright folds which, with further deformation, rotated to become overturned and even recumbent. At the same time the tectonically lower basinal strata were deformed to give recumbent folds with a well developed slaty cleavage. The thrust faults represented in the basinal area could have had a variety of origins; they may have resulted from continued deformation with a northerly sense of movement, or have been associated with the later overriding of allochthonous sheets from the south. A further possibility is that they too represent surface expression of deep-seated faults.

Matthews (1977b) noted that the change of fold style from upright to progressively overturned, observed southwards from Bude, could be associated with contrasts in rock competency and sedimentary thicknesses which might have been induced by an undetected rise lying north of the Bude Basin. The change in fold style northwards from the Start/Dartmouth Rise into the South Devon Basin can be similarly explained, with upright folds on the rise fanning to the north and becoming recumbent in the basinal area. The Teign Valley Zone would thus represent a comparable fan on the rise (described in Chapter Two) which is believed to flank the north side of the basin; the folds overturn towards the adjoining basinal areas.

In this context the Bridford Thrust would find satisfactory explanation as a late-stage structure tending to carry tectonic units from a tectonic high (Teign Valley Zone) north towards the Bude Basin.

Finally, it seems that already-folded Middle Devonian to Upper Carboniferous rocks from the area of the southern rise were shed northwards into the tectonic low of the South Devon Basin. Some of the earliest sheets shed, which would have included the highest stratigraphic units, appear to have travelled farthest; some came to rest on the upright folds of the northern (Teign Valley) rise. The northern limitation of these thrust sheets, between Bodmin Moor and Dartmoor, against the Rusey Thrust, can best be explained by introducing a late southerly downthrow on the fault. This would involve a throw reversal as, it has been suggested, it was downthrowing to the north during deposition.

The dynamic view of active thrusting and the more fixist interpretations of Matthews are by no means exclusive. Isaac et al. (1982) emphasised that the overall regime they recognised west of Dartmoor is one of thin-skinned thrust and nappe tectonics with the allochthon probably not exceeding 1 km in thickness. Thrusting on an Alpine scale was not envisaged. They interpreted the rocks involved in terms of a basin and rise model; this could well be linked to basement faulting. It is conceivable that persistent rises (geanticlines) later played a part in deformation, shedding tectonic units into adjoining tectonic lows. The distances of tectonic transport involved need not be great.

Chapter Five

The Dartmoor Granite and Later Volcanic Rocks

The granite mass of Dartmoor dominates the geological scene in Devon in a similar way to its topographic eminence: as in the latter it forms a great natural division in the county, so in geological history it marks a turning point. It is intruded into rocks that have been folded, faulted and, to varying degrees, metamorphosed, and is succeeded by rocks that show little deformation and no metamorphism at all. The intrusion of the granite was not the cause of this change in geological style, however, but the most obvious reflection of a major tectonic episode, the Variscan Orogeny. This was the culmination of the plate convergence that occurred during the Devonian and Carboniferous Periods in Devon, and when the resulting mountain range was formed, with granite at its core, a new phase of crustal rifting began accompanied by red-bed sedimentation. A short-lived volcanic phase followed the intrusion of the granite, giving rise to rhyolite and basalt lava flows.

Granitic rocks are common features of the Earth's continental fold belts and, by analogy, it is widely believed that the Dartmoor granite and the five similar plutons of Cornwall were formed and emplaced in a comparable tectonic environment. The evidence is that all six are linked at depth in one huge South West England batholith, but nothing is known directly of its deeper parts. Geophysical evidence, gravity and seismic, indicates the widespread occurrence of rock with densities characteristic of granite in a wide belt from Dartmoor to seaward of Land's End, the six plutons appearing at surface being merely the peaks of a continuous body of rock over 200 km long and up to 50 km wide. Part of this evidence is shown in *Figure 4.6* by the zero Bouguer anomaly linking the negative anomalies caused by the Dartmoor and Bodmin Moor plutons.

Parallel examples of batholiths in orogenic fold belts can be found in many parts of the world, for instance in the Sierra Nevada sector of the Western Cordillera of the United States, where a whole series of late Jurassic and Cretaceous granitic intrusions appear to be connected in a general way with the compressive forms generated by the oblique convergence of two crustal plates—the continental North American plate and the oceanic Pacific plate which has underthrust it for many millions of years. The bulk of the intrusions on the western side of the batholithic complex which, at 650 km length, is probably the largest in the world, are of quartz dioritic or granodioritic composition; those on the eastern side are predominantly leucocratic granodiorites and quartz monzonites.

THE DARTMOOR GRANITE

The Dartmoor granite is, more precisely, an adamellite (or a quartz monzonite, as it is better known across the Atlantic): this is a type of granite with approximately equal proportions of orthoclase and plagioclase in it, and quartz constitutes around 30 per cent. It crops out over an area of approximately 625 sq. km and is the largest single silicic pluton exposed in Britain (*Figure 5.1*). The granite gives rise to an elevated region of wide sweeping moorland dotted with numerous rocky tors, some of striking appearance. High Willhays (621 m), the highest point, lies about 5.5 km south of Okehampton and is within 2 km of the intrusion's northwestern boundary. The highest ground occurs, in fact, mainly in the north-central parts of the granite, around Black Ridge and Whitehorse Hill. Here a majority of Dartmoor's radially draining rivers begin their courses. Two, the Rivers Teign and Dart, pass from granite onto country rocks via impressive tree-clad gorges. However, despite the tors and the many scattered stream sections, less than one per cent of the granite is exposed and only recently has a reasonably detailed map of the pluton been prepared.

De la Beche published the first map showing an outline of the granite in about 1835. Revisions were made around the turn of the century when the Geological Survey was conducting the primary six-inch geological mapping programme in south Devon, but this programme was not continued into north Devon. As a result, no Survey map carrying a detailed trace of the granite's northern boundary was available until 1969, following reconstitution of the South West England Field Unit in 1962.

Many individual studies have been made on the granite since a brief account by De la Beche in 1839. Generally, observers have accepted an igneous origin for the mass, which Ussher (1888) suggested was probably laccolithic in form. Only Hunt (1894) ventured a metasomatic explanation: that is, genesis through *in situ* alteration of sedimentary material by silicic alkalic fluids.

The occurrence of two main rock types in the pluton, one coarse and the other fine, became well known from the earliest work. Further subdivisions were proposed by Reid *et al.* (1912), namely pegmatitic material containing hornblende, and dark fine-grained granite forming inclusions and *basic segregations*. This account commented on the presence of numerous small xenoliths in the coarse granite and also carried a good deal of mineralogical information. Later, in the 1920s and early 1930s, Brammall and Harwood examined the mass in much more detail, providing a series of papers on various mineralogical aspects (including one in 1926 on the occurrence of gold and silver) and an essentially new appreciation of the rock types (1923, 1926a and b, 1932).

They divided the coarse granite into two varieties: one characterised by abundant large feldspar phenocrysts ('giant granite'), the other by only a few large feldspar phenocrysts ('blue granite'). General observations led them to believe that most tors consisted of the 'giant' variety, while tracts of lower ground were underlain by 'blue' granite. At Hay Tor (SX 758771), a body of fine granite apparently intrudes the main mass of 'giant' granite. Its base is not exposed, but Brammall and Harwood reasoned that this material represented the chilled upper margin of 'blue' granite which crops out in lower

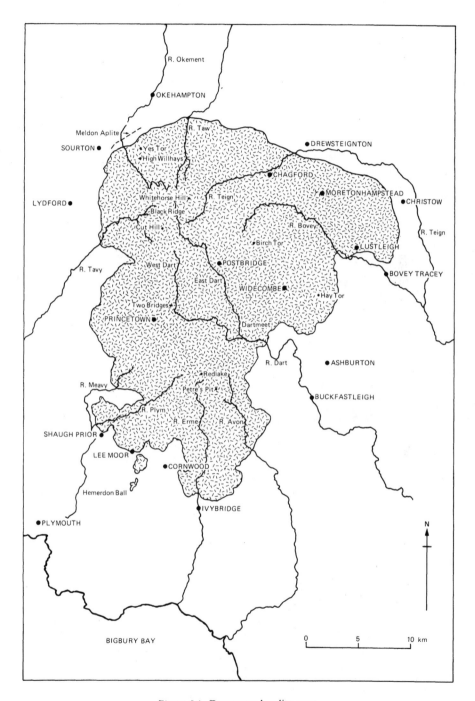

Figure 5.1. Dartmoor: locality map

ground to the west. They suggested that the dark inclusions and basic segregations could be relics of an earlier dioritic phase, broken up and incorporated by the intrusive magmas responsible initially for the 'giant' granite and then for the 'blue' granite. Fine-grained granitic rocks form only a very small proportion of the Dartmoor pluton, occurring chiefly as dyke-like and sill-like bodies. These were assigned to a fourth phase of intrusive activity, followed eventually by quartz-tourmaline veining, greisening and, locally, metalliferous mineralisation.

Apart from a wealth of mineralogical information, Brammall and Harwood also presented 78 chemical analyses of various rock types, including xenoliths, aureole specimens and individual minerals. Their principal conclusions were that the pluton was igneous, composite, probably laccolithic in form, and that the magmas were derived by melting of sedimentary rock at depth but had suffered considerable modification through assimilation of materials at higher levels.

Work then proceeding on the neighbouring plutons in Cornwall seemed generally to support these views. However, since that period many new studies have been initiated on the granites of South West England and a particularly useful review by Exley and Stone appeared in 1966. Evidence gained in recent years suggests that some modification of opinions expressed by Brammall and Harwood about Dartmoor is necessary, but the overall importance of their contribution should not be underestimated.

The Granite Boundary and the Form of the Intrusion

The irregular outline of the Dartmoor granite contrasts with the more compact, smaller ovoid shapes of the Cornish plutons (*Figure 5.2*). According to one interpretation, based on gravity data, the magma may have risen in the south and spread northwards as a laccolithic tongue several kilometres thick (Bott *et al.*, 1958). This model agrees broadly with previously held opinions about the pluton's general form. However, from the shapes of the Cornish granites, it seems possible that the magma actually rose through relatively resistant Devonian rocks in the south-central region (*Figure 5.3*) spreading out on reaching the Carboniferous–Devonian interface, both to the north, and to a lesser extent, southwards.

There is some evidence that, before emplacement, the upwelling magma caused vertical compression of overlying strata and attendant tectonic displacement (Dearman and Butcher, 1959; Edmonds *et al.*, 1968; Freshney *et al.*, 1972). Consequently, further structural studies in surrounding country rocks may help in understanding more clearly both the form of the upper regions of the pluton and the mechanisms of intrusion. A fair measure of agreement certainly exists on one matter: the general proximity of the present erosion surface to the original roof of the pluton. No exposures seem to be at any great depth in the mass. Commonly cited reasons for this are a relative abundance of partly digested shale xenoliths, and signs of contamination in the granitic rock. In addition, two round patches of contact-altered sedimentary rock rest on top of the granite on Standon Down (SX 555818) and near Sharp Tor, Lydford (SX 553846). A third body of country rock consisting of contact-altered dolerite occurs in the upper, north face of Tavy Cleave (at about SX 557837). This view is also supported by the spatial distribution of large feldspar crystals which will be described in more detail below.

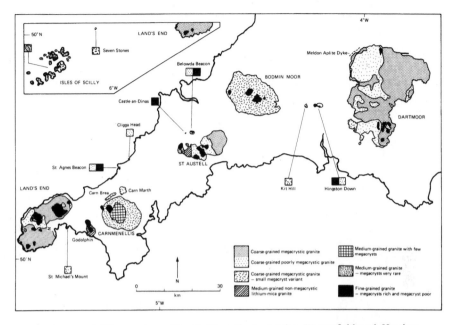

Figure 5.2. The granites of South West England (after Dangerfield and Hawkes, 1981)

Contacts of the granite with country rock are exposed only in about eight places and although slate and granite outcrops at several other localities show to within a few metres where the junction must lie, the general course of the boundary has been inferred mainly from rock fragments in the soil. Dips of the contact mostly appear to be outwards at relatively high angles. The best exposure is possibly at Whitstone Quarry (SX 814792) near Lustleigh. There, the contact plane is seen undulating along both strike and dip directions, plunging broadly to the southeast at 60–65°. Elsewhere exposure is too poor for direct measurement. In a few places the granite boundary crosses steep valleys without any marked deflection and so locally must be vertical or near

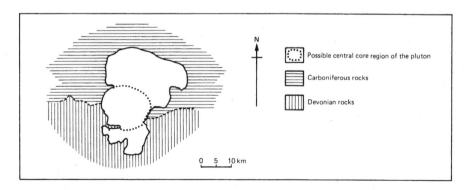

Figure 5.3. Possible position of the main feeder of the Dartmoor granite

vertical. One such example occurs in the Teign Valley near Hunters Tor; others, again to the northeast, in the valleys of the Shuttamoor and Beadon Brooks.

Contacts between the granite and country rocks are sharp, generally with little veining. Granite veins are fine-grained and most range from a few centimetres to 60–70 cm in thickness. The largest known extends north from the pluton boundary near Sharp Tor, Drewsteignton (SX 727898) for about 200 m and is approximately 10 m wide. Texturally the material of this dyke resembles adjacent 'giant' granite, but is finer grained. At the main contacts, the granite remains coarse grained to within a few centimetres of the junction with sedimentary rocks. This suggests that the latter were hot when intrusion took place, a point reinforced by the presence of a well developed metamorphic aureole 1–3 km wide, surrounding the pluton.

Joints and Faulting

The shapes of the majority of tor outcrops are closely controlled by major joint planes, which fall into two obvious categories: those that are inclined at high angles or are sensibly vertical, and those that are sub-horizontal and usually termed floor joints. The origins of the two categories are probably different. A third group of joints, less well developed and inclined broadly at angles of between 20° and 80° may be due to a third cause. However, all joints show one common feature: when viewed in quarry exposures or in the higher tors, it is evident that they occur in greater numbers per unit volume of rock towards the top. For example, in the upper parts of exposures, the spacing of open floor joints may be no more than a few centimetres; at the bottom separation may exceed 10 m. The respective figures for open high angle and vertical joints are around 20 cm and in excess of 10 m. Suitable data for the third group of joints are not available, partly because they are not so well developed, and partly because the labour intensive operation of measuring them has never been undertaken on a sufficiently large scale.

Jointing is commonly regarded as a strain phenomenon produced in response to stored stress, and the upward increase in the number of open partings as an expression of the hydrostatic relief due to removal of overlying rock by erosion. In the past, it was generally accepted that stored stress in plutons like Dartmoor related to the period of intrusion and subsequent cooling. This may have been so in the case of stress responsible for the floor jointing. Floor joints are approximately horizontal in exposures on ridges and hill tops, but in rocks cropping out on the flanks of hills, they are inclined towards neighbouring valleys or hollows at angles of up to 20–25°. In broad terms, the joints mirror topographic form which in itself may broadly indicate the original shape of the pluton's roof. If the latter assumption is correct, floor jointing could well be a reflection of stress built into the rocks by contraction during the primary cooling phase.

High-angle and vertical joints strike in all directions, but with marked maxima along north–south, east–west, northwest–southeast and northeast–southwest directions. These maxima are evident everywhere in the granite mass and show no sensible deflections with respect to the arcuate path of the pluton's boundary. This feature makes it unlikely that they result from an inherent stress field. It is more probable that the joints are due to externally imposed stress. Blyth (1957, 1962) was the first to suggest that some at least (the northwest–southeast and northeast–southwest) might be a reflection of horizontal forces which produced tear faulting in the northeastern portion of the granite mass. In fact, joints with the same general directions occur in each of the granite bodies of South West England. They may all relate to regional stress fields associated with the prolonged opening of the North Atlantic ocean basin.

The third group of joints merge at one extreme with the high angle and vertical group, at the other with floor joints. Strike directions are not yet well documented. However, it seems that quartz porphyry dykes and metalliferous lodes occupy planes which may equate with many of the higher angle joints of this type. One explanation for their origin possibly lies in the fact that a high proportion of the Cornubian crustal block is, according to geophysical evidence (Bott *et al.*, 1970), composed of relatively light granitic materials. This must mean that there was considerable isostatic imbalance with respect to adjacent denser crustal rocks during late Carboniferous times. Attempts to achieve equilibrium may have given rise, among other effects, to a field of gravitationally induced vertical compression within rocks of the Cornubian block. Joints of the third group conceivably relate to stress of this nature.

The explanations for jointing offered here are undoubtedly oversimplifications. Individual planes locally deviate as much as 20° either side of their mean strike direction and, in places, by more than 10° about the mean plane of inclination. In some tors north–south and east–west sets are well developed, in others it is the northwest–southeast, and northeast–southwest sets that are the most obvious. More rarely, only one direction from each of the sets is evident, with the result that outcrops weather to lozenge- rather than cube-shaped blocks. Low-angle joints may be clear in one part of an exposure, non-existent in another. In rare instances joints make arcuate traces in the granite. Because of features like these and the generally incomplete state of observational knowledge, it is perhaps wiser simply to state that jointing is an expression of an extremely complex and changing web of stress fields which has acted on the granite over a long period of time.

Surfaces on many joint planes are grooved (slickensided) and thus indicative of movement. This movement probably amounted only to small scale adjustment. Larger adjustments resulted in significant faulting. In the interior of the pluton, many rivers have carved out pronounced northwest–southeast and north–south valleys over parts of their courses where the granite had been weakened by fracturing. However, lateral or vertical displacements cannot be demonstrated. Clear evidence of faulting is restricted to the granite boundary. Small faults, mainly with northwest–southeast strikes, cause displacement at several localities along the northern margin of the pluton (*Figure 5.4*). Others will doubtless be detected in the future when country rocks adjacent to other regions of the pluton are investigated. Nevertheless, two zones of major dislocation are easily recognised. One running southeast from Sticklepath, through Chagford and the Bovey Valley is thought to consist of a plexus of subparallel and *en échelon* faults. It displaces the northern granite boundaries dextrally some 2,000 m in the vicinity of Sticklepath and Ramsley. At the southern end, the zone encompasses the Tertiary Bovey Basin and extends into Tor Bay. The dislocation passes northwest to the north Devon coast, causing dextral displacement of similar proportions in Permian rocks of the Crediton Trough. An accumulation of 2,000 m of fluvial

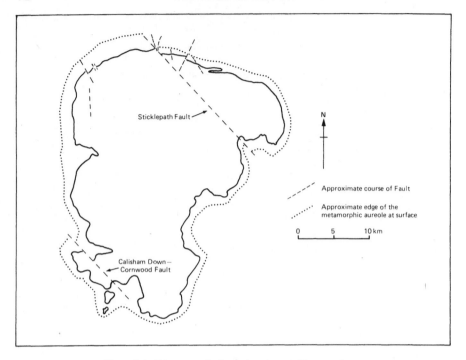

Figure 5.4. Dartmoor: faults and metamorphic aureole

Oligocene sediments in another basin near Petrockstowe reinforces evidence of instability during Tertiary times, but some observers believe initial movement may have occurred as early as Carboniferous time (Chapter 4). The second major fault zone strikes in the same direction, displacing the south-western tip of the granite between Calisham Down and Cornwood, again dextrally by about 2,000 m. There is no definite evidence here of vertical movement, although this cannot be discounted.

The Rock Types of the Dartmoor Granite

From a study encompassing all the Carboniferous granite intrusions of South West England, Hawkes and Dangerfield (1978) proposed a broad textural classification for the rock types based on a clearer understanding of their field relationships. The most important difference with respect to the divisions Brammall and Harwood recognised in the Dartmoor mass lies in an amalgamation of the 'giant' and 'blue' granites into a single megacrystic variety. This has been done on evidence gained from percentage counting of large feldspar crystals in the three granites, Dartmoor, St Austell and Land's End, where 'giant' and 'blue' varieties are particularly well developed. In each pluton, it transpires that the feldspar megacrysts range from a few scattered individuals amounting to less than one per cent of the rock by volume up to numbers locally constituting as much as 60 per cent. Increases or decreases between the large-feldspar contents of individual granite outcrops appear everywhere to be gradational, making it possible to construct feldspar megacryst distribution maps like the one for Dartmoor shown in *Figure 5.5*. There is no single exposure on Dartmoor large enough to demonstrate a significant range in feldspar megacryst content, but exposures in the coastal section of the Land's End granite between Porth Nanven (SW 355308) and Gribba Point (SW 355302) show complete gradation from rock with less than one per cent of megacrysts to rock containing about 15 per cent.

'Giant' and 'blue' granites form over 90 per cent of the Dartmoor pluton's exposed portion. Because field mapping shows them to be variants of a single intrusive phase, the mass should no longer be regarded as composite in the normally accepted sense. It is appropriate now to consider the rock types in more detail.

The principal textural materials recognised in the Variscan granites of South West England are:

1. coarse granite
2. medium-grained lithium-mica granite
3. fine-granite.

Coarse granite can best be defined as rock in which the mean grain size of ground-mass constituents is about 2–3 mm. In the case of Dartmoor, the coarse granite is characterised for the most part by large white feldspar crystals ranging in length from around 10 mm to 170 mm, and comprising between approximately one and 30 per cent of the rocks by volume. As the percentage of megacrysts increases, so does their mean length.

The general disposition of the megacrysts in the pluton, shown in *Figure 5.5*, indicates that rock with few feldspar megacrysts is not confined only to lower tracts of ground as originally believed by Brammall and Harwood. It actually forms the highest northwestern parts of the moor around High Willhays (SX 580892), Steeperton Tor (SX 618887) and Cut Hill (SX 598828). Another large area of poorly megacrystic granite occurs in the southeast bounded by Legis Tor (SX 571656), Rook Tor (SX 602616) and Collard Tor (SX 556621). Other small patches outcrop in the central parts of the pluton, to the north and west of Hay Tor, and where the River Dart flows from the granite onto country rocks (SX 696714). At Hemerdon Ball (SX 571582) a small granite boss isolated from the main mass by a thin sedimentary cover is also composed of poorly megacrystic material.

Exposures adjacent to areas of poorly megacrystic granite show gradual increases in the numbers of large feldspar crystals, with granitic rocks in the eastern, south-central and extreme western parts carrying between 10 per cent and 30 per cent by volume. It is as if this richly megacrystic material forms an incomplete capping to poorly megacrystic granite.

To the southeast of the Calisham Down-Cornwood fault, poorly megacrystic granite grades laterally into another variety (*Figure 5.5*). The mean grain size of groundmass constituents is about 2–3 mm, so the rock meets the definition of coarse granite. It differs, however, in the size of feldspar megacrysts which are numerous, relatively closely packed and with a mean length in the range 15–20 mm. Although numerous, the feldspar megacrysts, because they are small form no more than seven per cent of the rock. Poorly megacrystic granite commonly tends to grade locally into material of this sort, but nowhere else is it developed on any scale. Reasons for the textural difference are not understood. Nevertheless they are worth seeking because most of the

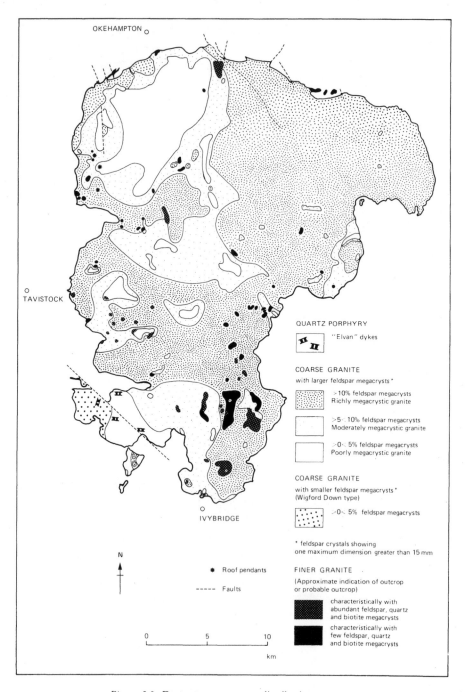

Figure 5.5. Dartmoor: megacryst distribution map

coarse granite forming the Bodmin, Carnmenellis and Isles of Scilly plutons (*Figure 5.2*) is also the smaller-feldspar type. Textural alternation on this regional scale must have some significance with regard to the complex physical conditions existing while intrusion and crystallisation of the six major granite bodies was taking place.

Mineralogically, the two coarse varieties of granite are very similar, their principal constituents being feldspars, quartz and biotite. Muscovite, tourmaline and chlorite are the most conspicuous of numerous accessory species, which also include apatite, monazite, zircon, garnet, fluorspar, ilmenite, magnetite, rutile, topaz and cordierite. Feldspars form up to 65 per cent of the rocks, quartz roughly 30 per cent, and biotite 5–7 per cent. In general accessory minerals account for less than one per cent by volume.

The main constituents are easily recognised in hand specimen. Feldspars typically are opaque and greyish white, although locally they may have a greenish hue, or are reddened where adjacent to mineral veins. Two types occur, but identification is difficult except by microscopic examination of rock thin sections. Plagioclase feldspar crystals vary from small grains less than one millimetre in diameter to well-formed prismatic crystals up to 25 mm long. Compositions are chiefly in the albite-oligoclase range. The other type, potassium feldspar, has developed by solid-state replacement of plagioclase in such a way that host crystals may show any stage between minor and complete alteration. Commonly, the largest plagioclases became sites at which replacement proceeded beyond the original crystal boundaries through mergers with matrix feldspar and the simultaneous expulsion of matrix quartz and mica grains. A majority of feldspar megacrysts were formed in this manner and, as most contain some exsolved plagioclase, they are properly termed perthites. X-ray studies indicate that potassium feldspar is the structural form orthoclase, and the exsolved plagioclase low-temperature albite.

Because of the secondary, replacive nature of the potassium feldspar, the amounts present in individual granite specimens vary between roughly 30 per cent and 40 per cent. Plagioclase constitutes 20–30 per cent of the rocks by volume.

Quartz occurs as irregular, translucent, grey crystals ranging from less than one mm in diameter to composite megacrysts up to 25 mm across. It seems from thin section examination that these composite structures also developed in the solid state, probably while the replacement of plagioclase was taking place.

Biotite mica, the only other major constituent of the coarse granites, forms dark brown flakes typically 0.5 mm to about 4 mm across. Many are ragged and they appear to have suffered corrosion. Most contain numerous inclusions of apatite, zircon and monazite. Monazite and some zircon are surrounded by pleochroic haloes. The haloes mark the extent of damage in the mica's atomic lattice caused by α-particles emitted during radioactive decay of uranium and thorium present in the two accessory minerals.

Partial alteration of biotite to colourless muscovite is common, notably in granite adjacent to mineral veining. In one southwestern area bounded approximately by Legis Tor (SX 571656), outcrops in the River Plym (SX 576654; SX 572652; SX 575653) and Trowlesworthy Tor (SX 577646), biotite shows extensive replacement by muscovite. It appears that iron expelled in this mineralogical transformation may have been dispersed as finely divided haematite which causes a general reddening of plagioclase in the granite thereabouts.

Most samples contain a little muscovite resulting from the alteration of plagioclase, and chlorite that has replaced biotite. Near mineral veining, chloritisation of both biotite and plagioclase is locally marked. For example granite, in a disused railway cutting (SX 787821) at Casely Court, has a distinct greyish-green colour due to chloritisation connected with the passage of mineralising fluids. The plagioclase there shows signs of kaolinisation as well.

Tourmaline is generally the most conspicuous accessory mineral in the coarse granites, occurring dispersed, or clustered as small, dark brown or black prismatic crystals up to a few millimetres in length. It also occurs intergrown with quartz in pods within the granite one to several centimetres across. None of the pods seem to be linked with tourmaline veining, which is another common feature of the pluton. The tourmaline in these patches, and the crystals present in the granite matrix look to have developed through a metasomatic process involving localised partial destruction of biotite, muscovite, plagioclase and, in places, orthoclase. Vein tourmaline is usually associated with quartz in sub-vertical or high-angle structures a few millimetres thick. Many can be seen in what are now open joints. Some major quartz veins, up to 1 m thick and generally striking east–west or east-northeast–west-southwest, carry tourmaline and, locally, cassiterite and haematite. These and the finer veins post-date the development of quartz and orthoclase megacrysts and result from mineralising processes which affected South West England over a lengthy period following the original unroofing of the Dartmoor granite possibly in early Permian times (Dangerfield and Hawkes, 1969; Hawkes, 1974).

Of the other accessory minerals, only garnet, cordierite and fluorite are clearly visible in hand specimen, but they are rare. Garnet forms imperfect, rounded, dark red crystals up to five millimetres in diameter and cordierite, which commonly has been replaced by aggregates of finely divided mica and chlorite flakes (pinnite), typically measures up to about 10 mm in length. Both minerals occur within the granite. In contrast, visible fluorite appears rarely as small colourless, or very pale green, crystals coating joints.

Medium-grained, lithium-mica granite characterises parts of the St Austell pluton and much of the Godolphin mass (*Figure 5.2*), but is not developed at surface in any of the other granites. However, to the northwest of Dartmoor, country rocks are cut by a granitic dyke made of a lithium bearing material similar to that found in the St Austell and Godolphin bodies. The dyke is approximately 12 m wide and dips southeast at 50°. It strikes northeast from the area of the old ice ponds on Sourton Tor (around SX 544901) to the 'Granulite' Quarries (SX 567922) near Meldon village. The presence of thin stringers in rocks exposed in the British Rail quarries (around SX 572927) suggests an overall length of at least 3.5 km. Because the present erosion level only just intersects the irregular upper extremities of the body, outcrop is sporadic. This point is exemplified in the smaller of the two 'Granulite' quarries where the dyke fingers out upwards in contact-altered rocks of the Meldon Chert Formation.

Over most of its width the dyke consists of a whitish aggregate of quartz, albite and orthoclase, with flakes of lithium-micas providing a pale-brown speckled appearance. The micas are variants of the species lepidolite, and topaz, and green and pink lithian tourmaline are abundant accessory constituents. The same minerals occur in the hanging wall and footwall zones of the body, but the rock there is fine-grained and resembles white marble. Mineralogical banding is commonly developed, and scattered small vein-like structures carry coarser crystals of quartz, feldspar, lithium-mica, tourmaline and, locally, pale pink petalite. In addition to those already mentioned, the Meldon dyke contains a rich variety of accessory minerals, including purple fluorite, apatite, garnet, various clay minerals, the zeolites heulandite and stilbite, prehnite, axinite, cordierite, also small amounts of amblygonite-montebrasite, spodumene, pollucite, columbite, and at least five beryllium-bearing species. Finding most of these minerals is difficult because of the fine to medium grain size of the rocks.

The fine granites show considerable variation in form, texture, modal composition and in groundmass grain size. Mean groundmass grain size ranges in fact from around 0.06 mm to 0.6 mm. Broadly, the rocks divide into those characterised by numerous megacrysts of feldspar, quartz and biotite, and those in which megacrysts are rare. Megacryst rich material typically contains between 10 per cent and 15 per cent of biotite: the biotite of megacryst poor types normally lies between one and five per cent. As a consequence, the former rocks are darker, and the latter paler in colour than the coarse granites.

Megacryst rich, fine-grained granite equates with the 'dark inclusions' and 'basic segregations' first recognised by Reid et al. (1912). The material occurs in the form of small rounded pods perhaps 5–20 cm in diameter, and as more irregular masses ranging from, say, less than a metre to as much as a kilometre in maximum extent. Thicknesses in the larger bodies may exceed 150 m.

Modal data for rocks of this sort are not available, but visual inspection indicates appreciable mineralogical variation. This feature makes it unlikely that such material represents the incorporated remnants of an earlier dioritic phase as suggested by Brammall and Harwood (1923). In a typical specimen, the fine-grained matrix consists of poikiloblastic quartz, subhedral plagioclase showing partial replacement by potassium feldspar, biotite, and numerous accessory minerals which include tourmaline, monazite, zircon, ilmenite, rutile, apatite, topaz and garnet. Quartz megacrysts, commonly rounded, are smaller than those in the coarse granites and measure no more than 10 mm across. Feldspar megacrysts are also smaller and very variable in numbers. The largest are about 60 mm long. Most are perthite, but some consist of partly potassium feldspathised plagioclase. Biotite megacrysts in this type of fine granite have no counterparts in the coarse granites. The conspicuous dark brown flakes are up to 3–4 mm across.

Contacts against granite are sharp and the smaller pods in particular have every appearance of being xenoliths. By analogy, the larger masses are probably inclusions, a point reinforced by the occurrence of megacrystic granite veining in one body exposed in the river valley known as Newleycombe Lake (SX 594700). Richness in biotite suggests that the material may originally have been argillaceous sedimentary rock. However, shale xenoliths derived from surrounding contact-altered rocks are common in exposures of the megacrystic granite, but they are quite unlike the megacryst rich fine granite. A sedimentary origin for the latter material would seem to imply some form of metasomatic alteration before incorporation within the granite magma.

The megacryst poor fine-grained granites show a greater variety of forms; sheets interlayered with megacrystic granite, more clearly defined sill-like and dyke-like bodies, small rounded pods and larger irregular bodies. All are characterised by a general paucity of megacrysts and biotite when compared with the coarse granites and fine-grained, megacryst-rich, material.

Larger, irregular bodies are exemplified by the one in the east face of Yes Tor (SX 581902) which is approximately 13 m long and at least 2.5 m thick. Its contacts are sharp and the margins locally finer grained than the interior. Rare feldspar megacrysts lie across the junction with surrounding granite. Similar features are evident in some dyke-like and sill-like bodies, although not all show fine-grained margins. 'Dykes' range from about 3 cm to 1 m in thickness, 'sills' from roughly one centimetre to 25 m or 30 m. Little work has been done on modal compositions, so the question of their origin remains unsettled. Superficially, they resemble igneous intrusions, but the occurrence of feldspar megacrysts lying across contacts indicates that present relationships with the granite were established before the period of large feldspar growth. Further doubts as to an igneous parentage are cast by internal variations in grain size and the local presence of biotite rich layers. A few sill-like bodies display finer, apparently chilled upper contacts, yet their lower margins grade imperceptibly into coarse granite. In others, upper contacts show sporadic developments of pegmatite. Coarse-grained perthitic feldspar and quartz are generally the principal minerals. Muscovite, tourmaline and biotite can occur as accessory constituents.

Fine-grained granite sheets interlayed with coarse granite range from a few centimetres to about 2 m in thickness. Rarely can their lateral extent be determined, but this is certainly of the order of several metres. Passage into granite above and below can have an appearance of being transitional. It is possible that this material and also the small isolated pods of fine-grained granite were originally sandstone prior to incorporation within the granite magma. However, no criteria have been observed to confirm this proposition. At present, the nature of all the megacryst poor fine-grained granites remains enigmatic.

Another enigmatic rock type is the hornblende bearing pegmatite described by Reid *et al.* (1912). It is found only as boulders on Bittleford Down (around SX 708756). Field relationships with the granite are unknown. The rock is composed mainly of relatively coarse-grained potassium feldspar and quartz, with subsidiary plagioclase, and sphene, apatite and a dark-green mineral as the chief accessory constituents. Originally identified as hornblende, X-ray study has indicated that the dark mineral has a pyroxene structure. There are no other records of pyroxene bearing rock among the Variscan granites of South West England.

Xenoliths

Shale xenoliths, typically from 10 to about 75 mm long and roughly ovoid in form, are common in many outcrops of megacryst rich coarse granite. Hay Tor (SX 757770) is a particularly good locality in which to study them. Most xenoliths are rich in biotite and may carry andalusite and/or cordierite in addition to quartz, plagioclase and potassium feldspar. Accessory minerals can include apatite, monazite, zircon, rutile, ilmenite, tourmaline, topaz, garnet, corundum and xenotime.

Although aureole rocks such as sandstone, limestone and mafic igneous material might be expected to occur as xenoliths in the granite, none is easily recognised. It is possible that the isolated pods of generally megacryst poor fine-grained granite were originally sandstone or silty sandstone, and certain biotite-plagioclase rich patches and schlieren may represent vestiges of included basalt or dolerite.

METAMORPHIC AUREOLE

Rocks surrounding the Dartmoor pluton were baked for horizontal distances of between 0.75 km and about 3.25 km, depending on the general inclination of the contact. From a chemical point of view and because of an original, relatively high-temperature origin, mafic igneous material responded most readily to the thermal stimulus. In the inner parts of the aureole, these rocks show stages in reorganisation towards hornblende-plagioclase hornfels, an assemblage characteristic of the hornblende-hornfels facies. The same degree of metmorphism is indicated by minerals like tremolite, garnet, diopside and hedenbergite which are developed in certain calcareous sedimentary rocks. Thus temperatures and water vapour pressures in the inner regions of the aureole may have been respectively of the order of 500–600°C and 1–2 kb.

It is unfortunate that water vapour pressures cannot be used to extrapolate the likely hydrostatic pressure at which the mineralogical transformations occurred. Knowledge of this particular parameter would allow deduction of depth of emplacement and hence the thickness of rocks originally overlying the pluton. However, the extent of the aureoles suggests that water vapour pressures in the range 1–2 kb were maintained for some time, and that during this period no open channel-ways to the surface existed. Confining pressure may therefore have been considerable implying, perhaps, a cover of at least 1–3 km.

Only an outline of the metamorphic changes can be given here. Detailed descriptions of the rock types and of locations where they can be seen are contained in the Geological Survey Memoirs for the Ivybridge (Ussher, 1912), Dartmoor Forest (Reid et al. 1912) and Okehampton (Edmonds et al., 1968) districts.

In the south, the Dartmoor granite penetrated folded sequences of Devonian rocks consisting chiefly of argillaceous and silty rocks, with scattered calcareous horizons and sporadic tuff, lava and intrusions of mafic igneous parentage. To the north, shale, siltstone, sandstone, lenticular limestone, chert, and intermediate and basaltic volcanic rocks of Carboniferous age were involved. All the rocks had suffered the effects of regional metamorphism before being baked by the granite, and belonged to either the high zeolite or low greenchist facies. This meant that the intermediate and basaltic rocks had been respectively downgraded to keratophyre and spilite.

Metamorphic changes seem to have been accomplished mainly by the rise in temperature and by circulation of hydrous fluid derived from the magma. However, acquisition of some potassium is indicated by the widespread occurrence of biotite in spilites and keratophyric rocks. It may be that the growth of this biotite correlated with the period of autometasomatic potassium feldspathisation of plagioclase inside the granite; the mica certainly post-dates the development of hornblende in the aureole mafic rocks. Patchy growths of tourmaline and axinite late in the sequence of metamorphic changes suggest also an influx of boron from the granite.

In the outer parts of the aureole, slates are spotted. Closer to the granite, they were converted to quartz-biotite-andalusite (chiastolite) and quartz-biotite-cordierite hornfelses. As the argillaceous material was generally richer in illite rather than chlorite, andalusite (chiastolite) bearing types are the more common. Silty rocks show similar mineralogical changes, but are richer in quartz.

Changes are less obvious in sandstones. Apart from recrystallisation of quartz, the chief response appears to have been a conversion of clay mineral impurities into flakes of muscovite. Locally, growth of biotite occurred as well.

Devonian sedimentary rocks on the west, south and east sides of Dartmoor contain interbedded argillaceous and silty calcareous horizons. Within the aureole they were converted to fine-grained, calc-silicate hornfelses bearing (in a variety of combinations) garnet, epidote, amphiboles, hedenbergite, feldspar, quartz, calcite, muscovite, biotite, sphene and, rarely, scapolite.

Similar assemblages were developed along the northern margin of Dartmoor in an interbedded sequence of Carboniferous chert, shale, lenticular limestone, and calcareous and siliceous mudstone. The hornfelses there are commonly much coarser grained, and in addition to the minerals seen in Devonian rocks, species such as grossular, andradite and spessartite garnet, wollastonite, diopside (salite), idocrase, bustamite, rhodonite and tephroite occur. Only the purer limestones in these sequences show little change other than the recrystallisation of calcite.

Prior to the baking, Carboniferous and Devonian spilitic rocks consisted of albite, relict clinopyroxene, uralitic amphibole, chlorite, epidote, sphene, ilmenite and probably prehnite. Textures in contact-metamorphosed rocks vary because the parent materials ranged from ophitic dolerite, through basaltic lava to tuffs, but the ultimate products are completely recrystallised hornblende-andesine-biotite hornfelses with and without diopside, and biotite-andesine hornfels containing accessory ilmenite, sphene and apatite.

The keratophyric rocks, chiefly agglomerates and crystal-lithic tuffs, are interbedded with argillaceous material. Contact metamorphism converted their zeolite/greenschist albite-chlorite-quartz assemblage to one consisting chiefly of albite, biotite and quartz. Cummingtonite, anthophyllite and cordierite occur locally in the more argillaceous layers.

Axinite and tourmaline can occur in any of the rock types, apparently as late-stage replacements of other minerals, or in irregular veins. Tourmalinisation is locally extensive. For example, Carboniferous siltstone and sandstone near the granite contact on Mardon Down (SX 775883) have been converted to banded quartz-schorlite hornfels.

Another interesting mineralogical development is the occurrence of sulphides in Lower Carboniferous rocks at various localities around the northern margin of the pluton. The sulphides appear chiefly in the calc-silicate hornfelses as disseminations and small irregular vein fillings. Pyrrhotite is the most abundant mineral species; others found in smaller quantities include pyrite, chalcopyrite, arsenopyrite, sphalerite and löllingite. Bornite, molybdenite, gersdoffite and bismuthinite have also been recorded, but are rare. The metals contained in these minerals probably came from both the granite and the country rocks.

THE AGE OF THE GRANITE

The youngest rocks cut by the Dartmoor granite are Namurian sandstones and shales of the Crackington Formation. However, fine-grained granite pebbles, xenolithic material and feldspar fragments derived from the granite are locally common in adjacent red-bed breccias now believed to be of Stephanian age. Unfortunately, direct dating of the granites in South West England by isotopic means has not been entirely successful. For instance, potassium-argon (K/Ar) data show an apparent spread of ages between about 303 Ma and 265 Ma (Miller and Mohr, 1964). Recent changes in the constants used to calculate K/Ar ages means that these figures should be revised to about 309 Ma and 271 Ma. The most reasonable way of interpreting this information may be to assume that the figure of about 309 Ma (late Westphalian) represents the approximate date of intrusion and that, following the temporary unroofing of the Dartmoor granite in Stephanian times, periodic heating of groundwater during subsequent volcanic episodes was sufficient to open the various radiometric decay systems (Hawkes and Dangerfield, 1978).

GRANITES IN FOLD BELTS

Large Cretaceous and Tertiary batholithic complexes, commonly consisting of diorite, granodiorite and quartz monzonite (adamellite), occur in several of the Western States of the USA: California, Colorado, Idaho, Montana, Oregon and Washington. The underlying tectonic situation is complicated though by a passage of en échelon strike-slip (transform) fault displacements of the East Pacific Rise beneath the Californian portion of the continental plate. Because of this complex interaction between the mantle and both continental and oceanic crust, the region is an excellent one in which to investigate the origins of granite.

Conventional hypothesis for the genesis of granite have developed from studies based primarily on older fold belts, particularly those in Europe, and they were made before the theory of plate tectonics was developed to explain the Earth's major crustal features. These hypotheses centred around two main propositions: firstly differentiation of basic, or possibly ultrabasic magmas and, secondly, the partial melting of sedimentary and volcanic materials which accumulated in unstable, sinuous geosynclinal belts. Neither view completely satisfied the requirements of all field and analytical observations made on granitic rocks.

Gabbroic and intermediate igneous rocks commonly occur in fold belts alongside granites. Chemical data from these rocks appear to demonstrate a general trend (the calc-alkali trend) which has been taken by some observers to indicate a common mafic igneous parentage. Direct evidence of differentiation is rare, however, and where mafic rocks are seen to grade locally into intermediate or silicic types as, for instance, in the Caledonian Insch mass of Aberdeenshire, the proportion of less dense material is relatively small. The overall ratio of mafic to silicic rocks is exactly the reverse in fold belts. It should be mentioned too that, in the Aberdeenshire example, assimilation of sialic country rock may have played an important part in the process.

There is more substantial evidence favouring partial melting in geosynclinal sequences. Experimental pressure-temperature data on mineralogical transitions suggest that release of liquids of granitic composition is possible when folding results in metamorphism, and the degree of metamorphism passes through the amphibolite facies into the granulite facies. Amphibolite and granulite-grade rocks are a common feature in older, more deeply eroded orogenic belts.

Metamorphic terrains bearing streaky granitic intercalations (the regions of gneiss and migmatite) are considered by many observers to be, or to be close to, the source regions of magma production. Once formation has begun, granitic liquids are thought gradually to coalesce and rise through crustal rocks because of their low density. Because of enrichment, particularly in sodium and potassium, it is assumed that the liquids are sufficiently reactive to cause partial assimilation of materials through which they pass. The enormous bulk of batholithic complexes is explained in this way.

Exposed batholiths show crystallisation to have been a complex multi-stage process. Individual groups of plutons may demonstrate an apparent compositional sequence of emplacement ranging from early dioritic types to later adamellites or leucocratic granites.

Looking afresh at the problem of magma production in the Western United States, Hamilton and Myers (1967) emphasised that the batholithic complexes there do not invariably occur in geosynclinal settings. For instance, Tertiary granitic rocks in Colorado invade the Precambrian basement and the Palaeozoic and Mesozoic sedimentary rocks which form part of a stable platform. Certain Mesozoic granites in California similarly appear in old Precambrian basement rocks. The origins of the Tertiary plutons of the Cascade Range present further difficulties in accepting a simple geosynclinal hypothesis. Plutons were mostly injected through pre-Tertiary crystalline rocks, but in the central Cascade region it seems the igneous activity was built on a basement of oceanic material. Likewise, Tertiary granites in some of the Aleutian Islands intrude andesitic, basaltic and dacitic volcanic rocks which constitute an island arc structure, where continental crust is unknown.

From these examples and others cited in their paper, Hamilton and Myers inferred that granitic magmas can arise by means other than the partial melting of the materials in geosynclinal sequences. They also inferred from initial $^{87}Sr/^{86}Sr$ isotope data that the ratios in granites are such as to rule out derivation either wholly from basalt-producing regions in the mantle or wholly from the melting of continental crust.

As an alternative to the two conventional general theories, Hamilton and Myers have put forward a series of suggestions which also took account of views expressed Gilluly (1965). They reasoned on the basis of petrological and seismic data that the upper mantle and lower crust are broadly of basaltic composition, with the dividing line between the two (the Mohorovičič discontinuity) probably marking a transition from basalt containing appreciable olivine below, to basalt with little or no olivine above. It has also been established, from experimental work on mineral phase relationships, that progressive melting of plagioclase and pyroxene (but not of olivine) can occur at pressures characteristic of the upper mantle and lower crust. Provided the appropriate temperature stimulus exists, and if ambient pressures are lower than those required by the basalt-eclogite transformation, plagioclase will melt preferentially to produce the high-alumina basalt, andesite and dacite magmas which, as Hamilton and Myers emphasised, typify island arc and some geosynclinal situations.

Where the crust is relatively thin, as in oceanic areas, this form of melting can occur at some depth in the mantle before pressures are at the level of the basalt-eclogite transformation. However, in continental regions, pressures corresponding to this transition are likely to exist near to or above the Mohorovičič discontinuity depending on the thickness of superincumbent sialic materials. Thus high-alumina basalt, andesite and dacite magmas originating in these regions may arise from basalt in the lower crust rather than in the mantle.

Aware that the more silicic portions of magmas generated in this way could hardly account for the sheer volume of many batholithic complexes seen in the Western States, Hamilton and Myers developed their theme by introducing a broad correlation, noted also by Gilluly: the correlation is between an increasing proportion of silica rich and potassium rich rocks and the eastwards thickening of the continental crust. A similar chemical correlation can be observed with respect to altitude of emplacement. Both Mesozoic and Tertiary igneous suites appear to show these relationships which may be evident too in some individual complexes.

To account for these observations, Hamilton and Myers reasoned that magmas derived either from the upper mantle where the Mohorovicic discontinuity is shallow, or from the lower crust where it is more deeply depressed, must be capable of modifying their composition by wholesale selective assimilation of materials through which they rise. How closely the more silicic components of the magmas approach highly leucocratic compositions could depend partly on the general composition of materials assimilated, and particularly on the length of their upward passage. The thicker the crust, the more leucocratic the residual magma composition.

The concept of zonal melting helps to explain the process in terms of energy transfer. Given that an initial magma will rise simply because it is liquid, and thus less dense than surrounding and overlying solid materials, it becomes subject at once to a pressure gradient which will increase the more it rises. A consequent upward fractionation of the most volatile components results in the progressive lowering of melting temperatures at the ascending boundary of the evolving magma system, compensated by an elevation of melting temperatures at the bottom. Set in motion originally by gravitational means, such a system will become more vigorous as selective fusion at higher levels incorporates increasingly larger quantities of low-melting sialic components, while exothermic crystallisation of mafic constituents proceeds at the base. Crystallisation in all parts of the system will be achieved when heat loss by conduction in the upper crust finally exceeds all further input through crystallisation itself.

There is another important aspect to the views expressed by Hamilton and Myers. They thought that when a batholithic system of this sort develops, a surrounding envelope of metamorphism causes crustal materials lying in its path to move aside, flow down and perhaps to close underneath as the detached igneous mass passes to high levels. These rocks, revealed by erosion, are the foliated gneiss and migmatite terrains considered by many observers to be the source regions of granite magmas. Instead they could, in some cases, be products of regional metamorphism related to batholithic complexes now largely or completely lost through erosion.

This approach to the genesis of granites in fold belts has a broad appeal and certainly reconciles more observations than any other single hypothesis. Not least among the observations are the $^{87}Sr/^{86}Sr$ isotope data which seem to favour a mixed origin. It is also possible to imagine circumstances in which particular plate configurations could provide the necessary deep-seated thermal stimulus for the formation of initial magmas, at the same time exercising overall tectonic control.

ORIGIN OF THE DARTMOOR GRANITE

Most observers have favoured a palingenetic origin, that is, derivation of granitic magma from the melting of crustal materials at depth (Brammall and Harwood, 1932; Exley and Stone, 1966). Compression during the Variscan Orogeny is thought to have caused partial melting in basement rocks and the evolution of liquids of ternary minimum composition. Following the assimilation of large quantities of overlying rock, it has been assumed that the bulk of the liquid phase crystallised at moderate depths, chiefly as granodiorite, leaving residual material enriched in water and potassium to migrate higher and eventually to crystallise as the adamellitic granite plutons of Dartmoor and Cornwall.

Geophysical data certainly indicate a large mass approximating to granite in composition underneath the peninsula and extending westwards beyond the Isles of Scilly. According to seismic information, the Mohorovičič discontinuity lies at a uniform depth of 27 km (Bott et al., 1970). Further evidence apparently supporting a crustal origin comes from the high $^{18}O/^{16}O$ ratios found in granite samples at outcrop (Sheppard, 1977).

There are certain factors which suggest caution in accepting this general picture. One concerns the way in which the Dartmoor granite crystallised. Ternary minimum theory envisages that, given steady pressure conditions and a falling temperature, granite magmas differentiate to produce a residual liquid with a composition that, on crystallisation, yields quartz, potassium feldspar and sodic plagioclase in roughly equal proportions. However, in the case of the Dartmoor magma, textural evidence indicates that primary crystallisation produced a crystal mush consisting of plagioclase, biotite and quartz, with the replacement growth of potassium feldspar (and muscovite) resulting later from interaction with contained aqueous fluid. This sequence of events suggests crystallisation under isometric or adiabatic conditions rather than those predicted by ternary minimum theory. Physical conditions of this sort imply a comparatively rapid ascent of the magma, possibly from deep crustal regions. Conceivably, it may have had to find its way through the large mass of granitic material believed to exist at depth.

A second factor stems from consideration of possible plate configurations. One view is that the South West England batholith resulted from crustal melting associated with the development of a long-standing shallow northward-dipping Variscan subduction zone in the region now marked very approximately by the Mediterranean sea. This simple collision model could provide for the granites on a palingenetic basis, but overlooks an absence of many features characteristic of perpendicular ocean-continent plate convergence. Perhaps the most important of these is the general lack of andesite-rhyolite-dacite volcanism in western Europe. The alternative plate model invokes complex lateral and rotational movements between west European microplates along major strike-slip (transform) faults, but an explanation for the granites of South West England is still problematical, mantle activity rather than Variscan subduction being a possible causative influence.

Uncertainty as to the origin of the South West England granite magmas arises naturally from a lack of detailed knowledge about the deeper geological environment. There is doubt, too, about a possible extension of the pluton chain eastwards. For example, granite intrusions may occur in the arcuate ridge of probable Variscan material connecting the Start–Torbay region with the Contentin peninsula, Normandy (Smith and Curry, 1975). If so, genetic theories for the South West England granites must also account for the Carboniferous masses at Barfleur and Flamanville, and indeed the similar intrusions of Brittany and the Vendée. It should be added that even the solitary experimentally determined $^{87}Sr/^{86}Sr$ ratio of 0.7067 (actually for the St Austell granite, and subsequently modified by a later computational method to 0.7086) conspires in this general uncertainty because it favours neither crust nor mantle as a source.

POST-OROGENIC VOLCANISM

Contrary opinions have been expressed as to whether volcanic activity accompanies the intrusion of granite batholiths. In the Andean Cordillera and again in the western United States, there is no doubt that magmatic material reached the surface. Dacitic lava, pyroclastic rocks and welded tuffs are the commonest types, which accord with the overall dominance of granodioritic plutons. Where adamellites and leucocratic granites occur, the associated volcanic materials are typically of rhyodacitic or rhyolitic composition. Andesitic rocks in some cases accompany dioritic intrusions.

Examples are not confined only to the more recent fold belts. Hamilton and Myers (1967) quoted the case of a Precambrian complex in Missouri where a roof of rhyodacitic and rhyolitic volcanic materials was invaded by adamellite and leucogranite. Contemporary volcanism also seems to have been fairly widespread in the Scottish Highlands during intrusion of the many newer Caledonian granites.

The same suite of rocks apparently marks post-orogenic volcanicity in many regions. From a survey of fold belts in eastern Europe and the USSR, Ustiyev (1970) concluded that andesite, dacite and rhyolite are the commonest volcanic products. He referred to outpourings of tholeiitic basalt as being characteristic of end stage activity in some instances. However, Ustiyev admitted that insufficient attention has been paid in the past to post-orogenic volcanism. In fact, in classical geosynclinal theory, high-level plutonic rocks such as the Dartmoor granite and many of the newer Caledonian granites in Scotland would also be assigned to this phase of igneous activity. Dividing lines are not necesarily clear cut and with the introduction and wider application of plate tectonic theory, it is becoming clear that sequences of igneous activity are not always the same.

In Devon, for example, there is no evidence of volcanicity of andesite-dacite-rhyolite affinity, nor of the tholeiite flood-basalt type, either accompanying the intrusion of the granite, or following later. Nevertheless, volcanism did occur in Stephanian and early Permian times, eruptions producing lamprophyric lavas and lesser quantities of potassium rich basalt and rhyolite. The trace-element content of these rocks differs strikingly from that of the granite, suggesting an entirely separate evolutionary path.

At least three other volcanic episodes occurred in South West England prior to development of the Tertiary igneous complex of Lundy (discussed in Chapter Ten). Only one can be recognised in Devon, however: the Dartmoor granite and country rocks are cut in a few localities by quartz porphyry dykes of Lower Permian age. Similar dykes appear in greater numbers to the west. Evidence of the subsequent outburst is at present restricted to lavas encountered in North Sea drilling and to the Fullers earth of the Bath district. This Middle Jurassic deposit was water-laid, but consists predominantly of altered tuff and vitric material originally of a mainly trachytic composition (Jeans *et al.*, 1977). Then, in Lower Cretaceous times eruption of phonolite occurred in the vicinity of Wolf Rock and Epson Shoal, a few kilometres south-southeast of Land's End.

Except perhaps in the case of the quartz porphyry (elvan) dykes, each of these volcanic episodes produced rocks arguably symptomatic of crustal rifting. They may represent localised expressions of igneous activity induced by complex tectonic events related to the opening of the North Atlantic ocean basin.

STEPHANIAN AND LOWER PERMIAN RHYOLITES, BASALTS AND LAMPROPHYRES

Several observers have suggested that intrusion of the Dartmoor granite resulted in penecontemporaneous volcanism. No evidence has been found to support this proposition, but equally there is no proof that rhyodacite volcanism did not occur above the highest parts of the pluton. All signs could have been removed by erosion, and if such activity had been largely ignimbritic, the glassy materials might not have survived transport and eventual deposition in surrounding red-beds. Nevertheless, it should be acknowledged that the existence of an extensive metamorphic aureole so close to the roof region of the pluton, coupled with evidence for the metamorphic growth of feldspar megacrysts inside the granite, make the possibility of a volcanic connection with the surface unlikely.

In a renewed search for signs of penecontemporaneous volcanism, Cosgrove and Elliott (1976) proposed as possible candidates the intrusive quartz porphyry material at Withnoe (SX 404518) and the rhyolitic lava at Kingsand (SX 435506), both roughly 18 km southwest of Dartmoor. They also suggested that the matching boulders and fragments occurring abundantly in various red-bed breccias might have come from comagmatic lava fields directly overlying the Dartmoor granite. An enrichment in potassium, rubidium and strontium which all these rocks show was taken to indicate continued differentiation of the granite magma, with the additional observation that as the detrital material in the breccias contains relatively more potassium than the Kingsand Withnoe rocks, correlation with the notably potassium-rich quartz porphyry (elvan) dykes seen throughout Cornubia might solve the problem of feeders for the proposed lava field.

However, it is important to realise that the potassium enrichment in the Kingsand Withnoe rocks and breccia debris is due to a richness in biotite, while that in the elvan material results from an abundance of finely divided muscovite. In mineralogical terms, the two groups of rocks differ significantly, and it seems unlikely that either could have been produced from the granite magma by simple differentiation. There are in addition stratigraphic grounds for separating the three rock types.

The biotite rich quartz porphyry rocks at Withnoe cut Lower Devonian Meadfoot Slate, so their exact age is unknown. Fortunately, the chemically and mineralogically related rhyolitic lava at Kingsand is interbedded literally with basal red-bed sandstone and breccia; an unpublished K/Ar isotope determination indicates an age of about 295 Ma (Stephanian). On this evidence, and bearing in mind the probable late Westphalian age of the granite, correlation with mafic and lamprophyric volcanic rocks developed near the base of red-bed strata to the north and east of Dartmoor seems much more appropriate, though it is likely that the porphyries were being eroded before these lavas were erupted, as they occur in lower strata as well as higher. The stratigraphic relationship of the elvan quartz porphyry dykes will be dealt with in the next section. Nevertheless, it is worth mentioning here that no detritus from these dykes has been recognised for certain among the breccias, which carry abundant silicic volcanic material of the Kingsand–Withnoe type.

Distribution and Rock Types

The present rather insignificant outcrops of volcanic material (*Figure 5.6*) belie the original extent of the activity. Associated and overlying red-bed breccias contain enormous quantities of the volcanic detritus including boulders up to 12 m. Sporadic dykes of lamprophyric composition occur westwards at least to Chyweeda (SW 612326), near Leedstown in Cornwall, and as far north as Fremington (SS 517338), near Barnstaple. According to unpublished K/Ar data, they also yield a Stephanian age of about 295 Ma. At Horswell House, about 15 km south of Dartmoor, Ussher (1904) described a poor quarry exposure (SX 688420) where 'mica-trachyte' lava is associated with 'Permian' strata and also with rhyolitic material.

The majority of outcrops occur in four areas: to the north and west of Tiverton, between Silverton and Killerton, in red-bed strata of the Crediton trough, and from Dunchideock northeastwards towards Exeter. The rocks are mostly vesicular and scoriaceous lavas varying perhaps from less than 1 m to about 20 m in thickness. In some places they are brecciated and contain included fragments and blocks of red sandstone. Locally, agglomeratic rocks are associated with the lavas; several vent agglomerates like those near Posbury (SX 815978) and Webberton Cross (SX 885872) are known. The quartz porphyry outcrop at Withnoe is roughly circular and may also be a volcanic neck. A few feeder dykes occur, the largest exposed concentration being in

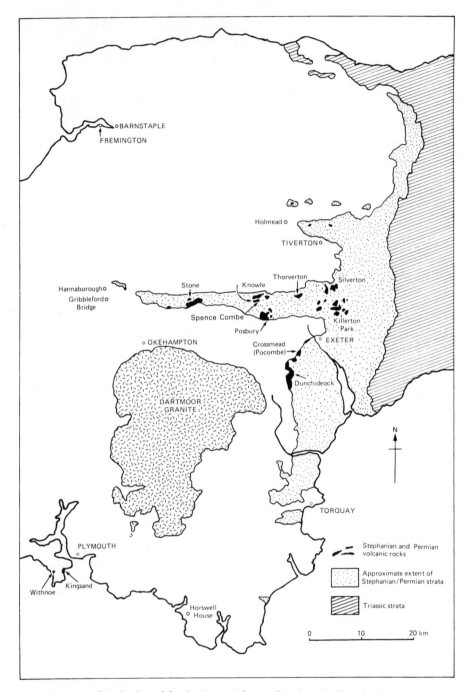

Figure 5.6. Distribution of Stephanian and Lower Permian rhyolites, basalts and lamprophyres in Devon

Culm rocks around Gribbleford Bridge (SX 526013). Similar volcanic material in the nearby Hannaborough Quarry (SX 529029) is associated with agglomerate and possibly marks the site of another neck. One small dyke merging into a lateral flow was temporarily exposed in red sandstones beside the M5 motorway near Killerton.

At some localities the lavas rest unconformably on Devonian or Carboniferous rocks; at others they are separated from this erosion surface by thin red sandstone or breccia. Thus it is clear that eruptions took place at about or soon after the time red-bed deposition began. Early K/Ar isotopic data from lava specimens indicated a basal Permian age of about 280 Ma (Miller and Mohr, 1964). However, a more recent determination suggests an age range from a Stephanian (around 295 Ma) to Lower Permian for the volcanism, and this is therefore the latest date for the onset of red-bed sedimentation.

One particularly interesting aspect of the volcanism is the variety of rock types involved. Teall (1902) was the first to attempt a systematic division, proposing four major groupings: basalt, olivine trachyte, biotite trachyte and augite minette; he pointed to similarities with 'Permian' igneous rocks in Germany and also with certain volcanic types found in Montana and Wyoming. Tidmarsh (1932) later revised these groupings, suggesting instead that the rocks formed two parallel series grading from basalt through trachyandesite to lamprophyre. After examining samples from each of the four main areas mentioned earlier, Knill (1969) thought Teall's classification the more logical, but decided to amalgamate the trachytic and minette rocks into a single potassium rich group, with the basalts constituting a group on their own; she emphasised the similarities between the potassium rich types and examples such as yogoites, shonkinites and prowersites from the western United States. Cosgrove (1972) has subsequently affirmed this general similarity, likening the Devon rocks petrographically to the absarokite-shoshonite-banakite series of Wyoming.

Clearly there have been points of disagreement amongst observers of the Devon rocks, due partly to locally extensive alteration and the consequent difficulties in determining primary mineral assemblages. However, the existence of three rock types—rhyolite, basalt, and minette—is beyond serious dispute.

The main outcrops of rhyolite are at Kingsand, in Cornwall. The rocks there are a pale reddish-buff colour and very fine-grained. They contain a proportion of glassy material and generally show flow-banding and spherulitic structures. Coarser grained silicic volcanic material at Withnoe, nearby, consists of quartz, feldspar and biotite megacrysts in a fine-grained matrix composed of the same constituents.

Reddened basaltic rocks can best be seen in quarries at Webberton Cross, near Dunchideock, and Stone (SS 681013), near North Tawton. Typically, they consist of phenocrysts of labradorite and olivine (commonly corroded and largely replaced by haematite and/or carbonate), set in a groundmass of labradorite, diopsidic augite, pyroxene which has been altered to bastite, iron oxides and zeolites. Some specimens may bear quartz xenocrysts. Coarser grained doleritic material, perhaps carrying calcite and quartz filled amygdales, occurs in several places, for example in Raddon Quarry (SS 909020) near Thorverton, and as veins at Dunsmoor (SS 955015) near Silverton.

Minette lamprophyres appear both as dykes and as lavas. Essential minerals are biotite, orthoclase, minor sodic plagioclase (oligoclase), and abundant accessory apatite. Most, however, contain at least some pyroxene (augite); others bear olivine in addition. Biotite, augite and olivine tend to form phenocrysts, with feldspars constituting the matrix. Alteration can be extreme in that many rocks now consist only of calcite, dolomite, chlorite, quartz and iron oxides. Examples of biotite minette and vesicular augite-biotite minette lavas can be found at Killerton Park. Olivine-bearing material occurs at Hannaborough Quarry. Neighbouring dykes are also olivine-bearing minettes, but most show a high degree of alteration.

Coarser grained, highly vesicular varieties of these rocks appear, for example, at Knowle Quarry (SS 783020), where altered phenocrysts of olivine and biotite, with smaller crystals of carbonated diopsidic pyroxene, apatite, sodic plagioclase, magnetite, quartz and zeolite are poikilitically enclosed in a groundmass of relatively large orthoclase crystals. On the basis of chemical data, Knill suggested that this kind of material has affinity with the yogoites and shonkinites of Montana.

Certain of the lamprophyric lavas in the Killerton Park district Knill found to be especially rich in potassium and low in sodium, indicating a resemblance to prowersite, another rock type first described from Montana. Other rare biotite and olivine bearing rocks which occur near Holmead Farm (SS 890160) contain very little potassium feldspar, carrying instead analcime and possibly leucite. One is close in chemical composition to orendite from Wyoming, but at present there is some uncertainty as to correct nomenclature.

It is appropriate to describe now a final group of lavas, because they seem to form some sort of chemical bridge between the basalts and the potassic lamprophyric material. Chemically, the rocks are trachybasalts and, although much altered, appear to have consisted of olivine and andesine/oligoclase phenocrysts, set in a fluxion textured groundmass composed of oligoclase, orthoclase, minor augite and accessory iron oxide. The best outcrops are at Posbury Quarry (SX 814978) and at Pocombe (SX 900913), near Crossmead, Exeter.

Origin

Tidmarsh (1932), Knill (1969) and Edmonds *et al.* (1968) all favoured the possibility of interaction at depth between basalt or periodotite and either solidified granite or magmatic phases of the Dartmoor pluton, to explain the highly potassic nature of many of the Stephanian and Permian volcanic rocks. However, isotope age data suggest that eruption probably occurred perhaps 10 Ma after intrusion of the granite (by which time erosion was cutting into its top surface), so this hypothesis must be considered tenuous.

More recent investigations by Cosgrove (1972) show that many of the rocks in question are enriched in so called 'incompatible elements' like phosphorus, rubidium, strontium, zirconium, barium, lanthanum, cerium and neodymium. Studies of similar potassium rich rocks elsewhere have led to a belief that their magmas originated at considerable depth in the mantle, either by the melting of oceanic material descending on subduction zones, or by less well understood processes connected with crustal rifting. In the case of the Devon rocks, Cosgrove proposed that magma genesis resulted from the melting of a

Tethyan oceanic floor descending on a low-angle (20°) subduction plane at a depth of about 300 km beneath a European continental plate. However, geological evidence for the alternative tectonic case—crustal rifting—appears far stronger.

The geological picture of Britain in late Carboniferous–early Permian times is one characterised by continental deposition in a series of subsiding, fault-bounded basins. In fact, Whittaker (1975) has emphasised the major role played by graben subsidence in the pattern of sedimentation throughout the Mesozoic. With regard to South West England, the unexplored thickness of red-bed and later Mesozoic deposits in the Bristol and English Channels, and those further afield in the Irish Sea and Western Approaches, testify to the development of very large graben structures.

The eruption of lava in Devon must have coincided, therefore, with the beginning of a period of pronounced crustal tension. This is certainly the view expressed by Smith and Curry (1975) in their structural appreciation of the developing English Channel. Genesis from subducted oceanic material cannot be ruled out, of course, but at that time active subduction seems unlikely to have been the cause. The tectonic setting of the potassic rocks of Wyoming and Montana referred to by Teall and by Knill offers further support for the tensional case. These rocks occur in regions peripheral to the hot-spring province of Yellowstone Park, where the activity is possibly related to a mantle convection plume and continental rifting.

The connections between potassic volcanism, a mantle plume and rifting in Wyoming may not seem particularly relevant to questions posed by the Stephanian lavas in Devon, unless South West England did subsequently become a hot-spring province as suggested by Dangerfield and Hawkes (1969); if this was the case, then these features may well be related to the origin of the lavas.

DYKES (ELVANS)

The granites and associated Devonian sedimentary rocks in Cornwall are cut by numerous quartz porphyry (elvan) dykes. Most strike east–west or east-northeast–west-southwest, and their thicknesses range from less than one metre to more than 20 m. Dips vary from about 45° to vertical. A few, like the intrusion on Praa Sands described by Stone (1968), are composite structures. Generally, wall zones are very fine-grained and may show flow-banding. The interior parts are a little coarser grained and commonly carry conspicuous megacrysts of quartz and feldspar. Fresh elvan material is a pale grey colour; in outcrop, the dyke rocks typically weather to buff or pale shades of brown.

For some reason, few elvans occur in the Dartmoor granite and surrounding country rocks. Only five are exposed at the surface; the presence of another two is suspected from debris.

Distribution of the Dykes

The most accessible of the elvan dykes occurs in the River Teign (SX 841845), near Christow (*Figure 5.5*). Two small exposures (SX 835843; SX 836844) roughly 500 m south-southwest of Christow Post Office are probably parts of the same intrusion. At the river outcrop, the dyke is vertical, approximately 1.7 m wide, and it strikes a few degrees north of east. Thin associated stringers cut the country-rock shales about 2 m south of the main exposure.

This particular dyke does not possess flow-banded margins, although each wall zone is finer grained than the interior. It consists of megacrysts of white and pink-stained feldspar, rounded quartz anhedra and scattered ragged flakes of chloritised biotite, set in a fine matrix of quartz, sodic plagioclase, potassium feldspar, chlorite and sericite. Accessory minerals include ilmenite, apatite and zircon. Some of the feldspar megacrysts are potassic, others are plagioclase showing extensive alteration to potassium feldspar and sericite. The chlorite and sericite in the rock matrix are developed mainly at the expense of plagioclase.

A fragment of similar material has been found on the west flank of Zeal Hill (around SX 670638). The rock contains more sericite and only small quantities of chlorite. In this respect it bears a close resemblance to a majority of Cornish elvans.

Fragments of very fine-grained silicic rock occur in Forestry Commission trackways on Soussons Down (SX 6779). They have come from dump material belonging to the Birch Tor and Vitifer group of mines and conceivably were derived from an elvan dyke encountered somewhere in those workings.

A hitherto unrecorded dyke striking about east–west cuts Legis Tor (SX 571656). The other three dykes crop out in china clay pits of the Lee Moor district. Two occurring respectively in the Shaugh Moor and Cholwich Town Pits strike approximately east–west, and may represent the same structure displaced dextrally by the Calisham Down–Cornwood Fault (*Figure 5.4*). Both are subvertical and have flow-banded margins. If they belong to the same body, it must vary from about 2 m to 7 m in thickness. The third elvan appears in the dormant Brisworthy Pit and may no longer be visible. Measurements taken in 1967 when the pit was being worked indicated a north-northeasterly strike, which is unusual. The intrusion is roughly vertical and about 3 m wide. Flow-banding occurs over its entire width.

The original mineralogy of the three dykes has been destroyed by the kaolinisation process that affected the granite hereabouts. They now consist of a fine-grained, very pale yellowish-green mixture of quartz, clay minerals and sericite mica, but flow texture has been preserved and it seems that their interior parts bore quartz and feldspar megacrysts.

Age and Origin

Stratigraphic evidence places the period of intrusion between the arrival of the granites and the process of kaolinisation. As no elvan detritus has been recognised for certain in basal Permian breccias, the dykes almost certainly post-date the rhyolite-basalt-lamprophyre volcanic outburst at 295 Ma. Rb/Sr isotope data actually indicate an early Permian age of about 270 Ma (Hawkes *et al.* 1975).

Because of a relative enrichment in silica and potassium, it has generally been assumed that the elvans represent a late-stage differentiate of the granite magmas. For several geological reasons this geochemical hypothesis seems unsatisfactory. The first and most important is that the Dartmoor granite was apparently being eroded before intrusion of the dykes took place. Other reasons stem from textural features which led Stone (1968) to propose emplacement in the form of a particle-fluid system where ion exchange between sodium and potassium was an active process. He still accepted the existence of a granitic source magma at depth, but certainly stimulated fresh thinking along textural and mineralogical lines. Henley (1972, 1974) took the concept of fluidised emplacement further by suggesting that the source material was solid granite mobilised at depth as a result of attack by aqueous potassic fluids derived originally from the assimilation of foundered pelitic xenoliths. Examination of elvan material shows that many feldspar megacrysts are, in fact, clusters of feldspar, quartz and, in some cases, partly chloritised biotite crystals which texturally resemble pieces of granite. These apparent granitic fragments provide some substance for Henley's view of the source material.

The fact that potassium enrichment in the elvans is due mineralogically to a widespread replacement of plagioclase by finely divided muscovite (sericite) certainly implies the presence during emplacement of a reactive aqueous fluid capable of promoting alkali ion exchange (potassium for sodium). It is also clear from field evidence that the elvan volcanism and primary metalliferous lode mineralisation were closely related. The mechanisms responsible for one must therefore have largely conditioned the other (Henley, 1974; Hawkes, 1974). This point is important because in rejecting the differentiation hypothesis for the elvans, the conventional idea of a direct connection between granite emplacement and the mineralisation is excluded as well.

Isotopic age data indicate a lapse of some 30–40 Ma between granite emplacement and dyke injection, sufficient time it would seem from recent calculations of granite cooling rates (less than 1 Ma to 7 Ma) to rule out any possibility of a direct magmatic connection. The time gap also poses problems in accepting foundered pelitic xenoliths as the source for Henley's reactive potassic fluids, because of the enormous energy losses implicit in the crystallisation processes that must have occurred within the magma column during that period. An unsubstantiated appeal to mantle activity for the required energy input to the crust round about 270 Ma is all that can be suggested in the light of current knowledge.

KAOLINISATION OF THE DARTMOOR GRANITE

There are several areas on Dartmoor, where plagioclase in both coarse-grained and fine-grained granite has been partly or completely altered to clay minerals, the chief of which is kaolinite. The alteration is confined mainly to the southern parts of the pluton. Smaller isolated patches like those at Petre's Pit (SX 658647) and Redlake (SX 645669) are approximately round or elliptical in plan, and range from perhaps 10 m to about 200 m across. Much larger irregular areas occur within a zone 1–3 km wide roughly marking the line of the Calisham Down–Cornwood Fault. Some workings for china clay in this zone are extensive and suggest that, at the surface, individual patches may exceed 500 m in diameter. There are no published figures for the full extent to depth, but similar kaolin deposits in the St Austell granite are known from borehole data to be greater than 250 m (Bristow, 1977).

Borehole evidence from St Austell shows that certain kaolinised regions in the granite are trough-shaped or funnel-shaped. Others form inclined pipe-like structures which widen towards the surface. However, no such definite information is available for the three-dimensional shapes of kaolin deposits on Dartmoor.

One particular feature of the alteration is the comparatively short distances between hard, relatively fresh granite and fully kaolinised material. These distances can be as little as 2–3 m. Exley (1959) has described the sequence of mineralogical and textural changes evident in the case of contacts in the St Austell granite: briefly, the changes seem to have involved the conversion of plagioclase to secondary mica and kaolinite-mica aggregates, and of some potassium feldspar to secondary mica. At the same time, alteration of biotite to non-hydrous secondary mica produced bleaching, and quartz may show signs of recrystallisation. Granite so affected can easily be crumbled in the hand.

Age and Origin

As elvan dykes were involved in the alteration processes, kaolinisation of the granite cannot have occurred before early Permian times. Unfortunately the dykes are the only reliable stratigraphic indicator. Subsequent magmatic events which might have been of service in putting a lower age limit on the kaolinisation occurred in regions distant from the granites, and are therefore of no direct help. Similarly, later sedimentary formations provide no positive evidence for dating the deposits. For example, kaolinite is certainly a common constituent in the Tertiary deposits of the Bovey Basin, but there are no obvious means of proving whether or not the mineral came from altered zones in the Dartmoor granite. Nevertheless, recent isotopic studies aimed at understanding the conditions under which the kaolinite formed suggest the need of a tropical or subtropical climate such as may have existed in Cretaceous or Tertiary times (Sheppard, 1977).

Sheppard's work on the oxygen and hydrogen isotope composition of the kaolinite implies a comparatively low temperature origin by weathering. This is an important contribution to the long-standing controversy between observers who have favoured genesis by hydrothermal means and those who considered weathering the dominant process. The evidence commonly quoted in support of each of these opposing hypotheses has been summarised by Bristow (1977). He thought that a compromise view might be nearer the truth.

Thus, Sheppard's isotopic data seems convincing enough, yet weathering alone cannot easily account for the sporadic nature of alteration to such great depths. Clearly the granite must have suffered preliminary degradation in selected areas prior to the growth of the kaolinite observed at present. Because of the shapes of the deposits, hydrothermal activity remains the most attractive agency for the initial alteration.

Circumstantial evidence appears to support the view of Dangerfield and Hawkes (1969) that South West England became a hot-spring province during the outburst of elvan volcanism. If so, the altered zones in the granites may delineate the root systems of geysers. However, as in the case of the elvans, the required thermal input to the crust needs an explanation. Bearing in mind the earlier potassic volcanism, one possibility is to recall the analogy between Devon and the Yellowstone Park district of Wyoming, and to speculate that each of the Stephanian and Permian thermal events may have related in some way to the activity of a mantle convection plume. Such a proposition will need rigorous testing in future research work, because it invokes the likely coexistence of other mantle hot spots under the Variscan crustal regions of Saxony and Bohemia. Like Devon and Cornwall, each of these regions is characterised by lithium rich and boron rich potassic granites, late Carboniferous or Permian volcanic rocks of lamprophyric and elvan type, and also by attendant metalliferous mineralisation and kaolinisation. Alternatively, it may prove that the events in both provinces were due to less dramatic tectonic causes such as general rifting associated with the breakup of Wegener's primaeval continent, Pangaea.

LOCALITIES

Dartmoor Granite

5.1. *Hay Tor* (SX 758771) and adjacent quarries (SX 760774; SX 755773; SX 750777): Sections in richly and poorly megacrystic granite, and also one apparently intrusive, megacryst poor, fine granite sheet are seen here. Xenoliths abound in the tor exposures. Jointing is particularly well developed in both the tor rocks and in the quarry granite. All exposures occur on open ground and can be reached on foot from the Hay Tor car park.

5.2. *Dewerstone Rock* (SX 536637), near Shaugh Prior: The smaller megacrystic variant of the coarse granite is best seen here. Access to this National Trust property can be gained via a footpath from the road bridge crossing the River Plym just by its confluence with the River Meavy. There are several outcrops in the River Plym.

5.3. *Between Legis and Trowlesworthy Tors* (SX 576654; SX 572652; SX 575653): Reddened muscovite bearing coarse granite crops out in the River Plym, upstream from Cadover Bridge. Exposures can be reached easily from the Cadover Bridge-Cornwood road by walking north along either of the river banks.

5.4. *Birch Tor* (SX 686815): The most accessible example of megacryst rich fine granite occurs as a raft several metres long in the south face of one of the coarse granite crags forming the tor. It can be approached by way of a poor moorland path leading south from a point on the B3212, approximately 1.3 km northeast of the Warren House Inn.

5.5 *Kestor Rock* (SX 665862): Coarse xenolith bearing granite contains a sill-like mass of megacryst poor fine granite, which bears several pegmatitic patches carrying quartz, potassium feldspar, biotite and tourmaline. The outcrop is reached by way of a lane and footpath leading southwest from Teigncombe Farm, Gidleigh.

5.6 *Two Bridges* (SX 610751): Partial kaolinisation of granite can be seen in a quarry just north of the road and nearly opposite the hotel. Dyke-like bodies of megacryst poor fine granite are also visible. However, the best exposures for kaolinisation are the working china clay pits. Access to these can be gained at the express permission of owners who are, perhaps, more conscious of dangers presented by potentially unstable material of this sort than a casual observer might be.

5.7 *Red-a-Ven Brook, Meldon* (SX 567920): Examination of the Meldon aplite presents a similar difficulty in that the most revealing exposures are in quarries which, in this case, are potentially dangerous because they are not worked. Nevertheless, the dyke may be examined in Red-a-Ven Brook between the two disused quarries. The locality can be approached along a rough road leading from the A30, near Fowley, to the British Rail ballast quarries.

Metamorphic Aureole

5.8. *Red-a-Ven Brook*: Several contact-altered rock types crop out in the brook. Keratophyric tuffs and cherts occur in the vicinity of the Meldon aplite. Further upstream slaty and more arenaceous rocks are exposed. Mine dumps beside the stream (at approximately SX 570918) contain calc-silicate hornfels debris in which garnet, pyroxene and wollastonite may be found.

5.9. *Sourton Tor* (SX 543898): A few kilometres to the southwest of Meldon, baked dolerite and keratophyric agglomerates are seen among the crags of the tor. They can be reached on foot from Sourton village.

Volcanic Rocks

5.10. *Kingsand, Cawsand Bay* (SX 435506): Spherulitic and flow-banded rhyolite is associated with Devonian limestone and basal sandy Permian strata on the beach north of Kingsand.

5.11. *Dunchideock* (SX 875871): Good exposures of basaltic material are found in the lower of two quarries where purplish-brown, calcite-veined and haematite-veined vesicular basalt can be found containing inclusions of red sandstone.

5.12. *Stone Quarry* (SS 681013) near Stone Cross, North Tawton: In the roadside a dark brown amygdaloidal and vesicular quartz basalt shows admixture with sandstone.

5.13. *Near Spence Combe Farm* (SS 795018) coarser, doleritic rock occurs in a quarry.

5.14. *Pocombe Quarries, Crossmead, Exeter* (SX 898916): These have been worked since Roman times. They expose a reddish-purple vesicular trachybasalt lava flow resting on Culm shale. The lava, which is cut by calcite veins and bears calcite geodes at the top, is overlain by soft red sandstone.

5.15. *Killerton Park quarries* (SS971005): Minette lavas.

5.16. *Hannaborough quarry* (SS 529029), near Hatherleigh: Agglomeratic lava accompanies minette lava and a crag in the quarry may be part of a volcanic vent.

5.17. *Quarry 1 km east of Knowle village* (SS 790022): Coarse-grained mafic syenitic rock of yogoitic affinity.

Dykes

5.18. *River Teign near Christow* (SX 841845): This dyke is the easiest to visit. Parking may be possible in a disused roadside quarry on the B3193. From there it is a matter of descending the steep west bank of the Teign at the grid reference point. The east–west dyke can be seen in a river bed outcrop where it cuts the country rocks.

Chapter Six

Metalliferous Mineralisation

Measured in terms of the number of workings and total output, Devon's metalliferous mining industry is dwarfed by that of neighbouring Cornwall; nevertheless it can boast that it possessed, in Devon Great Consols Mine, the largest single producer of both copper and arsenic in the two counties and, for a time, in the world. Much of that arsenic was exported to control the depredations of the boll weevil in the American cotton fields.

A variety of ores has been worked in the county although, due to the vagaries of reporting, their quantities are very imprecisely known. The figures given in *Table 6.1* are compiled from several published sources but are certainly incomplete.

There is little doubt that the streams which drain the Dartmoor granite were worked for cassiterite from very early times, although the earliest recordings of identifiable alluvial workings are from the first half of the fifteenth century (T.A.P. Greeves, personal communication). The ore was smelted locally, usually on site, in small furnaces dug into the valley slopes; during recent prospecting one of these hearths was located near Elfordleigh golf course (SX 545586) north of Plympton, and a small granule of metallic tin (the β phase) was recovered 2.4 m down in alluvium near Gale (SX 795715) northeast of Ashburton. The beginnings of underground mining are equally obscure, but silver-lead mines at Combe Martin and in the Tamar Valley were active in the thirteenth century (Dines, 1956) and, together with the Exmoor iron deposits (mainly in Somerset), they may have been mined by the Romans. Modern mining for base metals began at the end of the eighteenth century and developed rapidly thereafter, reaching its peak between 1855 and 1875. By 1910 only a few mines remained active and the last of these, Great Rock Mine near Hennock, closed in 1969.

The prospecting boom of the 1960s scarcely affected Devon, although some small alluvial tin operations were established in the Kester Brook (SX 791714) near Ashburton and at Pinsents Quarry (SX 866725) between Newton Abbot and Kingsteignton. Recent years have seen a renewed interest in copper around North Molton, in barite within the Teign Valley, in tin and tungsten on the banks of the River Tamar and in tungsten at Hemerdon Ball (SX 573585) northeast of Plympton.

With a few notable exceptions, the Devon mines were small operations the viability of which was constantly threatened by increases in working costs or decreases in ore grade or metal prices. In consequence, many have a history of repeated closure and restarting and only the larger or richer ones successfully weathered the economic vicissitudes of the nineteenth century. It is distinctly possible, therefore, that some deposits still possess a potential which has not been fully explored. Contrary to popular belief, modern improvements in extraction methods, bulk handling and separation techniques rarely permit the more economic exploitation of lower grade ores; the benefits from advancing technology are rapidly offset by inflationary labour and materials costs. Only in the cases of alluvial and opencast prospects is there hope of reworking the 'old men's' leavings.

Perhaps the most emotive issue about metalliferous mining is its environmental impact, and in this field the industry has an unfortunate legacy to overcome. Barren waste tips and slime ponds bear mute testimony to the insensitivity of former owners whose unsaleable, often biologically toxic, residues were strewn flagrantly over agricultural land, while insecurely fenced, open or poorly filled shafts remain a hazard to the unwary or adventurous. Paradoxically, there is also within this dereliction something to be preserved; at Wheal Betsy (SX 510812) on the Okehampton–Tavistock road it is the engine house, at Morwellham (SX 446697) the loading and shipping facilities, and at Birch Tor and Vitifer Mine (SX 682814) southwest of Moretonhampstead, the lines of surface workings.

In modern mining practice there is more scope for the disposal of waste underground, landscaping of surface dumps is becoming obligatory, and noise and dust levels are minimised by legislation. Impact there must be, but that impact is now much less unacceptable and less enduring.

Table 6.1   ORE PRODUCED IN DEVON

Copper	1,300,000 tonnes ore yielding 84,800 tonnes metal (1,070,385 tonnes ore = 69,642 tonnes Cu, 1848–1914)
Tin	10,750 tonnes smelted tin metal 7,000 tonnes dressed ore, yielding about 4,800 tonnes metal (6,370 tonnes dressed ore, 1852–1914)
Lead	69,000 tonnes ore yielding 44,100 tonnes metal (63,247 tonnes ore = 40,324 tonnes Pb, 1845–1890)
Zinc	(4,344 tonnes ore, 1856–1873)
Silver	34.284 tonnes metal from lead ores. (26.389 tonnes metal, 1854–1882)
Manganese	(62,730 tonnes of variable grade, 1858–1907)
Iron Ores	460,000 tonnes of oxide and carbonate ores (356,390 tonnes, 1855–1913)
Pyrite	24,100 tonnes concentrates
Arsenic	96,600 tonnes refined and crude oxide 73,600 tonnes unroasted ore
Tungsten	58 tonnes wolframite concentrates (36.3 tonnes concentrates, 1900–1913)
Barytes	497,000 tonnes high grade, white product (14,464 tonnes, 1855–1896)
Fluorspar	3,335 tonnes (2,370 tonnes, 1874–1886)
Ochre	8,192 tonnes of various pigment grades

Figures in parentheses are totalled from government statistics collected for computer bank storage by the Economic History Department of Exeter University.

TYPES OF DEPOSIT

Metalliferous mineralisation in Devon can be classified into five broad physical types: alluvial, vein, stockwork, replacement and bedded deposits. Each of these reflects a differing range of depositional controls and calls for different mining approaches. In each the ore minerals are present in markedly different concentrations, and are associated with a diversity of by-products, gangue minerals or wallrocks. Detailed consideration of these factors is reserved for the description of individual localities but a brief outline of the deposit types is given below.

Alluvial deposits, in the strict sense, are confined to the alluvium and terrace sediments laid down during establishment of the modern (post-Pleistocene) drainage network. For convenience this category is stretched to include all forms of surface detritus (eluvium, head, valley-fill, etc.) and the basal gravels of the Bovey Basin Tertiary succession (Chapter Nine). Cassiterite (tin oxide) is the only ore yet to have been won from alluvial workings, although in the Shaugh Prior–Plympton area wolframite (the major ore of tungsten) may also be present in considerable quantity and a little gold occurs in the River Dart sediments. Economic viability depends as much upon the length, breadth and depth of the deposit and the size distribution of its sediment as upon the ore mineral content. Locally the grade may exceed 3 kg cassiterite per cubic metre (0.15 per cent Sn) but exploitation at 0.45 kg per cubic metre (0.022 per cent Sn) has proved profitable. Most of the ore mineral is present in the form of free grains, although close to source or near former crushing plants there may be some composite grains. The gangue minerals comprise quartz, feldspar, micas, tourmaline and iron oxides together with rock fragments and clays; their size ranges from large cobbles to clay grade. Benefication involves washing, screening, sizing and simple wet gravity separation with removal of iron oxides and wolframite by electromagnetic means.

The greater part of base metal production has come from vein (lode) deposits in which the ores occur as irregular lenses, pipes or scattered patches. Workable veins usually have a strike length of at least 0.5 km (Devon Great Consols Main Lode is 4 km long), they vary in width from 0.2 m to 12.5 m and some have been followed to depths of 250 m or more. Most Devon mines, however, were less than 150 m deep. In several instances veins are arranged *en échelon* within a wider and much longer vein zone. Most dip at 60° or more, although flatter ones do occur, particularly southeast of the Dartmoor granite. Two main lode directions are apparent, an earlier set generally containing high temperature minerals trends roughly east–west and is cut by a later north–south set with lower temperature minerals. Within the Dartmoor granite, however, the veins have a more variable strike.

Mineralogically vein deposits may be simple or complex, with ore and gangue assemblages which largely reflect the depositional temperature range. Some of this complexity arises from repeated reopening of the vein fracture which allowed sequential deposition of progressively lower temperature ores during its geological history. In a few instances mineralisation is carried into the host rock where it replaces parts of the rock fabric; disseminations of this nature, where rich enough or large enough to mine, are known as *carbonas*. Near surface the lodes are commonly altered, parts being depleted and parts enriched in some metals, particularly copper, lead and silver. At surface, the weathered cappings or gossans consist predominantly of quartz and iron oxides and were aptly termed 'iron hats' by the miners.

One of the characteristic features of vein mineralisation is the development of alteration in the adjacent wallrock. This may toughen or soften the host rock, thus affecting the cost of mining of the vein or the stability of open workings; it can extend for a mere few centimetres to more than two metres from the vein walls.

The traditional method of mining is by shafts (vertical, or incined down the lode) and horizontal levels, the ore being extracted by upward or downward stoping between levels. Patchiness of ore mineralisation is reflected in irregularities of the stoping pattern and in many mines the overall extraction has not exceeded 35 per cent. Surface treatment has varied through the years but depended mainly upon ore type and complexity. The richer ores, mainly of copper and arsenic, were formerly hand cobbed and sold as unrefined high-grade ore; later all ores were crushed and concentrated by wet gravity or froth flotation methods to produce mineral concentrates. Complex tin ores were commonly calcined to remove unwanted sulphides and many arsenic mines roasted their ore to sell arsenious oxide.

Grades at which the ores were worked are not reliably known but from examination of abandoned workings and discarded waste it is possible to derive an approximate lower limit. Cut-off grades seem to have been about 0.7 per cent Sn, 2 per cent Cu, 2–3 per cent Pb, 4–8 per cent As_2O_3 and 20 per cent Fe with run-of-mine ores averaging 50–100 per cent higher than these figures. Recorded descriptions, however, indicate that some ore shoots carried extremely rich values, particularly of copper and arsenic.

Stockwork deposits are bodies of rock in which the mineralisation is contained in numerous veinlets of varying strike and dip which form a netted or interlacing mesh. Generally one set of fractures is preferentially developed and contains wider or richer veins. The physical form of a stockwork is rarely fully resolved, shape and size being determined by mining considerations. At Bridford Mine (SX 830864) a barite stockwork in the hanging wall of No. 1 Vein (Vipan, 1959) has maximum horizontal dimensions of 120 m by 30 m and has been followed for 160 m vertically. Its veinlets range from 7 cm to 30 cm in width. Dines (1956) quoted the surface expression of the Hemerdon tin-tungsten stockwork, northeast of Plympton, as 600 m by 145 m but current exploration indicates that the body is larger than this, persisting to depths of 400 m and containing a proven 45 million tonnes of mineralised rock with an overall grade of 0.18 per cent WO_3 and 0.025 per cent Sn. An additional 15 million tonnes of similar grade is also thought to be present.

Ore grades in stockwork deposits are always low, often around 0.15 per cent to 0.3 per cent Sn (or WO_3), and can only be worked profitably by open pit operations handling large daily tonnages. Current proposals for Hemerdon are to mine at a rate of 2.5 million tonnes per year. Underground stockworks are now exploited along similar lines by block caving mining but this technique has not been employed in Devon. Large excavations on Dartmoor near Crazy Well Pool (SX 582704), the workings of Whiteworks Mine (SX 613710) and parts of the Birch Tor and Vitifer complex (SX 680808) suggest that they may have been tin stockworks.

Replacement of original rock fabrics and minerals by metalliferous ores can be recognised from the microscopic scale through to ore deposit size and though always effected by the mechanism of percolating solutions, the replacements may have been due to hot or cold fluids moving upwards, downwards or laterally, and of markedly variable chemical composition. The location of replacement deposits may be controlled by an equally variable combination of factors, both chemical and physical.

In south Devon large replacement bodies of iron oxides are developed in the Middle Devonian limestones, particularly in the vicinity of intense fracturing. An example is the Sharkham Point Mine (SX 935548), southeast of Brixham. No surface treatment seems to have been practised so, presumably, only the high-grade ore was mined. At numerous localities manganese ores (mainly the oxides) replace either the volcanic rocks or associated cherts of the Lower Carboniferous succession. Many small mines were tried in the mid nineteenth century but the only large producer was Chillaton and Hogstor Mine (SX 433812), northwest of Tavistock.

Bedded ores, as the title implies, are confined to distinct horizons within a sedimentary sequence. Two subcategories are recognisable, namely ore minerals deposited with the sediments and ore minerals introduced after consolidation. The former, *syngenetic ores*, are poorly represented in Devon, and in known deposits the metallic content is nodular, suggesting that there has been some redistribution during, and perhaps after, compaction. The only worked example seems to be the concretionary ironstones of the Smallacombe (SX 777766) and Atlas (SX 779762) mines near Haytor Vale. At Haytor Mine itself (SX 771771), on strike with Smallacombe, the ore is magnetite with some sulphides in a hornblende-garnet host, and seems to be a thermally altered equivalent. More spectacular are the pyrometasomatic bedded sulphides of the Sticklepath area (SX 640940), on the northern margin of the Dartmoor granite. Two mines have produced small tonnages of high grade copper ore (estimated at about 17 per cent Cu) from four beds of garnetiferous rock, one of these being 30 m thick. At both Haytor and Sticklepath the ore was worked by normal mining techniques and most of it was hand dressed for sale.

NORTH DEVON

The mineralised area of Exmoor (*Figure 6.1*) forms a self contained entity within the metallogenic province of South West England; it contains only low temperature mineral deposits, which are difficult to relate spatially to the Cornubian granite batholith. This raises questions as to the source of the metalliferous ores, the fluids which transported them, and the temperature environment in which they were deposited. It may be that we need to look to the Devonian sedimentary rocks and their metamorphism to provide these requirements. If so, the Exmoor mineralisation must pre-date the granite emplacement and, therefore, the main phase of Cornubian mineral deposition. Such a relationship is suggested by recent isotopic age dating (Ineson *et al.*, 1977), although this is not the conclusion of those authors.

It is probable that Exmoor iron ores were worked in Roman times, but modern production dates from about 1850 and continued into the present century. Some of the mines received attention in World War I, but no production resulted, and they were again examined during World War II. Argentiferous galena production is first recorded from the Combe Martin area in 1293 (Dines, 1956), since when the mines have closed and reopened several

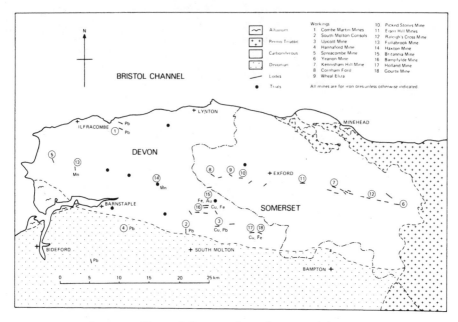

Figure 6.1. The north Devon and west Somerset mining area

times. The earliest workings for copper near North Molton date from the thirteenth century, but records of output begin in 1696 and end in 1888 (Rottenbury, 1974). Recent geochemical, geophysical and drilling exploration has failed to reveal any further significant copper concentrations.

Argentiferous Lead Deposits

The principal district of production for silver-lead ore was around the town of Combe Martin. Records of mining extend from the Middle Ages to the end of the nineteenth century. Tristram Risdon (1811) a seventeenth-century historian, whose work was published posthumously, stated that the Combe Martin mines were first found in the 22nd year of King Edward I (1293), at which time 337 men were moved from Derbyshire to work them. Exploitation continued spasmodically thereafter, with the last major phase of activity in the reign of Queen Elizabeth I. Attempts to re-open the workings or to find new deposits since that time have proved largely unsuccessful. Work in the nineteenth century resulted in small returns of ore and most of the operations were little more than trials.

Little direct evidence is available concerning the precise location and form of the ore bodies; the early workings appear to have been concentrated close to the present main street of the town which roughly lies along the line of the Combe Martin Fault. It would seem from study of the coastal section and from examination of specimens in museum collections and from mine dumps that typical Cornubian-type vein structures are not present. It is considered that the mineralisation takes the form of discontinuous lenses of ore which are particularly well developed in the cores of folds. These bodies are concordant with the strongly developed local cleavage. Evidence of this may be seen in the debris around Knap Down Mine (SS 597467) where irregular pods of white quartz with siderite and galena are enclosed in sheared and crumpled grey slate.

The mineralisation appears to be stratigraphically controlled in that it is restricted to that part of the Ilfracombe Slate referred to the Lester Slate and Sandstone. At outcrop these rocks commonly bear traces of sulphides as streaks along the cleavage planes of the mudstones. Dating of the mineral deposits by Moorbath (1962) gave a model lead isotope age of 370 ± 50 Ma, and results from isotopic work on clays by Ineson et al. (1977) suggested an age of about 304 Ma. Both ages indicate pre-Cornubian granite mobilisation, and the deformed nature of the ores supports pre-folding deposition. An originally syngenetic origin of the deposits cannot be ruled out, with redistribution and concentration effected by the Variscan earth movements. Late movement (during the Tertiary) of the Combe Martin Fault, which trends northwest–southeast, may have been responsible for further remobilisation of the ores.

Galena is the principal ore mineral at Combe Martin; it occurs either in a relatively pure crystalline form or as very fine-grained material intergrown with sphalerite. Both varieties of ore exhibit evidence of deformation, the former having curved crystal faces and the latter being sheared. In addition, minor amounts of chalcopyrite and pyrite are found and Dines (1956) noted that covellite, malachite and azurite are present in specimens of gossan. The minerals stibnite and millerite, although rare, have been recorded in the ores. Dines also stated that the silver content of the ores has been quoted as ranging up to 5.14 kg per tonne (168 oz per ton). Filaments of native silver in the gossan are the only recorded evidence of the occurrence of that element in a separate mineral phase. The gangue minerals are white quartz and pale brown siderite which is usually altered to a limonitic weathering product.

Other lead deposits have been worked in north Devon, mostly trials on small mineralised structures in the Pilton Shale. Rottenbury (1974) listed several mines which produced ore, among them Hannaford (SS 607297) and South Molton Consols (SS 701285). The assemblage at South Molton Consols included lead, zinc, silver, antimony, arsenic and iron. Lead deposits in the Pilton Shale are known to be crosscutting veins, in contrast with those at Combe Martin. Small lead mines in the Lower and Upper Culm Group have been located at Yarnscombe (SS 541241) and Brushford (SS 675081) but, although Dines (1956) referred to documentation of the latter mine in 1498, nothing else is known of the workings.

Iron Ore Deposits

By far the greater part of the Exmoor iron ore deposits (*Figure 6.1*) lie within northwest Somerset; only a handful of workings and several prospecting pits are located in Devon. Most of these sites have been recorded by Rottenbury (1974), the major workings were listed by Dines (1956) and an account of the development of iron mining in Exmoor Forest, between Exford (SS 853384) and the Devon–Somerset border, was given by Orwin (1929).

The iron mines fall into two groups, a northern belt stretching from Spreacombe (SS 481420) in the west to Yeanon (ST 059334) in the east, and a southern group scattered around North Molton (SS 736297). In the northern belt most of the veins have an east-southeast trend, conforming to the strike of the enclosing rocks (Ilfracombe Slate, Morte Slate or Pickwell Down Sandstone) although a few, including the Devon occurrences, have a general northerly strike. The southern veins are less regular in attitude and occur only in Pickwell Down Sandstone.

Exploitation seems to have begun in open trenches and, in some favourable instances, developed through adit levels, deep pits or shafts. Records of the Devon mines are scanty and the best known workings are those around Brendon Hill (ST 027343) where production began in the mid eighteenth century and persisted until 1911. Nearly all the sites are within enclosed land and are largely overgrown, although several adits are still visible and at Kennisham Hill Mine (SS 963362) the engine house stands amid a forestry plantation. Just west of the road junction at Brendon Hill can be seen the remains of the railway incline down which the ore began its journey to Watchet. The momentum of descending loaded trucks was employed to haul up the empty ones.

Some dump material is preserved at most sites but the ores can be best seen at Cornham Ford (SS 749386), Wheal Eliza (SS 784381), Picked Stones (SS 797378), Eisen Hill (SS 908371) and Raleigh's Cross Mine (ST 025342). The ore types are highly variable, even at individual mines, and include buff to near white siderite (spathose iron ore), friable, spongy and earthy limonite or goethite, hard, massive and occasionally mamillary dark red-brown to black haematite, and dark brown, fine grained, siliceous haematite. Rarely, there may be plates of specular haematite or narrow bands of soft micaceous haematite. Black manganese oxides (psilomelane and wad) coat many of the joints and occur less commonly as pockets and botryoidal masses in the iron ores; at Fullabrook (SS 515399) and Haxton (SS 655371) manganese seems to have been the main ore (Dines, 1956). Traces of copper are recorded at many mines; green secondary copper minerals and some primary sulphides are scattered through the iron ores at Kennisham Hill Mine and Wheal Eliza. At the latter locality barite is also recorded (Dines, 1956). Grains of gold were reported in haematite from the Britannia Mine (SS 746335) (Pattison, 1865). Soft and earthy ores largely represent near-surface alteration products of primary siderite and intermediate stages in this conversion can be recognised in many dump specimens. It seems certain that the siliceous, specular, micaceous and mamillary to massive forms of haematite are also in part primary mineralisations and were probably formed at the same time as the siderite. When altered they give rise to a soft brick-red ore.

The ores occur together with quartz, often interbedded with it, in veins up to 8 m wide, usually following the dip of the enclosing strata or the dip of the slaty cleavage, steeply south. Although apparently continuous over much of the Brendon Hills, it is probable that the worked veins are disposed en échelon and that individual ore bodies are of limited lateral and vertical extent. The maximum recorded mined strike length was 335 m and depth was 250 m. Available descriptions suggest that the veins narrowed downwards and many split into several branches.

Various authors have offered differing theories for the formation of these iron deposits. Pattison (1865) saw the gold at Britannia Mine (and Bampfylde) as genetically related to intrusive mafic rocks and, by implication, must have considered the iron ores to have been deposited under their influence, if not actually derived from them. However, no geologist since this time has been able to confirm the presence of mafic igneous rocks in the North Molton area. Rottenbury (1974) treated the deposits as hydrothermal ores suggesting, on geological grounds, a post-folding, pre-New Red Sandstone age. But in a later note (Rottenbury and Youell, 1974) the veins were referred to as Permo-Triassic fissure fillings, the deposition presumably resulting from lateral and downward transportation from a ferruginous sedimentary cover.

On close examination there can be little doubt of the hydrothermal nature of the deposits, of their emplacement within an early structural fabric and involvement in later, minor movement phases. It is difficult to visualise a close genetic connection between the ores and the distant Dartmoor granite and the writers favour the concept of derivation from Devonian sedimentary rocks with transportation and deposition into open fracture systems by deeply circulating reactive waters; these were partly generated and wholly heated by the metamorphic effects of the Asturic (Stephanian) folding. This would imply an age of about 300 Ma, a figure in close agreement with the 296 Ma, date derived by Ineson *et al.* (1977). These authors related the mineralisation to the Dartmoor granite via low-angle thrusting (the Exmoor Thrust), the existence of which is a matter of continuing debate.

Copper–Iron Deposits

Although many of the iron ore veins report only small copper contents (less than one per cent) there are a few near North Molton in which copper occurs in concentrations sufficient to have supported limited exploitation. The major mines were those of Bampfylde or Poltimore (SS 738327), Molland or Bremley (SS 819283) and Gourte (SS 823282). Compared to Cornish mines these workings are small; they show larger recorded outputs of iron ore than of copper ores (Dines, 1956). Rottenbury (1974), however, questioned these records and derived for Bampfylde a minimum output of 15,750 tonnes of 15 per cent copper ore; much of the recorded iron ore production he ascribed to nearby Stowford and Crowbarn mines. All three mines lie within the outcrop of the Pickwell Down Sandstone though the immediate wall rocks are slate. Mining began in the late seventeenth century and continued intermittently until the 1880s, but Bampfylde may have been tried for copper as early as 1250 (Rottenbury, 1974).

The veins all trend east–west and dip steeply to both north and south. They are cut by small, apparently unmineralised, north–south cross-courses but the Poltimore Lode of Bampfylde Mine has been followed for 430 m west and at least 280 m east of the River Mole and the Molland-Gourte veins seems to have a similar strike length, though less well developed. At Bampfylde the workings extend 200 m below the valley floor and at Molland to 145 m. According to Collins (1912), interpreting Pattison (1865), the worked veins were in fact zones of interlaced narrow veinlets following the bedding of the slate host. Such descriptions were perpetuated by Dewey (1923) and Dines (1956) but neither they nor Collins saw the underground exposures. Rottenbury (1974), who has penetrated underground and who opened up the higher workings on North Lode, reported that the veins are typical fissure fillings with sharply defined walls between which the vein material consists of varying widths of quartz, iron oxides, quartz with contained slate fragments or ferruginous cemented slate breccia (*Figure 6.4*). The copper ores are distributed as discrete leaders, patchy lenses or interlacing veinlets. Rottenbury confirmed that the mineralised veins follow the steeply dipping cleavage planes rather than the more gently dipping bedding. The veins are highly variable in width, from 0.48 m at their poorest to 2.44 m at their widest, and average about 1 m. In Bampfylde Mine the richest ore occurred in a shoot on Main Lode which extended 270 m down dip and 90 m along strike.

Extensive dumps along the eastern bank of the River Mole show typical vein material. Vein quartz with massive, specular and occasional micaceous haematite or siderite carries crystals, spots, veinlets and small lenses of copper sulphides. These consist of primary chalcopyrite and bornite altered in places to covellite and chalcocite and eventually to malachite and azurite. Some of the iron ore is soft goethitic material, probably a gossan from the upper levels. A similar mineralogy, but with predominant siderite, is seen on the small dumps at Molland. At this mine the usual depth sequence from oxidised chalcocite ore to primary chalcopyrite seems to be reversed (Rottenbury, 1974); the cause of this unusual relationship is not understood.

Gold was reported from the gossan at Bampfylde Mine (Pattison, 1865) and led to unscrupulous attempts to attract investment capital; Rottenbury separated microscopic grains from the North Lode gossan and recorded assays of up to 85 ppm. This mineral is not known from Molland or Gourte.

THE DARTMOOR AREA

For descriptive convenience an area comprising the whole of the Dartmoor granite outcrop and most of its thermal aureole is considered here. To the southwest it merges into the highly mineralised Gunnislake-Tavistock district from which it has been separated by a line joining Tavistock and Sampford Spiney and then following the valley of the River Meavy (*Figures 6.2 and 6.3*). A wide range of mineral associations and deposit types is represented within this area, some of them unique (so far) in Devon. Of these, the manganese deposits and the late lead-zinc and uranium lodes are not described in this part of the account but are treated separately.

The beginnings of mining in Dartmoor are lost in the mists of history but tin streaming almost certainly dates from pre-Roman times, since when it has been practised periodically through to the present day. Abundant evidence of these activities remains in the form of hummocky valley floors, deeply sliced gullies and scattered blowing houses. Underground mining probably began in the Elizabethan era with the search for argentiferous galena and copper, became widespread in the late eighteenth century and reached its peak in the latter half of the nineteenth century. Despite a general mining decline in South West England since 1890, several small workings in Dartmoor have been active during this century and an important new discovery, the Hemerdon stockwork, was made as recently as 1916.

Most of the mines are of small to medium size with outputs of only a few thousand tonnes of ore and, with a few exceptions, their ore has not been outstandingly rich. It has become customary, therefore, to regard the deposits as being both small and of patchy or low grade; on either count they were of doubtful promise! Such conclusions presuppose that adequate development was done to establish the true reserve potential, a proposition which is not

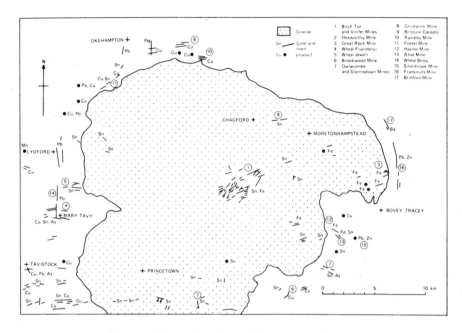

Figure 6.2. The north and central Dartmoor mining area

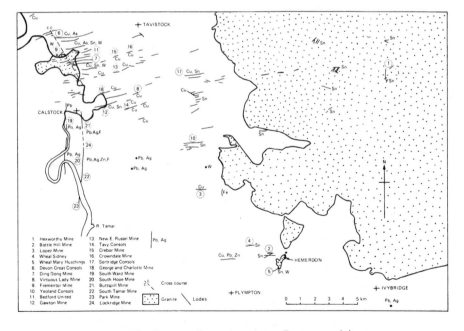

Figure 6.3. The Tamar valley and southwest Dartmoor mining area

convincingly borne out by available records. Unfortunately, recent exploration, except at Hemerdon, has been discouraged by environmental opposition to mining close to, and within, the National Park.

Spatial distribution of the deposits clearly suggests a genetic association with the Dartmoor granite and in some instances this can be amply demonstrated from available geological and mineralogical evidence. As yet there have been few sophisticated modern scientific studies of the Dartmoor mineralisation; the conditions of deposition and ages of mineral formation are deduced from comparison with the better studied Cornish occurrences and from reported geological observations. Neither micro-zoning nor macro-zoning of metalliferous minerals can be recognised with confidence: the complex veins consist either of intimate mixtures of both higher and lower temperature species or distinctly separate introductions of such species during sequential opening of the fissures. The depth, continuity of mineralisation and accompanying changes in mineral content, therefore, are matters of subjective speculation.

Placer Tin Deposits

Erosion of the Dartmoor granite and its metamorphic aureole has produced stanniferous gravels which have been worked from early times. Virtually every tract of alluvium in the district has been turned over by the tinners, and the now overgrown piles of waste along the valley bottoms form a familiar feature of the Dartmoor landscape. Production was generally carried out on a small scale, the ores being smelted locally in primitive blast furnaces after crushing and hydraulic separation. Although the annual tonnage of tin from Devon was generally much less than that from Cornwall, an independent stannary charter was granted to the county in 1305 (Lewis, 1908). Records indicate that a considerable production of metal was achieved in the twelfth and sixteenth centuries with a decline in the seventeenth century as the deposits were depleted.

The deposits were present in the alluvium of much of the upland area, particularly in the southward draining catchments. They were not, apparently, more favourably distributed in the area where vein complexes are known to outcrop at present, but were derived from ore bodies in the roof zone of the granite which have largely been removed by erosion. The valley of the River Walkham to the north of Merrivale Bridge (SX 550751), east of Tavistock, is an example of a tract of extensive former stream working which is remote from significant vein structures. Removal of the tin-bearing gravels from the valleys of the moor has been so thoroughly effected that little evidence of the original nature of the deposits remains. In many areas the river deposits have been removed to bedrock and the present-day streams flow in channels which are lowered commonly to an extent of several metres, indicated by scarp-like features at the edge of the alluvial tracts. A striking example of the tinners' activity may be seen in the valley of Newleycombe from the vantage point of Down Tor (SX 580694), east of Burrator reservoir, looking towards the north.

Examination of the tinners' waste piles at various localities shows that the discarded material, other than the sand and silt fractions which were washed away, consists of large, subrounded fragments of quartz-tourmaline rock together with a lesser fraction of the harder, generally finer grained granite varieties. Fragments of coarse, megacrystic granite, which is the most common host rock of tin mineralisation, are notably deficient; their inability to survive transport gives some clue to the conditions in which the stanniferous gravels were deposited. During much of the Tertiary, the Dartmoor granite was exposed and subjected to warm temperate climatic conditions which produced a deeply weathered profile. The presence of kaolinitic clays derived from the granite in the Oligocene Bovey and Petrockstow beds suggest that the kaolinisation of feldspars was widely developed. This altered granite was extensively eroded in the succeeding glacial epoch; while full-scale glaciation did not affect Devon, there was high seasonal precipitation under a periglacial climate and permafrost activity. Coarse granite debris was rapidly broken down and removed, while residual deposits of resistant veinstone filled channels along the valley floors.

In addition to the strictly alluvial deposits described above, some tin ore was worked from the ground between the fluvial gravels and lodes at outcrop. This type of residual material was known as *shode*, and represents an arrested stage in the development of the stream-tin deposits.

The areas surrounding the granite and its aureole have also supported stream-tin operations. During periods of high precipitation of the periglacial regime the rivers draining the upland areas were rapidly cutting back the valleys in which they flowed, causing pre-existing stanniferous gravels to be swept away and redeposited in the alluvium of the lowlands. This process operated to a very marked extent in the case of the River Dart, which has an extensive catchment over southern and central Dartmoor. Alluvial tin at Colston (SX 751649), south of Buckfastleigh occurs in a buried channel of the Dart (Scrivener and Walbeoffe-Wilson, 1975) and shows a close resemblance to the lowland stream tin deposits of Cornwall described by De la Beche (1839) and Dines (1956). The tin-bearing sediment rests directly on slate bedrock and is up to 80 cm thick. It consists of locally derived material grading from small angular boulders to clay without any degree of sorting, mixed with a fraction of well rounded, granite-associated gravel and sand. Detrital cassiterite is common in the latter, though the grade varies considerably over short distances. A thickness of about 2.54 m of barren alluvium overlies the deposit. It is considered that transport of the dense vein fragments took place in conditions of violent flooding, accompanied by scouring of the channel. Deposition of the barren sediment followed when the periglacial regime declined and precipitation moderated.

Some exploitation of alluvial tin deposits has taken place in recent years. The valley of the Kester Brook at Longstone Bridge (SX 796714) near Bickington yielded ore from a strip of alluvium up to 30 m wide and between 2 m and 3 m deep. Values were low, but the operation was viable because considerable upgrading of the ore could be achieved by screening and classifying the poorly sorted sediment. This working ceased in 1975 and the ground was restored and returned to agricultural use. Until recently, a small plant near Newton Abbot separated tin concentrates from the alluvial overburden which was stripped from a ball clay working. Sand and gravel were also produced and sold.

Cassiterite has been noted in the Tertiary Aller Gravel (Scrivener and Beer, 1971). The ore grade, however, is low and no commercial exploitation has been attempted.

Simple Hydrothermal Veins

Metalliferous veins are of widespread though scattered occurrence in the Dartmoor granite and its metamorphic aureole, and this section deals with those ore bodies which formed as fracture fillings during the post-emplacement cooling history of the intrusion. The distribution of the veins is shown in *Figures 6.2 and 6.3.*

Within the granite, metalliferous veins are preferentially developed in those types rich in potassium feldspar megacrysts which may, as at the Birch Tor and Vitifer complex or at Hexworthy Mine (SX 656718), be intruded by sheets of microgranite. The megacryst deficient granite of the high ground of northwest Dartmoor and the southern part of the intrusion are relatively lacking in mineral veins though, as previously stated, the presence of extensive alluvial workings in these districts indicates that ore bodies were formerly present but have since been removed by erosion. Scrivener *et al.* (1977) considered that the mineralised zone was thin and lay close to the roof of the pluton. This is borne out by the location of the workings and by their shallow depth; within the granite a maximum of 119 m (65 fathoms) at Birch Tor Mine was quoted by Henwood (1843). Most of the workings were shallower than this, a typical example was Wheal Cumpston (SX 672723), south of Dartmeet, where nineteenth century exploration of the ground beneath an old open working showed that the tin vein died out within 35 m of the surface.

The most important economic mineral of the veins is cassiterite, which is generally found in a gangue assemblage of quartz and tourmaline (*Figure 6.4*). In a typical specimen of ore, the cassiterite is moderately coarsely crystalline, of subhedral habit and showing striking zoning when seen in thin section. A blue acicular or finely prismatic variety of tourmaline accompanies the cassiterite and can commonly be shown both to pre-date and post-date the formation of that mineral. Several generations of quartz are usually present, the most recent of which may be vuggy and accompanied by cavity fillings of earthy haematite or clay minerals. The depositional history of a typical Dartmoor tin vein may be summarized as follows:

Fracturing is initiated and a first generation of massive tourmaline with minor quartz is deposited. This early tourmaline phase may invade the wallrock.

A second phase of fracturing occurs, brecciating the first generation of tourmaline which is recemented by the deposition of blue acicular tourmaline, cassiterite and quartz. The relative amount of tourmaline deposited declines.

Late quartz is deposited in the final phase of fracturing.

Inclusions of wallrock are common in the veins and, at some localities near to the granite contact, small fragments of metasedimentary rock have been noted (Scrivener *et al.*, 1977).

Sulphides are almost absent from veins in the granite but become more common in ores from sites close to the contact with the country rock. Thus arsenopyrite and sphalerite are to be found at Crownley Parks Mine (SX 760764) near Hay Tor, while only traces of pyrite and chalcopyrite have been noted at the Birch Tor and Vitifer complex. Veins in the aureole, however, are relatively rich in sulphide minerals.

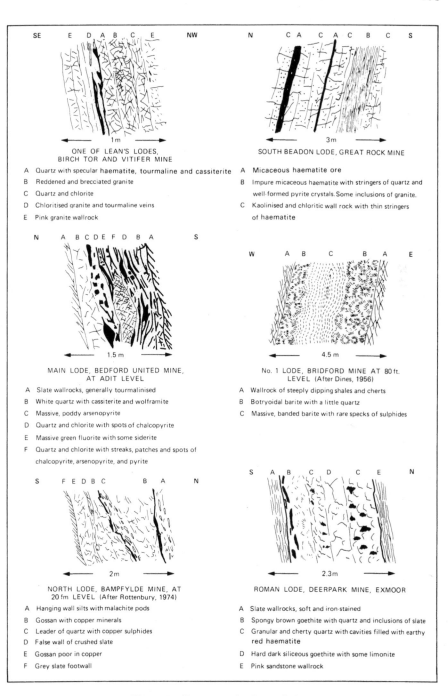

ONE OF LEAN'S LODES,
BIRCH TOR AND VITIFER MINE

A Quartz with specular haematite, tourmaline and cassiterite
B Reddened and brecciated granite
C Quartz and chlorite
D Chloritised granite and tourmaline veins
E Pink granite wallrock

SOUTH BEADON LODE, GREAT ROCK MINE

A Micaceous haematite ore
B Impure micaceous haematite with stringers of quartz and
well-formed pyrite crystals. Some inclusions of granite.
C Kaolinised and chloritic wall rock with thin stringers
of haematite

MAIN LODE, BEDFORD UNITED MINE,
AT ADIT LEVEL

A Slate wallrocks, generally tourmalinised
B White quartz with cassiterite and wolframite
C Massive, poddy arsenopyrite
D Quartz and chlorite with spots of chalcopyrite
E Massive green fluorite with some siderite
F Quartz and chlorite with streaks, patches and spots of
chalcopyrite, arsenopyrite, and pyrite

No. 1 LODE, BRIDFORD MINE AT 80 ft.
LEVEL (After Dines, 1956)

A Wallrock of steeply dipping shales and cherts
B Botryoidal barite with a little quartz
C Massive, banded barite with rare specks of sulphides

NORTH LODE, BAMPFYLDE MINE, AT
20 fm LEVEL (After Rottenbury, 1974)

A Hanging wall silts with malachite pods
B Gossan with copper minerals
C Leader of quartz with copper sulphides
D False wall of crushed slate
E Gossan poor in copper
F Grey slate footwall

ROMAN LODE, DEERPARK MINE, EXMOOR

A Slate wallrocks, soft and iron-stained
B Spongy brown goethite with quartz and inclusions of slate
C Granular and cherty quartz with cavities filled with earthy
red haematite
D Hard dark siliceous goethite with some limonite
E Pink sandstone wallrock

Figure 6.4. Representative Devon lodes

The mineral lodes over a considerable area of central and northeastern Dartmoor are characterised by the presence of specular haematite. At the Birch Tor and Vitifer complex this mineral occurs in association with the cassiterite-quartz-tourmaline assemblage in such quantities as to have hindered efficient upgrading of tin concentrates. Haematite crystallised somewhat later than cassiterite and is commonly seen replacing earlier quartz or growing along cleavage planes in biotite and feldspar of the wallrocks.

The haematite at Birch Tor and Vitifer is hard and granular, often formed as aggregates of splendent crystals several millimetres across. In contrast, the lodes of the Hennock and Lustleigh district in the northeastern part of the granite, bear a soft, fine-grained variety known as micaceous haematite. This was formerly mined for use in the preparation of anti-corrosion paints in which capacity it is superior to red oxide pigment. De La Beche (1839) noted the extraction of this material, and the industry continued throughout the nineteenth and much of the twentieth centuries. The last operation, Great Rock Mine (SX 827815), north of Hennock, closed in 1969 due to low reserves and rising costs of working.

Micaceous haematite occurs in lodes which vary greatly in width along their strike length, from fine strings only a few millimetres across to pods up to 1.2 m wide. The mineralised zones dip at a steep angle to the north and commonly show evidence of movement as vertical slickensides on the vein walls. The vein filling is predominantly fine-grained silvery haematite together with some coarser material. Quartz and tourmaline are commonly present in small quantities together with minor amounts of pyrite. Cassiterite is absent at Great Rock Mine, but small amounts are present in the haematite veins at Lustleigh. The granite wallrocks enclosing the haematite veins are locally impregnated with micaceous haematite and chlorite, which imparts a greenish colour to the rock. Kaolinisation is also common in the vicinity of the lodes (*Figure 6.4*).

Hydrothermal Sulphide Veins

Veins with predominant or significant contents of sulphide ore minerals are confined essentially to three areas outside the Dartmoor granite: around Mary Tavy, Plympton and Ashburton (*Figures 6.2 and 6.3*). All three areas have recorded productions of copper, arsenic and tin but the Mary Tavy district has by far the largest known output, and that mainly from one mine, Wheal Friendship (SX 506794). The beginnings of mining are uncertain, though Wheal Jewell (SX 528813) and Wheal Friendship were working in the late eighteenth century (Barclay, 1931) and continued periodically until 1925 together with Devon United Mines (SX 517789) (Dines, 1956). The deposits near Mary Tavy and Ashburton lie within the Dartmoor National Park and this has deterred exploration in recent years; at Plympton smaller lode deposits have been overshadowed by the attractions of stockwork mineralisation at Hemerdon. Waste dumps at Brookwood Mine (SX 697681) west of Ashburton, however, are presently being retried for fluorspar.

Production from Mary Tavy is dominated by the output from Wheal Friendship which according to Collins (1912) and Dines (1956) was at least 160,000 tonnes of copper ore, 18,000 tonnes of arsenopyrite, 720 tonnes of refined arsenic, 165 tonnes of tin and 7,053 tonnes of pyrite, in addition to significant quantities of lead and silver, and some zinc and scheelite. At various times this mine amalgamated with nearby Devon United and Wheal Jewell; the latter has no output recorded, but the former produced about 15,000 tonnes of copper ores, 1,500 tonnes of refined arsenic and 373 tonnes of tin concentrates. The extensive waste dumps of Wheal Friendship are well known to mineral collectors for their abundant specimens of arsenopyrite and chalcopyrite, many with associated scheelite and cassiterite. The gangue minerals are quartz, chlorite and dolomite.

Four major east–west lodes were worked in Wheal Friendship, but the two northern ones, which are separated by only 35 m, dip 45° to the north; they are joined by vertical 'dropper' veinlets, and numerous branches invade the cleavage of the enclosing slates. The resultant mini-stockwork has been followed to a depth of 450 m. A similar flat-dipping lode is recorded in the north of Wheal Jewell. Most of the other veins in this district also trend east–west and commonly dip steeply to the north, with only rare southerly dips. In Wheal Friendship the ore shoots appear to pitch westwards, away from the granite, and a similar pattern may be assumed for the other mines. Cassiterite and scheelite were best developed in the lower levels and their distribution suggests vertical thermal zonation of the ore minerals similar to that recognised in Cornwall. It is distinctly possible, therefore, that a significant tin resource could exist at depth below the Mary Tavy mines, an hypothesis which could only be proved by deep drilling.

In addition to the three major mines there are several small workings around Mary Tavy, most of which were developed little beyond the trial stage. Several of the copper-arsenic-tin veins are cut or heaved by north–south cross-courses parallel to, or extensions of, the Wheal Betsy Lode which is described later.

The outstanding feature of this small mineralised district is the ubiquitous and rich development of arsenopyrite in association with copper ores. Similar mineralisation on an equal or greater scale is seen only in the Gunnislake and Botallack areas in Cornwall. Comparing these three areas, it is tempting to deduce that the controlling factor for rich copper-arsenic deposition is the presence of mafic intrusive rocks (epidiorites) within thermally metamorphosed sedimentary rocks on the moderately gently dipping flank of a granite mass.

Less is known about mineralisation in the Ashburton and Ilsington area though there are several mines, one of which, Owlacombe and Stormsdown (Ashburton United) Mine (SX 770734) was worked until 1909. The deepest working is that at Brookwood Mine, 275 m deep, and this has recorded the largest output, some 30,000 tonnes of copper ore, a little pyrite and arsenopyrite.

The lodes of the area are variable in strike, from north–northeast to east–southeast, and the more important ones dip southwards at 60–70°. They seem to have been well defined structures but no descriptions exist as to their width, grade or nature. Two types of mineralisation are suggested by waste on the dumps. At Druids (SX 742717), Ausewellwood (SX 727709), Queen of the Dart (SX 734687) and Brookwood mines, all in the Ashburton district, the veinstuff consists of quartz and chlorite with pyrite, arsenopyrite and chalcopyrite; at the last named locality some specimens display a banded structure in which green and purple fluorite with siderite forms the final infilling. In contrast, the veinstuff from Sigford Consols (SX 774750), Owlacombe and Stormsdown, and Holne Chase (SX 722715) mines is of fine-grained quartz-chlorite-tourmaline with pyrite, arsenopyrite and scattered cassiterite in host rocks of tourmalinised slate. The lodes commonly seem to be brecciated. Again, in this area arsenopyrite appears to be an important lode constituent though less rich than in Mary Tavy and, apparently, only worked in a small way.

None of the Plympton mines is developed to appreciable depth and all were closed by 1880, though some have been re-examined in this century. Indeed, it was during work at Hemerdon Consols Mine (SX 565580) that the wolframite stockwork was discovered. Production statistics are almost certainly incomplete but outputs were generally small, the best from Bottle Hill Mine (SX 563589) known to have sold 2,032 tonnes of copper ore, 650 tonnes of tin concentrate, 30 tonnes of arsenopyrite and 13 tonnes of refined arsenic.

Two distinct forms of mineralisation are again evident. Lopez Mine (SX 515632), which lies west of the River Meavy, and Borrington Consols (SX 532585) north of Plympton, worked steep east–west lodes of composite type with early quartz, pyrite and arsenopyrite followed by chlorite veining carrying chalcopyrite in a brecciated slate host rock. At Lopez the lode was deeply weathered, yielding copper carbonates and native copper to a depth of 130 m and primary ores only in the bottom level (150 m); the lode is 1–4 m wide (Dines, 1956). In both mines galena and sphalerite are recorded and Borrington Consols sold lead, zinc and silver, but it is not clear whether some of these ores came from cross-courses. At Wheal Sidney (SX 550595), Wheal Florence (SX 575593), Bottle Hill Mine, Wheal Mary Hutchings (SX 564581) and Hemerdon Consols the wall rocks are of tourmalinised slate and may carry some tin. The lodes, varying in strike from east-northeast to east-southeast and usually about a metre wide, are of quartz and chlorite with cassiterite, pyrite and arsenopyrite, sometimes with local clusters of wolframite. Chalcopyrite and siderite are recorded from Bottle Hill Mine (Dines, 1956) where the lode reaches widths of 3.6 m. Wheal Sidney is unusual in that one of the lodes dips at only 40–45°; both here and at Bottle Hill late north–south lead veins cut the lodes but there is no record of lead production.

The isolated Wheal Emily (SX 542498), on the western bank of the River Yealm north of Newton Ferrers, which reputedly produced a little lead and antimony, and was reported to carry small amounts of silver (Dines, 1956), should also be mentioned here. This metallic association is suggestive of lower temperature deposition though, probably, these veinlets were not formed as late as the north–south lead lodes.

Stockwork Deposits

It seems probable that several small tin deposits within the Dartmoor granite which were formerly worked by open pits may have been of stockwork type. In addition to the examples previously quoted, descriptions by Dines (1956) of Gobbet (SX 649728), near Hexworthy, Crownley Parks (SX 760764), near Haytor, and Greatweek (SX 715876), near Chagford, suggest that these are also of stockwork type. Firm classification is prevented by the lack of contemporary descriptions and their present poor condition. If comparisons with better known Cornish examples are valid, their ore grade was low (about 0.2 per cent Sn) and their individual outputs probably less than 1,000 tonnes of concentrate. Mineralogically these deposits are simple; narrow tourmaline veinlets, with or without quartz, bear fine to coarse cassiterite, some pyrite and rare arsenopyrite or chalcopyrite. The host granite is commonly altered around the veinlets, usually ironstained and sometimes kaolinised.

By far the largest and now the most explored stockwork deposit in South West England is the wolframite-cassiterite body at Hemerdon. Discovered and initially developed during World War I (Terrell, 1920), it was again prospected just before and during World War II but never achieved full production. Demand for tungsten and concomitant price improvement encouraged renewed exploration in 1977. On the basis of wartime sampling, Cameron (1961) quoted reserves to a depth of 18.3 m of 4.57 million tonnes of ore carrying 0.14 per cent WO_3 and 0.04 per cent Sn. Recent work has indicated a proven 45 million tonnes of mineralised rock with an overall grade of 0.18 per cent WO_3 and 0.025 per cent Sn. Two boreholes which reached a depth of 400 m showed continuing mineralisation.

The stockwork is confined to the northern part of the Hemerdon Ball granite which has now been shown to have dyke-like form, some 120 m wide and dipping steeply eastwards. Its outcrop has been traced farther north than shown on the published geological map (Sheet 349) to within 150 m of the Crownhill granite outcrop. Texturally the granite is uniformly medium-to-coarse grained with large, generally anhedral feldspars. Within the stockwork the granite is always altered, variably kaolinised and iron stained. The southern, wider part of the outcrop is of harder, fresher and poorly mineralised granite.

Three intersecting quartz vein sets constitute the mineralised stockwork. The first strikes east-northeast–west-southwest, dips northwesterly at 45° and is commonly bordered by several centimetres of greisen; this set is particularly conspicuous in the south end of the body where the veins are closely spaced into sheeted zones. The other two sets are usually narrower and lack greisen borders; one trends northeast–southwest, the other north-northeast–south-southwest, and both have steep to vertical dips. Some of the northeast-southwest set are quartz-feldspar veinlets. Wolframite, cassiterite, and minor arsenopyrite (altered near the surface to scorodite) and haematite occur in all three sets but are best developed in the east-northeast–west-southwest set. Some of these veins persist a short distance into the surrounding Devonian slates, their mineralisation weakening.

The stockwork seems to have developed where the granite dyke cuts across an anticlinal crest within the country rocks; drilling results suggest that the orebody has a northerly pitch. Metal values are patchily distributed throughout the deposit.

Stratiform Deposits

At Belstone Consols (SX 632944) and Ramsley (SX 651931) mines, near Sticklepath, *bedded* copper deposits were worked intermittently from about 1860 to 1909 with a recorded output of some 13,000 tonnes of copper ores (Dines, 1956). There is reason to believe, however, that they may have been worked earlier than this and that the production figures are incomplete. From records of sales it can be deduced that the ore sold had a copper content of 12–17 per cent. These deposits form but part of a mineralised belt which flanks the northern margin of the Dartmoor granite and is marked by a pronounced magnetic anomaly. Trials elsewhere along the belt located similar mineralisation but, apparently, not in commercial grade or quantity.

The mineralised horizons are confined to the Lower Carboniferous outcrop and lie within, or just below, the Meldon Chert. Host rocks are shale and chert with some volcanic horizons, all of which carry disseminated pyrite, pyrrhotite and arsenopyrite, but the worked beds are usually garnet rich calc-silicate skarns which, at Ramsley, are locally intensely chloritised. The Lower Carboniferous rocks are disposed in a tight, southerly overturned anticline; at Belstone Consols and Ramsley the ore beds occur in the northern limb of this fold, at the Ford (SX 643935), Ivy Tor (SX 627934), Meldon (SX 570918) and Forest (SX 561912) mines (the last now beneath the Meldon Reservoir) the same beds were sought in the steeper-dipping overturned limb.

Ramsley Mine was 370 m deep and worked three beds, which were highly disturbed by the nearby Sticklepath Fault; Belstone Consols reached a depth of 180 m on Main Lode, which was 30 m thick (Smith, 1878), and investigated three other beds (*Figure 6.5*). Recent geochemical sampling suggests that additional unworked mineralised horizons may exist near Belstone Consols.

The only description of these deposits was by Smith (1878) who observed that copper sulphides occur both as widespread disseminations in the garnet rock and as small veinlets crossing the bedding, presumably in tension fractures. Unfortunately, the waste dumps at Belstone Consols contain very little ore material but pyrite, pyrrhotite, arsenopyrite, sphalerite, chalcopyrite and gersdorffite have been found, usually as disseminations or small patches. The Ramsley dumps, on the other hand, contain abundant ore specimens of both primary sulphides and secondary copper minerals. Kingsbury (1966) recorded large scheelite crystals from here and the same mineral from Ford, Ivy Tor, Meldon and Forest mines. Bismuth is reported from Ivy Tor Mine (Dines, 1956) though the mineral species is not identified. Forest Mine was quoted by De la Beche (1839) as a tin working, though dump specimens showed only abundant banded arsenopyrite and sphalerite. It is possible that the tin mineral present here was malayaite (a complex tin silicate) as has been described from the nearby Meldon (Red-a-Ven) Mine by El Sharkawi and Dearman (1966).

The unusual tin occurrence at Meldon may be indicative of an emanative centre and, by implication, suggest that part of the mineralisation is granite-derived. However, while this may be true for the higher temperature tin, and perhaps tungsten phases, there is less certainty about the sulphide mineralisation. Certainly some of the pyrrhotite was formed by the remobilisation of original sedimentary pyrite (Beer and Fenning, 1976) and the proximity of volcanic rocks in a condensed Lower Carboniferous shale sequence suggest strong possibilities that some part, at least, of the copper, zinc and arsenic content may have been derived from the sedimentary rocks by thermal reworking during emplacement of the granite. That granitic volatiles were involved at this stage is apparent from the formation of axinite, actinolite, datolite and bustamite in the skarns, and some of the ore constituents may have been introduced from the granite along with the volatiles. Cross-cutting

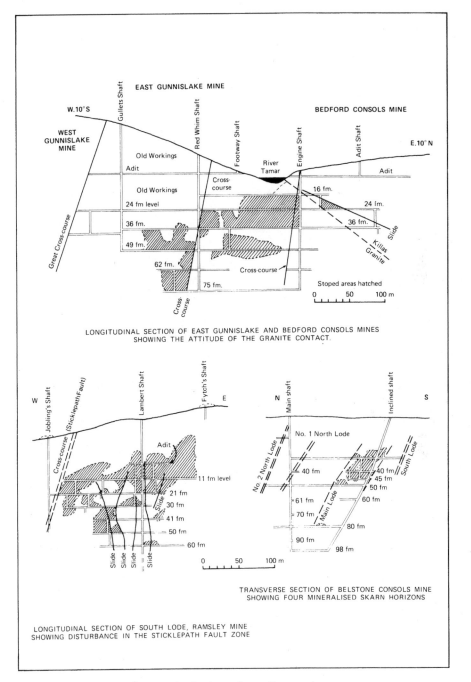

LONGITUDINAL SECTION OF EAST GUNNISLAKE AND BEDFORD CONSOLS MINES
SHOWING THE ATTITUDE OF THE GRANITE CONTACT.

TRANSVERSE SECTION OF BELSTONE CONSOLS MINE
SHOWING FOUR MINERALISED SKARN HORIZONS

LONGITUDINAL SECTION OF SOUTH LODE, RAMSLEY MINE
SHOWING DISTURBANCE IN THE STICKLEPATH FAULT ZONE

Figure 6.5. Sections of some Devon mines

veinlets of chalcopyrite bear witness to continuing post-metasomatic movement and deposition of ore minerals, presumably during a hydrothermal
phase which may be coeval with normal lode development.

To the south of the Dartmoor granite, at Haytor Mine (SX 771771), magnetite was worked from stratiform deposits in thermally metamorphosed Carboniferous sandstone and shale. Dines (1956) stated that *lodestones* from the
district were known in the sixteenth century and that the main period of
working was in the nineteenth century when some 12,300 tonnes of iron ore
were raised. There are three beds of ore 3 m, 4.3 m and 1.8 m thick, dipping
about 40° to the northeast.

Mineralogically the beds consist of an intergrowth of magnetite and hornblende, the latter altered to actinolite. Garnet is abundant and other skarn
minerals, notably axinite and apatite, are recorded. The ores also carry disseminated pyrite, chalcopyrite and arsenopyrite. Chalcedony is present in the
ore beds and the altered sedimentary rocks which separate them; this mineral,
pseudomorphous after datolite, was erroneously erected as a new species
(haytorite).

Apparently similar magnetite beds were worked nearby at Smallacombe
(SX 777766) and Atlas (SX 779762) mines. In the former working three beds of
nodular limonite, up to 0.75 m thick, lie several metres above the magnetite;
in the latter much of the ore was haematite and may have been from an
altered nodular horizon. The presence of unmodified hydrous iron oxides
suggests that the magnetite has not been derived by metamorphism of sedimentary ferruginous layers as considered by Collins (1912); an association of
skarn rocks and minerals supports a pyrometasomatic genesis, the iron perhaps remobilised from igneous rocks in the sedimentary succession. Magnetite from the volcanic outcrops occurs widely in south Devon stream alluvial
deposits (J. H. Walbeoffe-Wilson, personal communication) and impregna-
tions of magnetite are seen in volcanic tuff at Bulkamore Mine (SX 749631)
south of Buckfastleigh (Dines, 1956).

THE GUNNISLAKE–TAVISTOCK AREA

The largest mines of Devon are situated in the Gunnislake-Tavistock area,
and their output of copper, arsenic and pyrite dominates the county's mineral
production. By far the biggest producer was Devon Great Consols Mine (SX
435736), north of Gunnislake, in which the main lode was followed to a depth
of 550 m and traced over a strike length of 4 km. So important was the
mineral output that a shipping quay, storage bins and lime kilns were constructed at Morwellham and linked to the mine by an inclined railway track.
Incomplete returns for the major mines are listed in *Table 6.2*. Only three
other mines in the area produced more than 10,000 tonnes of copper ores.

Most of the Gunnislake mines began working in the early 1800s, producing
copper until 1880 and arsenic from about 1865 to 1910. Some continued in a
small way until 1925. Mines farther east seem to have started later although
Virtuous Lady Mine (SX 474697) probably began working in the sixteenth
century (Dines, 1956). Several properties were re-examined for tin around
1920–1925 and for wolframite around 1940. Ding Dong Mine (SX 436717)

Table 6.2 MAIN ORE PRODUCTION IN THE GUNNISLAKE–TAVISTOCK AREA

	Devon Great Consols (SX 435736)	Bedford United (SX 441726)	Wheal Crebor (SX 463723)	Gawton (SX 452688)
Copper-ore	776,080	67,100	41,860	22,230
Copper-metal content	49,975	4,570	2,720	1,000
Arsenopyrite	550	13,720	25,225	3,915
Arsenic (crude and refined)	72,970	2,145	—	15,905
Pyrite	9,735	510	3,100	1,460

and some of its neighbours are being explored at present and in recent years the calcined dumps at Devon Great Consols have been retreated for their fluorite, cassiterite and wolframite contents.

Mineralisation is essentially confined to an east–west zone, some 5.5 km broad, between the latitudes of Tavistock and Buckland Monachorum, and centred over the concealed granite ridge postulated by Bott *et al.* (1958) as linking the Dartmoor and Bodmin Moor outcrops. Only one mine penetrated this ridge; Frementor Mine (SX 425725) in the Gunnislake outcrop and, surprisingly, there are no records of granite veining in the deeper workings of any of the others. The increased importance of tin in lodes east of the River Tavy, however, suggests that the ridge culmination is nearer surface in this area and, therefore, that it probably plunges westwards.

The lodes all have strikes orientated between east-southeast–west-north-west and east-northeast–west-southwest, as do the associated elvan dykes, closely following the trend of the granite ridge. Most dip fairly steeply, both north and south, but in rare cases dips as low as 45–50° (all northward) are recorded and at Virtuous Lady Mine there is a unique dip of only 10–15° north. Lode widths differ considerably, ranging from 30 cm to 10 m, and individual structures may pinch and swell repeatedly. The most extreme variation is again found in Virtuous Lady Mine, where the lode width ranges from a few centimetres to 9 m. In some instances, as at Yeoland Consols Mine (SX 517662), south of Yelverton, closely spaced narrow veinlets within altered, sometimes mineralised, slate constitute a lode zone up to 4 m across; usually the ore is then bunchy.

Many of the lodes are cut by late northwest–southeast trending cross-courses, some displacing the earlier veins by several metres. In Devon Great Consols, however, the Great Cross-course heaves the Main Lode some 225 m in a right-lateral (dextral) sense. Most of the cross-courses dip westwards (*Figure 6.5*) at moderate to steep angles, and some carry a little galena and sphalerite mineralisation.

Only in a few mines is the pitch of the ore shoots definable and this is invariably towards the east and gentle, at angles of 15–45°. In the richer mines, such as Devon Great Consols, more than 60 per cent of the developed lode may have been removed by stoping.

Throughout this extensively mineralised area four major phases of ore emplacement can be recognised though not all are necessarily represented in any individual lode. The earliest phase caused intensive tourmalinisation of the slates around the main fractures and the introduction of disseminated fine cassiterite. This was followed by a fracture infilling of quartz and a little chlorite with cassiterite, wolframite, some pyrite and abundant arsenopyrite. After re-opening the fractures were again filled by a third phase, depositing chlorite with some quartz, pyrite, chalcopyrite, further arsenopyrite and a little galena, sphalerite and siderite. The final episode was the infilling of all fractures and cavities by fluorite and siderite with a little pyrite and chalco-pyrite.

Relationships of these phases are best recorded from mines along the banks of the River Tamar, especially at Frementor, Devon Great Consols, Bedford United (SX 441726) and Gawton (SX 452688) mines in which some of the lodes are banded (*Figure 6.4*). At Devon Great Consols and Bedford United the central parts of the lodes, rich in copper, were worked first and the outer parts were later stripped for their arsenopyrite or cassiterite-wolframite con-tent. Dines (1956) recorded that some of the arsenopyrite occurred in large lensoid bodies up to 2 m thick and areas of the unworked outer lode zones assayed 8–10 per cent As_2O_3. In Ding Dong Mine the chlorite-sulphide phase is emplaced between a quartz-cassiterite-wolframite vein and its hangingwall slates but, followed eastwards, the two structures separate. Available assays suggest that neither the tin nor tungsten content of the Tamar lodes exceeds one per cent.

In the mines to the east, less is known about the disposition of ores in the veins, but it appears from the few descriptions available and from the dump material that the tin and sulphide depositional phases are less obviously separated as zones in the lodes but appear to be more intimately mixed; their age relationships are still determinable, however. In this area, the unusual lode at Virtuous Lady Mine shows an equally unusual texture with massive and fractured vein quartz cemented by chlorite in which the ore minerals cassiterite, chalcopyrite, arsenopyrite and pyrite—occur as disseminations and lensoid patches. Siderite and fluorite infill the late cavities, the former in large crystals with curved faces (Dines, 1956). The vein is also noted for its spherical masses of hydromuscovite.

Surprisingly few exotic mineral species have been noted from this area though there are several secondary copper minerals and scorodite locally replaces arsenopyrite. Gold and silver have been assayed in some of the copper ores and gossans but are not recorded in mineral form. Surface alter-ation of the veins is a marked feature of this area and may extend to consider-able depths. Dines (1956) recorded gossans down to depths of 75–110 m in Devon Great Consols and at New East Russel Mine (SX 450710) the primary copper ores are leached out to depths of 215 m. Barclay (1931) noted that rich tin values were found in the gossans of Bedford Consols Mine (SX 457691) and Dines mentioned that at Tavy Consols Mine (SX 468668) tin and arsenic ores were found above copper ores, an apparent reversal of normal mineral zoning but explicable if the upper levels were all in gossan.

The question of vertical mineral zonation is a vexed one. Barclay (1931) assumed that the rich copper shoots of the Gunnislake mines would give way in depth to tin rich bodies. This has only been tested at the 550 m deep Richard's Shaft of Devon Great Consols Mine where a well defined lode of quartz, chlorite and tourmaline carried only traces of tin (Dines, 1956). Bar-clay argued that insufficient development had been carried out to prove or disprove the existence of tin-ore shoots. In the banded lodes, however, the high temperature cassiterite-wolframite-arsenopyrite phase is well represented in the outer bands and doubts must be entertained about its continuity at depths greater than that explored in Devon Great Consols. There is, in fact, little concrete evidence in the Tamar mines to suggest vertical mineral zonation on the pattern seen in southwest Cornwall. This is not to say, however, that such a feature may not exist further east, where admixture of the oxide-sulphide mineralisation phases is apparently more common. Here, in mines such as Wheal Crebor (SX 463723), Crowndale (SX 471725) and Sortridge Consols (SX 511707), there are vague pointers to the possibilities for improved tin grades beneath old copper workings.

The adits which penetrate the Devon bank of the River Tamar have long been a challenge for the local caving clubs and recently part of the adit system of the George and Charlotte Mine (SX 453699) has been opened to the public as a museum of nineteenth-century mining.

LATE MINERALISED CROSS-COURSES

Several of the east–west tin or sulphide lodes are cut, dislocated or even terminated by mineralised cross-courses which trend generally north–south. Such structures are widely represented throughout the Dartmoor and Gunnislake areas but the best known, most continuous and highly productive are those of the Tamar Valley, Mary Tavy and the Teign Valley. Argentiferous galena, sphalerite, fluorspar and barytes are the usual products but pitchblende occurs in one of these veins at Kings Wood Mine (SX 713665) west of Buckfastleigh.

In the Tamar Valley two parallel north–south veins, some 1.2 km apart, have been worked intermittently since the thirteenth century; they can be recognised to the north where they cut the Gunnislake tin-tungsten-copper lodes. The western vein, apparently 1.2 m wide (Dines, 1956) has been worked for lead and silver between South Ward Mine (SX 425687) and South Hooe Mine (SX 425657), in part beneath the River Tamar. Workings were deepest at the southern end, reaching 465 m. The eastern structure, mined almost continuously from Buttspill Mine (SX 438680) to South Tamar Mine (SX 436646) and probably traceable further southwards, has a strike length of at least 5 km. Again the deepest workings are in the south, at South Tamar (270 m) where they were flooded when levels beneath the river were breached.

Most waste dumps show quartz, fluorite and calcite gangues with rare traces of barite, abundant sphalerite and common pyrite but only scattered amounts of galena. Buttspill Mine dumps are notable for their content of green and colourless fluorite and this mineral was worked from the dumps of Lockridge Mine (SX 438665) in 1942.

Production statistics are woefully incomplete but the recorded output of the valley mines is some 25,400 tonnes of lead ores (16,150 tonnes Pb) and 18,980 kg of silver, of which more than half is attributed to South Hooe Mine. The early, near surface workings were particularly rich in silver but the content diminished with depth, clearly indicating enrichment by the action of percolating surface waters.

A similar north–south vein near Mary Tavy was worked in Wheal Betsy (SX 510812) to a depth of 310 m and in Wheal Friendship, but can be traced further north and south over a strike length of 5 km. It is also a narrow structure, up to a metre wide, and in Friendship has a westerly dip of 50°, shifting the earlier veins by as much as 55 m in a left-lateral (sinistral) direction. Total recorded production between 1820 and 1877 was 10,360 tonnes of lead ores (7,415 tonnes Pb) and 3,085 kg of silver, with minor amounts of zinc.

The protected Wheal Betsy engine house chimney is a well known landmark on the Okehampton–Tavistock roadside and the dumps a favourite hunting ground for mineral collectors. These dumps bear abundant samples of quartz, siderite and a little fluorite with pyrite, sphalerite and galena. Similar ores occur at Wheal Friendship, where Kingsbury (1966) recorded cobaltite intergrown with the galena.

Farther north, around Lydford (SX 510849) comparable lead veins have been worked in a small way. At Silverbrook Mine (SX 790758) near Ilsington the vein trend is northeast–southwest with a southeast dip of only 40–50°. Dines (1956) reported the vein as 30 cm to 1.5 m wide, carrying pyrite, sphalerite and galena in a quartz, calcite and barite gangue. The main product was zinc ore. Ivybridge Consols Mine (SX 646551), southeast of that town, worked an easterly dipping north–south lode to a depth of 140 m. Galena, sphalerite and pyrite occurred in a gangue of quartz, fluorite and pink calcite. At Loddiswell Mine (SX 721516) south of South Brent, argentiferous chalcocite occurs with galena in quartz and barite.

Between Dunsford (SX 812891) and Hennock (SX 831808), in the middle Teign Valley, a wide lode zone parallel to the granite contact has been worked for lead, zinc and spar minerals over a strike length of 8 km. Recent studies (Beer and Ball, 1977) have shown that the mineral veins splay out to the north and continue for 1.2 km beyond the southernmost workings. Lead mining probably began in about 1800 (Schmitz, 1973) and ended by 1880; barytes was first sold in 1855 but constant production dates from 1875 to 1959. The largest lead and silver producers were Frankmills Mine (SX 836819) and Wheal Exmouth (SX 838827), respectively 265 m and 195 m deep and the only barytes mine was Bridford (SX 829864), 182 m deep.

The vein zone contains several lodes (five were worked at Bridford) but in the south the main lead lode seems to be a continuous, well mineralised structure with only one or two significant branches. On average the lead lodes were some 1.3 m wide and consisted of quartz, barite and lesser fluorite and calcite with galena, sphalerite, pyrite and minor tetrahedrite. Cerussite occurs in small amounts and at Frankmills there is much siderite in the gangue and small amounts of stibnite are reported (Schmitz, 1973). The barite veins vary from 30 cm to 12 m in width but average about 2 m. They are predominantly of banded or botryoidal barite (*Figure 6.4*) with a little quartz, pyrite, galena, sphalerite and tetrahedrite.

There is a suggestion that the wall rocks influence the lode content, slaty host rocks favouring lead and cherty rocks favouring barite. If this is so, further lead and barite resources may be definable in the Teign Valley.

The mineralised cross-courses are clearly younger than the east–west sulphide veins but it is generally agreed that their ores, always of a lower temperature suite, are probably derived from waning hydrothermal emissions from the granite. Lead isotope determinations yield an age of about 260 Ma (Moorbath, 1962) but uranium-lead dating of the Kings Wood pitchblende shows that this uranium-lead-copper-cobalt-nickel vein is of somewhat younger age, 206 Ma (Darnley et al., 1965).

MANGANESE

Manganese ores are widespread in Devon, but the deposits are small and patchily distributed and their grade is generally low. In spite of this a significant tonnage of ore was raised during the period from the late eighteenth century to the end of the nineteenth. Initially the ores were used in the production of glass; small quantities added to the melt acted as a decolouriser, while larger amounts imparted colour—the distinctive lilac hue of 'Egyptian Ware' is due to manganese. From the early years of the nineteenth century an important use was as a reagent in the manufacture of bleach. Later on, demand for manganese grew with the development of the steel industry. High-grade foreign ores were imported on a large scale and the workings in Devon were unable to compete.

Three districts provided the bulk of production and these are listed below, together with the dates of working provided by Russell (1970).

Huxham, Upton Pyne and Newton St Cyres (Exeter district)	1788–1849
Christow, Ashton and Doddiscombsleigh (Middle Teign Valley)	c1810–1875
Milton Abbot and Brentor districts (West Devon)	1815–1896

The quantities raised at other localities and at times later than 1896 were insignificant, the workings being little more than trials.

In the Exeter district, deposits were worked from the Permian rocks close to the southern margin of the Crediton trough. De la Beche (1839) considered that the ores were associated with a fracture which splayed off from the fault separating the Culm rocks from the New Red Sandstone. Most of the workings are obscured at the present day and there is little to indicate the nature of the deposits. Some material, however, has recently been collected from a trench at Upton Pyne (SX 911977) and this consists of nodular masses of pyrolusite and manganite, with traces of rhodochrosite, barite and kaolinite (B. V. Cooper, personal communication). It was stated by Dewey and Bromehead (1916) that the manganese minerals form a cement for the sandstone host, swelling out locally into little pockets of ore.

The deposits of the Middle Teign Valley and of the Milton Abbot and Brentor districts are similar in that they occur in Lower Carboniferous chert beds which are invariably associated with contemporaneous igneous rocks— tuffs, spilitic lava and intrusive dolerite. The ore bodies are very irregular and were described by Dines (1956) as consisting of patchily developed replacements of the chert beds, usually linked by mineralized cracks. The enclosing cherts are commonly impregnated with fine disseminated pyrite, and in the West Devon district, ramifying, thin veins of quartz cut the mineralised ground. The principal mineralogical constituents of the ores are the oxides and hydrated oxides of manganese. Dines stated that at Chillaton and Hogstor Mines (SX 432812) the main mineral was pyrolusite, which enclosed kernels of rhodonite or of rhodochrosite. Pyrolusite and psilomelane enclosing rhodonite are found at other localities in the West Devon district and it may be that the chert was originally replaced by the silicate (rhodonite) which was later altered to the oxide assemblage. At Hill Copse Mine (SX 840843), in the Middle Teign Valley near Christow, oxides and hydroxides of manganese including pyrolusite, psilomelane and wad impregnate pyritic chert. Dines noted psilomelane as a constituent at other mines in this district.

The origin of these ores is to some extent a matter of speculation, but the lithological control of deposition, and the ubiquitous presence of mafic igneous activity, suggest that they may have formed from contemporaneous metal-rich solutions associated with the volcanism. It is unlikely that there is any direct genetic association with the emplacement of the Cornubian granite, as they are earlier than the intrusion, though later partial remobilisation of the deposits has undoubtedly occurred.

MINERAL EARTH PIGMENTS

Included under this heading are the various deposits of ochre, umber, red oxide pigment and mineral black which have been worked from time to time in Devon. The use of micaceous haematite as a pigment has already been discussed. Definitions of the substances listed above are as follows:

Ochre—contains finely divided limonite ($2Fe_2O_3.3H_2O$) in admixture with clay. It may contain from 17-60 per cent Fe_2O_3 and provides a range of yellow pigments.

Umber—similar in composition to ochre, but contains 10-20 per cent MnO_2. It yields a deep brown pigment, burnt umber, on being calcined.

Red oxide pigment—contains from 15-65 per cent Fe_2O_3 as haematite disseminated in a siliceous or clayey base. The red colour is very variable in hue.

At Combe Martin in north Devon, umber was formerly worked from deposits in rotted Devonian limestone. The manganese content was probably derived from the weathering of manganiferous siderite, a mineral commonly present in that district. Dines (1956) stated that old quarries at Muddiford (SS 566387) 'seem to have been openworks on lodes of soft iron oxide with some manganese, which may have been worked for umber and ochre'. Other localities from which these substances have been worked undoubtedly exist. In a different geological setting at Ashburton, the Devon and Cornwall Umber Works (SX 762705) produced some 7,000 tonnes of umber in the later part of the nineteenth century. The product was stated by Dines to contain 28 per cent MnO_2 and to occur in dolomitic Devonian limestone that had been altered in contact with underlying dolerite. Ochre and umber were also worked from Smallacombe Mine (SX 777766) where beds of limonitic nodules occur in thermally metamorphosed Carboniferous rocks. Decomposed and altered greenstones furnished ochre at several localities in west Devon, one of which, the Whitstone Ochre Works (SX 463818) was in production until 1942.

The principal source of red oxide pigment was the area around Sharkham Point (SX 936546) near Brixham. Deposits of haematite, occurring as irregular replacement bodies in limestone of Middle Devonian age, were worked for use as a pigment and for smelting. The earthy variety of haematite was employed for the former application, and kidney ore for the latter. Deposits of limonite and ochre were also worked in the vicinity.

Mineral black is not a metalliferous species, being composed of carbonaceous clay, but as it has been extensively worked in north Devon, it is appropriate to include it with other mineral pigments. Seams of this material, consisting of about 33 per cent carbon in a clayey matrix, have been worked for many years from Culm rocks in the Bideford district. The product, known as Bideford Black, found its principal application in the paint industry where it served as a cheap substitute for lamp black. After crushing, the run-of-mine material was upgraded by froth flotation prior to drying and packing. The last workings closed in about 1960.

OTHER IRON ORES

In south Devon there are various iron ore deposits that have been worked, mostly on a small scale, and which do not justify a separate genetic or geographical classification. Such are the iron mines of the South Hams, which were described by Jenkin (1974). Virtually no geological evidence is available about these workings, but it may be assumed that they are not unlike the iron ore veins of north Devon.

The iron mine at Bulkamore (SX 749631) near Rattery is of interest for its unusual geological setting. Devonian slate is in contact with volcanic tuff along a plane which dips steeply to the north. Near surface, this contact consists of a mass of limonitic material which produced 4,470 tonnes of ore in 1874–1875. Below the gossan zone, pyrite and magnetite impregnate the pyroclastic rocks while veins of quartz and siderite with chlorite are seen in the slates. Details of the mineralisation are given by Dines.

The gossan zone of a vein deposit was worked at Shaugh Iron Mine (SX 533633), north of Plympton, which produced 4,745 tonnes of limonitic ore between 1870 and 1874.

LOCALITIES

North Devon

6.1. *Lester Rock, Combe Martin beach* (SS 575476): lead mineralisation. Lead mineralisation in the Combe Martin district is best seen in the coastal exposure on the seaward side of Lester Rock. Access is from Combe Martin beach, following the line of the sewage outfall; the site can only be reached at low tide. Galena and sphalerite form irregular streaky masses along the cleavage of the Lester Slates and Sandstones in the core of an overturned anticline.

6.2. *Cornham Ford, Exmoor* (SS 749386): haematite ore. Adit workings and waste dumps of haematite ore can be conveniently examined down river northeast of Cornham Ford reached by tracks from the Challacombe–Simonsbath road (via Cornham Farm), the North Molton–Simonsbath road or along the River Barle from Simonsbath Bridge. The locality is in enclosed moorland grazing but the tracks are public rights of way.

6.3. *Raleighs Cross Mine, Brendon Hills* (ST 025342): sideritic iron ore. Sideritic ore is best seen on the site of Raleighs Cross Mine, 275 m west of the Watchet–Bampton–Wheddon Cross road fork at Brendon Hill.

The mine site and dumps lie a few metres south of the Bampton road in enclosed farmland; the farmer's permission is required for access. The incline of the old iron ore railway can be viewed on the north side of the Wheddon Cross road, 500 m west of the same road fork.

6.4. *Bampfylde Mine, North Molton* (SS 740327): copper-iron ores. Large dumps of copper–iron ores from the Bampfylde mine extend along the east bank of the River Mole where its valley turns eastwards some 600 m north of Heasley Mill, itself 2 km north of North Molton. Immediately west of the road there are two adits to the mine, and poorly mineralised rock dumps up the valley side.

Dartmoor

6.5. *Hexworthy* (SX 656718): stream tin. Evidence of former stream tin working is widespread on Dartmoor along the valley floors. Particularly good localities are along the Newleycombe Brook and the O Brook Valley near Hexworthy. The former site can be reached from a track which leads northeast from Norsworthy Bridge (SX 568693) at the head of Burrator Reservoir along the left bank of the stream. The valley of the O Brook is accessible to the north and south from Saddle Bridge (SX 665719), on the road from Hexworthy to Holne.

6.6. *Birch Tor–Vitifer complex* (SX 682814): tin lodes. The Birch Tor and Vitifer complex provides the best example of tin lode workings on Dartmoor. Extensive gullies cut across the open moorland, and the spoil heaps provide examples of vein material and altered wallrock. Access is by walking south from Bennett's Cross (SX 680816) on the Moretonhampstead–Plymouth road or from the track which leads southeast from a point 200 m northeast of the Warren House Inn (SX 684811). The workings occupy a tract of country about 1.5 km wide to the south of the road.

6.7. *Wheal Friendship* (SX 506794): copper mineralisation. Copper and arsenic minerals are well represented on the waste heaps of Wheal Friendship, and cassiterite, scheelite, sphalerite and galena may also be found. The dumps are widespread, some in enclosed land. They are best examined near the bridge (SX 506794) in the lower part of Mary Tavy village, reached from the Okehampton–Tavistock road by any of three lanes which run east from the roadside part of the village.

6.8. *Bottle Hill Mine dumps* (SX 563587): tin-tungsten-copper minerals. Lode material can be seen on the dumps of Bottle Hill mine, though these are within private land and permission to visit is needed. Access is from the Plymouth–Cornwood road, turning off north at Newnham along the road signposted to Lee Moor and forking right about 1.6 km along this road. The dumps lie either side of this lane, some 350 m from the turn off.

6.9. *Hemerdon Mine* (SX 573585): tungsten mineralisation. The open pit in the Hemerdon stockwork lies a short distance east of Bottle Hill and can be approached via narrow lanes from that site. The best route, however, is from the Plymouth–Cornwood road turning off towards Hemerdon village, taking a righthand turn in the centre of the village, posted to Drakeland. About 1.5 km along this lane is a narrow track forking off to the right which leads to the mine area. Permission to examine the open pit must be sought at the office and protective wear (hard hats, etc) is obligatory. The veining and kaolinisation are well exposed and the mineralisation is abundantly evident in surface spoil.

6.10. *Ramsley Mine, Sticklepath* (SX 649930): bedded copper-zinc-arsenic mineralisation. The minerals are conveniently displayed in the large dumps of Ramsley Mine; these dumps are in process of removal and good specimen material is continuously exposed. The dumps lie beside the A30 Okehampton–Exeter road, close to the bridge beside the Owls-foot Garage in South Zeal. Access is via the lane running south towards Throwleigh following the stream and emerging close at the bridge.

Gunnislake Area

6.11. *Devon Great Consols Mine* (SX 425735): sulphide mineralisation. The complex sulphide mineralisation of west Devon and east Cornwall is best displayed in the widespread dumps of Devon Great Consols Mine. These are on private land and may be in the process of being worked; if so, permission is required of the operators. Access is from the Gunnis-lake–Tavistock road turning off at Gulworthy towards the Horn of Plenty Restaurant. Some 500 m along this road there is a narrow lane towards the left advertised to Wheal Josiah. Followed for its full length it leads into the area of dumps and arsenic flues.

6.12. *Morwellham* (SX 446697): old mine workings and buildings. Conducted tours of the reinstated parts of the George and Charlotte Mine can be combined with visits to the old quays, storage bins, limekilns and rail-way incline at Morwellham. Access to the latter is from the Tavistock–Gunnislake road and is well signposted from Gulworthy and the inn to the east (SX 455727). Details are obtainable from the Morwellham Quay Recreation Centre, telephone Gunnislake 766.

6.13. *Bere Alston* (SX 432664): quartz-fluorite-sphalerite-galena mineralis-ation. This mineralisation from late cross-courses is well represented on the dumps of Lockridge Mine which lies beside the lane from Bere Alston to Lockridge Farm immediately west of the old railway track. Similar material occurs on dumps at Furzehill Mine at the minor cross-roads (SX 437655) immediately northwest of Cotts, some 1.5 km south-west of Bere Alston.

6.14. *Wheal Betsy* (SX 511812): sulphide mineralisation. Quartz-siderite-pyrite-galena-sphalerite minerals from a late cross-course are abundant on the dumps of Wheal Betsy, just east of the Tavistock–Okehampton road, 2 km north of Mary Tavy. A very rough track leading to the dumps and engine house leaves the road just after the cattle grid mark-ing the edge of the unfenced moorland section of the road.

6.15. *Teign Valley* (SX 842830): mixed mineralisation. Quartz-barite-sphaler-ite-galena-pyrite vein material occurs on the dumps of Wheal Exmouth (SX 842830) which line both sides of the road following the River Teign between Dunsford and the A38 dual carriageway at a point 1 km south of the Lower Ashton crossroads. An exposure of barite veining in a disused quarry (SX 829865) is reached farther north along the valley road by turning off to Bridford at Bridfordmills and forking left some 650 m uphill. The quarry, which is privately owned, is about 100 m along this lane.

Chapter Seven

The New Red Sandstone

In Devon, as in many other parts of Britain and Europe, the transition from the plate-convergence/orogenic regime of the Palaeozoic to the stable shelf conditions of the Mesozoic was marked by the deposition of a major terrestrial red-bed succession, the New Red Sandstone. Its deposition can be regarded as the last chapter of the Variscan Orogeny in the region, recording the degradation of the upland areas and dispersal of the eroded material to infill surrounding basins; or it can be seen as the commencement of the shelf-type sedimentary regime that lasted in the Cretaceous and beyond. In Devon it offers perhaps the best example of a red-bed succession to be found anywhere in Britain, and the finest and best-exposed non-marine Permian rocks in western Europe.

Although less than one quarter of Devon is underlain by the New Red Sandstone, its presence imparts a warm red coloration to much of the landscape; to most people, Devon is a predominantly red county. The south coast section displays an aggregate thickness of some 3,000 m of gently dipping strata, from Tor Bay northwards and eastwards: the breccias, aeolian and fluvial sands, conglomerates and mudstones ('marls') that make up the sequence form a coherent palaeoenvironmental picture of sedimentation under semi-arid climatic conditions on the flanks of a mountain range. Along this coast, where exposure and accessibility are generally excellent, most of the sedimentary features that characterise red-bed formations around the world can be seen in context; climate and scenery combine to make this a favourite location for geological field excursions.

The outcrop of the New Red Sandstone, which includes uppermost Carboniferous, the Permian and nearly all the Triassic in Devon, is confined mainly to the east of the county, linking to the Triassic basins of the English Midlands through Somerset, and dipping eastward beneath the Jurassic of Dorset into the Wessex Basin (*Figure 7.1*). The highly irregular western margin of the outcrop reflects the uneven topography of the land surface upon which the red beds were deposited: this was a desert landscape of mountain ridges and valleys, trending east–west and cut by faults. A number of semi-elliptical cuvettes, smaller basins which fringed the western side of the Wessex Basin, can be distinguished today from Tor Bay (itself a half-submerged cuvette) northwards into Somerset; of these, the Crediton Valley cuvette is the most outstanding, projecting westwards some 40 km from the main outcrop. The cuvettes contain coarse alluvial fan deposits which grade into finer grained sedimentary rocks formed in an aeolian and flood-plain/

148

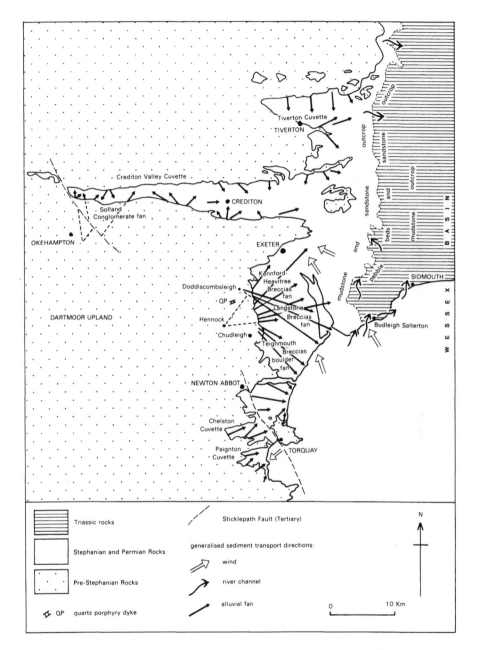

Figure 7.1. Depositional basins and sediment transport trends in the New Red Sandstone (younger formations omitted).

channel-complex depositional environment. These are all coloured in varying shades of red, with pale green, yellow and grey spots, layers and streaks occurring in a few fine-grained beds; the predominant colouring matter is ferric oxide, mainly as haematite.

Topographically, most of the outcrop is lowland, 'good red earth for the plough', less hilly than the older Palaeozoic rocks and displaying deep steep-sided valleys only where capped by Cretaceous rocks in east and south Devon. Nevertheless, the basal breccias and the conglomerate fomations higher in the succession give rise to more resistant escarpments and hill forms, and in many places the unconformity is marked by irregular high ground or more definite escarpments formed by the red-beds.

STRATIGRAPHY AND GEOLOGICAL HISTORY

Following the magmatic and orogenic events that culminated during the late Carboniferous in the Variscan Orogeny, sedimentation recommenced in Devon under semi-arid climatic conditions with coarse clastic deposits forming in intermontane valleys and wider low-lying desert basins. Sedimentation continued, apparently without interruption apart from minor tectonic upheavals, until the end of the Triassic: marine conditions were then re-established, ushering in the subsequent shelf-sea regime of the Jurassic and Cretaceous. The major non-marine red-bed sequences formed between these two events record much of the terrestrial history of the region, but virtually nothing of the faunal transition from Palaeozoic to Mesozoic. In fact, the position of the Permian–Triassic boundary, set conventionally at the base of the Budleigh Salterton Pebble Beds, is a matter decided by comparison with other successions rather than by any indigenous evidence, for fossils are almost completely lacking in this part of the succession and below.

This lack of fossils has meant that stratigraphic correlation in the New Red Sandstone has had to rely heavily upon lithological similarities. It has also meant that, compared with the marine successions above and below it, the New Red Sandstone strata have been unattractive to research workers until sedimentological and radiometric dating tools were developed to aid interpretation. Nevertheless, early accounts attributed Permian as well as Triassic ages to the succession (Murchison, 1854), largely on the basis of comparison with successions elsewhere and the presence of the Exeter Volcanic 'Series', mafic lavas which occur at and near the base of the sequence in the Exeter, Crediton and Tiverton areas. The new name New Red Sandstone was applied much earlier (Conybeare and Phillips, 1822), to distinguish it from the Old Red Sandstone which, as was recognised in other areas, lay beneath the 'Coal-Bearing Series' (Carboniferous), while this succession lay above. The name represents a lithostratigraphic supergroup in modern terminology (Laming, 1968), composed of a number of formations and groups defined on lithological rather than chronostratigraphic criteria; some of the units are definitely diachronous, but most appear to be roughly time-consistent and fit into the chronostratigraphic framework established elsewhere.

Age criteria

Radiometric dating of mafic lavas from Dunchideock (Miller and Mohr, 1964) and Killerton (Miller et al., 1961) provides the only age bench-mark in the lower part of the sequence: regrettably, the stratigraphic position of the Killerton lava mass is unknown and that of Dunchideock uncertain. The results were 279 Ma and 281 Ma respectively, very close to the accepted date of 280 Ma for the boundary between Carboniferous and Permian Periods. The basal part of the sequence, on this evidence, is thus late Carboniferous (Stephanian) in age (Laming, 1965) but how far back into the Carboniferous the red-bed deposition began is less easy to say. With the date of the main intrusion of Dartmoor granite possibly 300 Ma, it is unlikely that they date back before the early Stephanian, but the lavas at Kingsand, interbedded with red breccia beds just above the unconformity, have been dated at about 295 Ma which suggests that red sedimentation began very soon after the intrusion of the granite.

The youngest Culm rocks in the Exeter area are G_2 (early Westphalian) but there are younger rocks involved in Variscan folding in the Radstock area, which put the date of orogeny also at early Stephanian, at least in that area. Conceivably, folding and the onset of red-bed deposition could have occurred almost simultaneously at Radstock, indeed the overlap may have been a considerable length of time as neither can be thought of as an 'instantaneous' event; the stratigraphic hiatus between mid-Westphalian and early Stephanian strata in Devon is explainable on the basis of the intervening strata being removed following uplift. The causes and results of the orogenic/intrusive events preceding the New Red Sandstone are not the subject of this chapter, however: it is sufficient to remark that the silicic plutonism associated with the Variscan Orogeny subsided before the end of the Carboniferous; that after the plutonism there were extensive volcanic outpourings in the Dartmoor area; that a late phase of volcanicity occurred in the Lower Permian, producing large quantities of flesh-pink feldspar crystals; and that the Exeter volcanic rocks were from a different source and erupted during the early Stephanian at or soon after the beginning of the Permian.

Datable fossil evidence is confined to rare finds of Triassic rhyncosaurs and labyrinthodonts in the Otter Sandstones, and sparse miospores and ostracoids in the Mercia Mudstone Group of Sidmouth and vicinity. All ages indicated by these fossils are early 'Keuper'. The labyrinthodont *Mastodonsaurus lavisi* found on the coast west of Sidmouth in the last century came from near the top of the Otter Sandstones, and indicates an early Ladinian (Middle Triassic) age (Paton, 1974) while *Rhyncosaurus* (*Hyperodapedon*) from the same formation at the mouth of the River Otter (Whitaker, 1869) has been judged to be of early to mid-Ladinian age (Walker, 1969). Among miospores obtained from mudstones at Higher Dunscombe Cliff, Warrington (1971) identified *Camerosporites secatus* which is known only from the mid to late Carnian; and ostracods including *Euestheria minuta* were obtained from a sandstone in the Mercia Mudstone Group, suggesting a direct correlation with the Arden Sandstone of Worcestershire, of late Carnian age. Dating as precise as this is not possible on any beds below the Otter Sandstones, but the data confirm the Middle Triassic age of that part of the sequence and suggest that the Budleigh Salterton Pebble Beds are of 'Bunter' (Lower Triassic) age. The dawn of the Mesozoic Era in Devon might, therefore, have been celebrated by a few scattered amphibians, though whether this coincided precisely with the immediate appearance in the area of a river bearing large quantities of quartzite pebbles enclosing fossils from French Ordovician rocks will remain a matter for conjecture. For the present, however, most people involved are content for the boundary between the Permian and Triassic Systems, and thus the Palaeozoic and Mesozoic Eras, to be drawn at the base of the pebble beds.

In the past, comparisons of the Devon succession with red beds elsewhere, initially with Germany but more specifically with the Midlands, provided general dating and part of the terminology. Thus the upper mudstones of the Triassic were commonly known as the 'Keuper Marl', and the Budleigh Salterton Pebble Beds were correlated with the Bunter Pebble Beds of the Midlands—though the Lower Mottled Sandstone (Bunter Sandstone) beneath the latter is not represented in Devon. Fewer foreign names were available for the Permian part of the sequence, so 'Lower Marls', 'boulder breccias' and 'breccio-conglomerate' were applied to distinct formations, with the Watcombe Clay as the basal unit. The last-named resembles a clay only on weathering, so the name was changed to Watcombe Beds (Laming, 1966) and subsequently to Watcombe Formation (Smith *et al.*, 1974). Numerous stratigraphic terms were introduced after extensive field work by Laming (1966, 1968), and the more important sequences are presented in *Table 7.1*, showing

Table 7.1 STRATIGRAPHY OF THE NEW RED SANDSTONE IN DEVON

| | Stratigraphic Division | New Red Sandstone Succession at: | | | W.A.E. Ussher's Coast Section Stratigraphy (1876 1913) |
		South and East Devon Coast Section	Exeter District and Northeast Devon	Crediton Valley	
TRIASSIC	Rhaetian	Westbury Formation Blue Anchor Formation			
	'Keuper'	Mercia Mudstone Group	mudstones Nynehead Sandstones		Upper Marls
	'Bunter'	Otter Sandstones	White Ball Sands		Upper sandstones
		Budleigh Salterton Pebble Beds	Uffculme and Milverton Conglomerates		Conglomerates and pebble beds
PERMIAN	Upper	Littleham Mudstones Exmouth Formation	Aylesbeare Group		Lower Marls
		Langstone Breccias†			
	Lower	Dawlish Sands # #	Clyst Sands #	Clyst Sands 'VV	Dawlish sandstones and breccias
		Coryton Breccias # # #	Heavitree and Kennford Breccias # # #	St Cyres Breccias # # #	
		Teignmouth Breccias VVV	Alphington Breccias VVV	Crediton Breccias VVV	Boulder breccias
CARBONIFEROUS	Stephanian	Ness Formation			
		Oddicombe Breccias		Knowle Sandstones Bow Breccias VVV	Breccio-conglomerates
		Watcombe Formation		Cadbury Breccias	Watcombe clays and sandstone (shale paste)
		Chelston Paignton Breccias			
		Livermead Formation			
		Tor Bay Breccias			

Volcanic rocks are indicated by VVV, murchisonite feldspar by # # #, and unconformities by = = = =.
† Shown as "Exe Breccia" on the Newton Abbot IGS 1:50,000-scale geological map.

their relationship to the traditional terminology and their probable chrono-stratigraphic correlation. The occurrence of volcanic rocks is indicated at their likely stratigraphic levels.

Lithological correlations

Correlation on the basis of lithological similarity, although valid in most cases over short distances, may be subject to question when carried from one region to another. If the Variscan Orogeny had occurred much earlier in South West England than in Midlands, similar successions of coarse to fine red-bed strata would most probably have resulted in each area, due to the purely physiographic influences such as the infilling of basins and reduction of relief in the source area; in such a case, chronostratigraphic correlation of similar lithological units would not have been appropriate. Nevertheless, the time correlation of the pebble beds mentioned above seems soundly based, on the evidence of the vertebrate fossils found in the overlying sandstones; their deposition in Devon and the Midlands at the same time could have been due to a widespread climatic change such as increased rainfall in the (different) upland source areas. A similar climatic control may earlier have terminated aeolian deposition in the Dawlish and Clyst Sands, again due to increased rainfall, ascribed to the expansion of the Zechstein Sea in the Upper Permian of the North Sea area (Smith *et al.*, 1974). This latter idea is interesting, putting the base of the Upper Permian at the top of the dune sands; but the Zechstein Sea was some 300 km away, and aeolian deposition continued in a minor way at later times, including the Lower Triassic sandstones at Budleigh Salterton. It may have been that increased rainfall in Devon and elsewhere caused the Zechstein Sea to expand, rather than the other way round.

One undisputed lithological correlation, within the region but linking for-mations in different areas, is based on the occurrence of flesh-pink feldspar crystals, 2 cm long or more, in breccias and sands in the upper part of the Lower Permian in the Dawlish-Exeter-Crediton area. Their appearance in the succession, in the Coryton, Kennford, Heavitree and St Cyres Breccias, as well as in the Dawlish and Clyst Sands, may be counted the least-unreliable criterion of time correlation in this otherwise barren Permian succession. They are sedimentary particles, in that they have been deposited by water and show small amounts of abrasion; they appear to have originated in the area of northwest Dartmoor. One has to visualise a volcanic outburst showering the landscape with crystals of feldspar, quartz and mafic minerals: the last would have weathered quickly, but the feldspars retain their clear crystal faces to this day, almost immune to weathering.

A more general correlation of breccia formations is possible on the basis of the other fragments they contain. Limestone from Devonian outcrops is com-mon south of the Teign estuary, and from a Carboniferous source at West-leigh; quartzite, vein quartz and shale fragments derived from Carboniferous and Devonian sources are even more common and widespread. Igneous rock fragments appear at various levels from the very basal beds upwards, mainly red-weathered quartz porphyries that occur up to boulder size—the largest being over 4 m across. The source of these is, on the evidence of sedimentary indicators of current directions, the roof of the granite dome, and their erup-tion may have been linked to the Withnoe volcanic event which post-dates the granite intrusion by possibly only 5 Ma (as discussed in Chapter Five);

but the original outcrops above the Dartmoor granite have long since been removed by erosion. Moreover, porphyry fragments occur in abundance in the earliest Stephanian red beds in the Tor Bay area, which may be significantly older than the Kingsand-Withnoe event. Nevertheless, it is clear that a huge volume of porphyry detritus was supplied to the breccias from volcanic and subvolcanic sources on the granite roof. Few occurrences of comparable rock can now be found *in situ*, but the dyke near Christow (SX 841845, locality 5.18) may be one example. Many of the quartz porphyries have a granitoid appearance, though fragments certainly identifiable as Dartmoor granite are few in number: xenoliths and aplo-granite fragments are common in some breccia horizons, however. Evidently erosion cut deep enough into the granite roof in Permian times to reach the main body of the pluton (Merefield, 1981), though larger amounts of granitoid dyke rocks from higher levels were affected, along with numerous quartz porphyry flows and feeders.

A fragment succession is evident in most parts of the Lower New Red Sandstone breccias, changing from one or a few locally-derived lithologies at the base to polymict widely-derived assemblages higher in the sequence. This arises from the infilling of cuvettes by alluvial fans, which built up until the bedrock divides between adjacent cuvettes were overspilled. Material from more distant sources was then able to spread over each area and mix with indigenous lithologies as a piedmont fan complex or *bajada* was formed; at this stage wind-blown sand is a common constituent, either as separate aeolian horizons or as water-borne sandy matrix and interbeds. Thus a *palaeoenvironmental succession* is also evident—a major upwards-fining cycle that starts with bouldery alluvial fans and ends in the muds and sands of a river-channel/flood-plain depositional regime. In fact, there are two major cycles in the New Red Sandstone of Devon, the second commencing with the Budleigh Salterton Pebble Beds. The first was initiated by tectonism: the second probably by climatic change.

THE BASE OF THE SUCCESSION

A variety of older rocks underlie the red beds, almost all of them of sedimentary origin. The oldest are in south Devon, where the Lower Devonian Dartmouth Slate is overlain by a New Red Sandstone outlier at Slapton (SX 820450); the age of the outlier is not known, but a smaller one at Kingsand (SX 875871) on the Cornwall side of Plymouth Sound, also overlying Dartmouth Slate, has red breccias associated with a rhyolite body dated at between 280 Ma and 295 Ma (described in Chapter Five).

Progressively younger Devonian strata occur beneath the New Red Sandstone unconformity as it is traced northwards. In Torbay, Lower Devonian Staddon Grit forms a horseshoe-shaped margin to the Paignton Cuvette, while north of Newton Abbot Upper Devonian slate, limestone and igneous rocks give way to Carboniferous as the foundation. The youngest Culm Measures strata beneath the New Red Sandstone are G_2 near Clovelly, but in west Somerset the red beds rest on Devonian again, lying on Hangman Sandstone (Middle Devonian) in the Minehead and Quantock areas. The red beds here, however, are Upper Permian and Triassic: clearly the major structural framework of the older rocks was complete before the red beds appeared, and the onlap of progressively younger red beds, from Stephanian in Tor Bay to

Triassic in Somerset, was influenced mainly by topographic effects, and by subsidence of the large Wessex Basin to the east. Foundering and infilling of fault-bounded basins is a common feature of post-orogenic red-bed successions world-wide, and there is some evidence that in Devon a number of east–west normal faults influenced the formation of some cuvettes, including the Crediton Valley Cuvette: this continues westward as a fault zone and associated reddening effects (including an outlier) to the north Cornwall coast.

The Unconformity

The surface on which the red-bed sequence was deposited was entirely terrestrial, as far as evidence is available, and predominantly composed of bare rock mountain sides and canyons. The higher mountainous areas of the Dartmoor dome, underlain by the granite but with large amounts of volcanic rocks at the surface, formed an important source area that was possibly never entirely covered by red beds.

Detailed examination of the unconformity is possible at only a few localities, the two most significant being Waterside (Oyster) Cove, Goodrington (SX 895587) and Petitor Cove, north of Torquay. At Goodrington, purple Devonian slates show a distinct reddened zone for about 3 m below the coarse limestone breccia which forms the base of the succession there (Tor Bay Breccias). At the actual contact, the underside of the breccia carries, in places, a cast of the former bedrock surface which shows it was clean of soil and superficial detritus at the moment of covering. Streamflood scour before deposition may have created this condition, now preserved by the well-cemented breccia beds above; these, as indicated by their composition, came from elsewhere—on the evidence of roundness of fragments, possibly several kilometres away.

In Petitor Cove (SX 926664, *Figure 7.2*) the Watcombe Formation overlies a bed of Devonian limestone in which bedding is not discernible: in fact, where seen it is thoroughly shattered into angular fragments and boulders typically from 20 cm to 1 m across, many of them still in their original positions relative to their neighbours. The fragments have not been rolled or

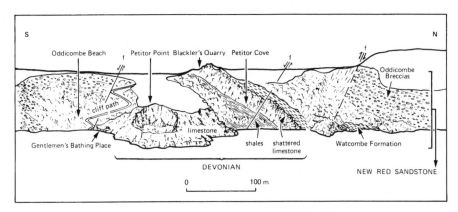

Figure 7.2. Cliff section at Petitor and Oddicombe

abraded, but between them are infillings of red sand with limestone particles, red silt and grey limy silt, accumulated apparently as downwashed material from what thin soil once overlay the limestone. The slate-fragment sandstones of the Watcombe Formation, which overlie with an angular discordance of 20°, show very few included limestone fragments, so the unconformity is again mantled with material brought in from elsewhere.

Associated with the unconformity, at the top of the cliff (Blackler's Quarry, SX 924664), are some infilled cavities in the limestone. Similar features, more easily observed, are present at the Gentlemen's Bathing Place on Petitor Point, at the north end of Oddicombe Beach. Here, a number of cavities can be seen partly or completely filled with stratified red silt and fine sand, cemented by calcite which also appears as fine nodules and crystal linings. The cavities occur in massive limestone, just beneath where it is overlain by a well-cemented talus breccia, grey in colour, but the red-bed unconformity is not present at this point. The south side of the Petitor limestone inlier is bounded by a large normal fault, against Oddicombe Breccias, but it may be judged that the non-red talus deposit formed on the side of Petitor mountain before it was buried by Watcombe strata. Pre-existing solution cavities were infilled at that time, and the northward dip of the bedding within them is consistent with the tilting of the whole block of Devonian rocks that is evident on the north side of Petitor.

Lavas

Volcanic activity possibly connected with the intrusion of the Dartmoor granite occurred before the earliest red beds now seen were deposited, and may have continued intermittently until the feldspar-effusive episode recorded in the Coryton, Kennford and St Cyres Breccias of the late Lower Permian; but, apart from the Kingsand lava and a quartz porphyry dyke in the Teign valley, all evidence of this is indirect, from the fragments and boulders incorporated in the breccias.

Olivine basalt and trachybasalt, lamprophyre and minette extrusions constitute the Exeter Volcanic 'Series', at and near the base of the sequence in the Exeter area and the Crediton and Tiverton Cuvettes. These are dealt with in Chapter Five, where their petrology and petrogenesis are discussed. Two of the lava occurrences, however, have yielded radiometric dates close to the Permian-Carboniferous boundary (280 Ma) and the Kingsand rhyolite gives an age somewhat older, though still younger than the granite. The association of these lavas with basal red beds, therefore, has some stratigraphic significance, though there appears to have been more than one episode of lava extrusion in the Lower Permian (*Table 7.1*). The association of the Dunchideock and other lava bodies with the Alphington Breccias puts that formation at the base of the Permian, as with the Teignmouth Breccias and Crediton Breccias: olivine basalt and trachybasalt are the main types present. All lower formations are, on this basis, of Carboniferous age; but there is no other evidence to confirm this. Further northeast, however, the minette and trachytic lamprophyre outcrops around Killerton, one of which also gave a basal Permian date, occur in association with the basal Clyst Sands (top of the Lower Permian): however, it is probable that the volcanic rocks were erupted earlier and stood as a topographic feature when the sands were deposited.

DEPOSITIONAL FEATURES: THE BRECCIAS

Thick breccia formations make up most of the lower part of the New Red Sandstone in Devon, and are very well exposed along the coast sections from Tor Bay to the mouth of the River Exe. Some are well-cemented rocks, such as the Oddicombe Breccias: their beetling cliffs north of Petitor defy the most intrepid geologist to negotiate them. Others, such as the Langstone Breccias, can be crumbled in the hand but nevertheless give rise to a prominent headland and an escarpment inland. The variation in cementation is due to the rock types contained in them; the Oddicombe Breccias have abundant limestone fragments, the Langstone none, but quartzite and other resistant types are present in the latter that give it some resistance to erosion. Cliff-falls are common on Langstone Rock, however, and it has diminished appreciably in size within the last decade.

Breccias are usually defined as coarse-grained clastic rocks with angular fragments, though a more soundly based criterion is the amount of matrix present between the fragments. Both criteria indicate, however, that the sediment was deposited relatively close to source by a powerful but unselective transporting agent; in the case of the New Red Sandstone breccias, the agent was sheet-flooding, resulting from occasional violent rainstorms in an otherwise mainly dry desert. Deposition took place as sheets of muddy or sandy gravel, spread wide over alluvial fans fringing upland areas, each fan fed by one or more canyons down which storm-fed debris-choked torrents poured. Reaching the fan surface, the floods spread out and lost velocity, the sediment load being dumped as layers of gravel, mud and sand spread fairly evenly over the fan surface; the latter quickly settled in between the larger fragments, giving rise to a sediment with a wide range of grain sizes. Many alluvial fans in present-day deserts are noted for the regularity of their slopes, generally less than 1°, and the evenness of their shallow conical form; channels radiating from the apex create very minor irregularities. In the ancient fans, fairly regular plane bedding and low-angle trough cross-bedding can be seen, formed by the same processes.

Boulders

In the past, questions have been raised about the origin of the breccias because of the large size of some of the included fragments. Rounded boulders of purple quartz porphyry up to 1 m across are common among fallen blocks of breccia at the foot of the Ness Cliff, Shaldon (SX 939718). There are smaller boulders of limestone, quartzite and rock types from the granite aureole, but the porphyries are the remarkable ones. The largest so far discovered was on the slopes of Great Haldon; it was 4 m long by 3 m wide and was estimated to weigh 45 tonnes.

That such large boulders can be moved by water is evident from the size of boulders that occur on present-day alluvial fans in semi-arid regions. They are usually partly buried, having been washed down half-submerged in a flood of water charged with large quantities of mud, sand and gravel—the density of the mixture, it is thought, aiding the transport of the boulders. The infrequency of rainstorms and flooding in these regions means that large amounts of debris can accumulate on slopes in between storms, so that when the rain comes plenty of material of all sizes is available to be swept down into canyons and carried away by the flood.

Trains of boulders have been mapped in the breccias outcropping on the slopes of the Haldon Hills, derived from a source lying to the northwest. Limestone boulders well over a metre in size appear to have had their origin in the Devonian limestone outcrop at Chudleigh—or rather the limestones that once existed above the present-day outcrop—but the source of the porphyries is less obvious though, as seen in *Figure 7.1*, the dyke at Christow lies upstream of the boulder train. The huge volume of porphyry in the breccias argues a source equally huge, and the only reasonable explanation is that advanced by Shannon (1927), that they came from a large supra-granitic lava field on the Dartmoor upland, long since eroded away. Volcanicity occurred after the granite intrusion, and rhyolitic flow-banding present in some of the porphyry fragments links these boulders to such a source; unfortunately, formerly perched presumably on the roof of the granite mass, the original outcrops are no longer preserved.

The boulders are well rounded, which could mean that they had travelled a great distance. It is more likely, however, that they achieved this rounding from spheroidal weathering at the original outcrop. Igneous rocks in Northern Nigeria and other parts of the world have been observed to weather into rounded boulder shapes, so the boulders apparently had that shape before they began to be transported. Other features of the climate of that area link it to the red-bed conditions of Devon, and it is easy to visualise these boulders rolling down a hillside into a canyon and being swept downstream. Their roundness would assist their rolling, of course, once they were sufficiently undermined to start moving.

Breccia fragments

Although quartz porphyry fragments are widely distributed in the red breccias, a variety of other rock types are found, reflecting the diverse geology of the Palaeozoic source area—from Brixham to Exeter, to Hatherleigh and Exmoor, and beyond. Volumetrically, quartzite is the most abundant, derived from the Culm Measures in mid Devon and from the Staddon Grit and other Devonian formations in south Devon. These fragments are commonly block-shaped, having come from more strongly jointed beds than other constituents, and more angular because of their relative toughness.

Varieties of quartz porphyry and related igneous rocks are found in nearly all breccia formations, though their abundance lessens northwards and they are mainly absent north of Tiverton. Their link with volcanicity on the Dartmoor upland seems undisputed, but a thorough study of the petrological variations might reveal valuable additional data on that volcanic province. Geochemical and petrological evidence links these rocks, many of which show rhyolitic texture, to the Kingsand and Withnoe outcrops in east Cornwall (Cosgrove, 1973), and with an extensive rhyolitic capping of the granite mass. Moreover, considerable enrichment of the trace element caesium in clays in the red bed succession was attributed to this volcanic capping (Merefield, 1981): high levels were analysed in the rhyolitic rocks whereas granite and mafic lavas had much less, while the decline in caesium content from base to top of the succession indicated that the source was most actively eroding at the beginning of New Red Sandstone time.

Lava fragments from the Permian Exeter Volcanic 'Series' are not common in breccias except close to actual sources; probably they weathered too easily in the harsh hot-wet-and-dry climatic conditions then prevailing. As mentioned earlier, true granite fragments are rare, though 'granitoid' porphyries suggest that erosion had cut down to the top of the pluton, probably with granite exposed in the bottoms of canyons.

Fragments of Devonian limestone are abundant in the lower breccias, particularly between Torquay and Teignmouth where the Oddicombe Breccias form prominent cliffs and escarpments. These fragments show a much greater degree of rounding than the quartzites, not from a greater distance of transportation but due to their less-resistant lithology. A statistical study on the rounding of these pebbles (Laming, 1966) has shown that contours of equal rounding elegantly outline the depositional basins and match up with other indicators of transport direction (*Figure 7.3*). The average rounding at most localities exceeds 0.40, and because of this degree of rounding the term breccio-conglomerate was formerly applied to these rocks; but on other criteria these are still breccias.

Less noticeable but still important are the fragments from Culm shale and Devonian slate source areas, which provided much of the matrix of the breccias as well as vein-quartz and slate fragments. In the Watcombe Formation, comminuted flakes of red slate make up most of the beds; in this case the material in the 'sand'-stones can be compared directly with Devonian slate that underwent red weathering on the former land surface before being covered by sediments—the purple-brown of the fresh rock has been changed to a dull haematite-red colour for several metres below the unconformity surface. This can be seen very well at Waterside Cove (SX 895587) where the unconformity is exposed, but the comparison with the Watcombe slate-fragment sandstones is best seen in Petitor Cove (SX 926664) where slate is also exposed (*Figure 7.2*).

Rock types attributed to the metamorphic aureole of the Dartmoor granite are common in some formations. They include a variety of green and black metavolcanic types and black fine-grained quartzites ('lydian stone') that are now only minor constituents of the aureole; presumably they were more abundantly represented on the top of the granite mass.

Finally, feldspar should be mentioned as an important constituent. Flesh-pink murchisonite, a variety of sanidine, has been mentioned earlier, and its remarkable resistance to weathering; other, slightly more reddish, varieties can be found in small quantities below the level where murchisonite first appears, and here—notably in the Teignmouth Breccias—it is associated with quartz porphyries. The centre of volcanicity which is presumed to have been the source of the murchisonite was, on evidence of distribution, located somewhere over northeast Dartmoor and erupted only for a short time. By the time the Langstone Breccias were being deposited, very little of the feldspar remained to be included.

Fragment succession

The succession of fragments in the breccias throws an interesting light on the erosional history of the surrounding area. Early-formed deposits contain few fragment types while later ones contain many, reflecting the change from local to more widespread sources.

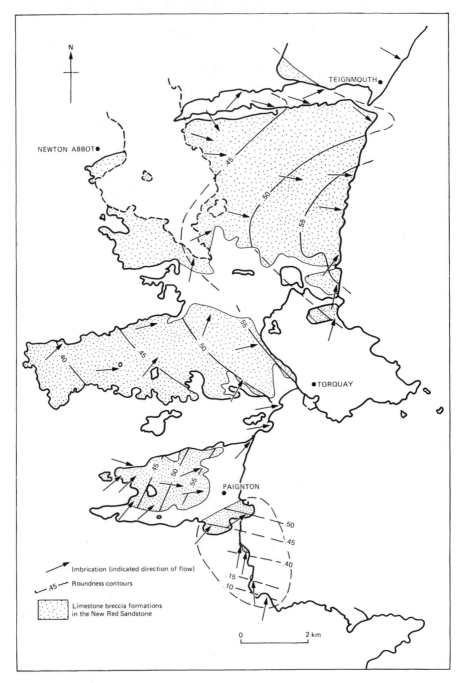

Figure 7.3. Imbrication directions and contours of pebble roundness in limestone breccia formations in the New Red Sandstone of south Devon

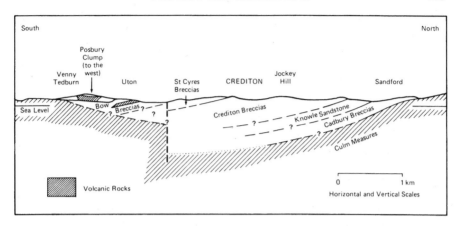

Figure 7.4. Geological cross-section of the Crediton cuvette

This is well illustrated by the succession in the eastern part of the Crediton Valley Cuvette (*Figure 7.4*) where the Cadbury Breccias at the base have a virtually monomict composition of Culm quartzite fragments, with a few pieces of vein quartz. Also, a very few fragments of fossiliferous sandstones from the Pilton Shale have been found, indicating derivation from the north Devon area as well as from the Culm terrain of mid Devon.

The Bow Breccias which follow, above and to the westward, show Culm quartzites as in the underlying formation but with the inclusion of trachytic lava fragments; these have been matched to some of the contemporaneous lavas (Hutchins, 1963). Above these come the Crediton Breccias, in which tourmalinised 'granitoid' porphyry fragments are found in addition, which distinguish them from the Bow Breccias: whether simply due to the widening of the source area to include the Dartmoor upland, or whether their appearance marks the intersection of 'granitoid' dykes and the roof of the granite itself by actively eroding canyons, is not entirely clear, but comparison with other nearby areas suggests the latter as the main reason for this increase in fragment content.

At the top, the St Cyres Breccias show a polymict assemblage similar to the Crediton, but are distinguished by the appearance of murchisonite feldspar. This development is of a different kind to the previous ones inasmuch as the new type has not been reached by erosional downcutting but by eruptive action; nevertheless, the catalogue of fragment types given by Hutchins for the St Cyres Breccias demonstrates that the downcutting was still opening up new sources. Culm, untourmalinised volcanic rocks, tourmalinised lavas, dyke rocks and slate, feldspar, quartz-tourmaline vein rocks, microgranites, 'granitoids', chert and a pebble of limestone were recorded. The 'granitoids' increase in abundance in the higher beds, as do the feldspar and the proportion of sandy matrix. To the east, this formation passes into wind-blown sand deposits of the Clyst Sands.

Boulders of Devonian (Upper Givetian) limestones are present in one bed of the Bow Breccias (Hutchins, 1958), at Solland (SS 615021). These appear to have been derived from the northern rim of the Dartmoor upland, on evidence of sedimentary flow directions, suggesting a former Devonian outcrop possibly in the vicinity of Okehampton, taking displacement on the Sticklepath Fault into account (*Figure 7.1*). Decalcified limestone fragments with ghost fossils can be found in the breccias in Bow village.

Elsewhere, similar but less well-known 'unroofing' successions are evident, and if the incoming of the murchisonite feldspar is taken as a valid time-correlative event, the successions are coincident. Thus, the St Cyres Breccias correlate with the Coryton-Dawlish-Kennford-Heavitree breccia formations south of the Exeter Culm ridge, all of which contain this feldspar; the Crediton Breccias beneath correlate with the Alphington and Teignmouth Breccias, in which it is absent (*Table 7.1*).

Lavas occur at the base of the Alphington and Crediton Breccias and are probably correlative; this puts both of them near the Carboniferous–Permian boundary on the basis of the Dunchideock radiometric date. In the fragment succession, this is the level where 'granitoid' fragments appear, suggesting that erosion had cut down to close to the granite itself at the beginning of the Permian. Likewise, the fading out of these fragments records the burial of that particular source by sedimentation, probably by the end of the Lower Permian.

Formations below the base of the Permian show little or no correlation since they represent more local derivation, particularly in the Tor Bay area. The Oddicombe Breccias present an interesting transition northwards from Oddicombe, however; from a thickness of some 350 m at the type section at Oddicombe Beach, these massive limestone breccias thin northwards and split into several beds separated by a different fragment facies, fine-grained breccia with a quartzite-porphyry-aureole rock suite typical of the overlying Teignmouth Breccias. This formation—for convenience in mapping, it includes the massive limestone breccia interbeds—is known as the Ness Formation, after the Ness at Shaldon (SX 940718), and it represents the interdigitation of limestone detritus derived from the Torquay area to the south and of porphyry detritus derived from the Dartmoor upland to the west (Laming, 1966).

Sedimentary structures and palaeocurrent indicators

When coarse sedimentary material is deposited by a strong current, elongate and blade-shaped fragments tend to orient parallel to the flow, while roller-shaped ones orient across the flow. Moreover, the former types tend to adopt a position of upstream dip on finally coming to rest, as this attitude is the most resistant to renewed movement. In conglomerates, pebbles can be seen overlapping one another, a feature termed *imbrication* from an analogy with tiles on a roof. An additional process, removal of fine material from the upstream 'heel' of each fragment (Laming, 1966), is also operative; however produced, though, imbrication is a most useful indicator of palaeocurrents, indicating directions of derivation and current flow in coarse-grained rocks of many types.

Imbrication directions mapped in the limestone breccias match remarkably well with the roundness contours (*Figure 7.3*), and, on a wider basis, demonstrate the general eastward flow of material from the Dartmoor upland and high source areas in mid and north Devon towards the Wessex Basin: indeed, the presence of the Wessex Basin in the earlier depositional history of the New Red Sandstone is inferred from this palaeocurrent pattern.

On a more detailed scale, cuvette infilling is illustrated north of Torquay in a convergent pattern made by fans coalescing in a semi-elliptical cuvette fed from two major and probably several minor canyon mouths (*Figure 7.5*). Higher in the succession, however, the Kennford Breccias (porphyry-quartzite-feldspar) have a fan-shaped imbrication pattern with a presumptive (geometric) apex in the southern Teign Valley, around Hennock, while the Langstone Breccias (quartzite-aureole-porphyry) appear to originate from further north, geometrically from Doddiscombsleigh (*Figure 7.1*). Both apices fit the fragment composition moderately well, but not all palaeocurrents travelled in straight lines so actual sources may have been different.

In Tor Bay, two directions of infilling are indicated (*Figure 7.3*). In the Paignton Cuvette the flow was generally eastward, but in the Goodrington area a northerly flow is evident. Indeed, some breccias are very close to source at Broadsands, as very low roundness values were recorded there (0.10 and 0.17, values close to those typical of untransported debris).

Other sedimentary structures in the breccias include cross-bedding and a few channels, but these are subordinate to the pervasive fabric and particle-shape features described above; directional indications of cross-bedding are parallel to imbrication, though channels appear more random in orientation.

Fossil burrows occur at a number of localities, notably Goodrington South Beach and Watcombe Cove. Originally ascribed to the activities of annelids (Ussher, 1913; Scrivenor, 1947), these structures have been the subject of much discussion and scepticism, for their dimensions—up to 18 cm across—would suggest worm-like organisms of quite frightening dimensions. The contents of the burrows include breccia fragments up to 3 cm which could hardly have been ingested willingly. A more soundly-based explanation (Ridgeway, 1974) was that they were formed by primitive reptiles or amphibians which made temporary burrows to escape the heat of the desert sun, as present-day species do.

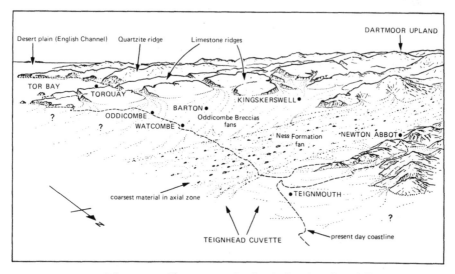

Figure 7.5. Palaeogeographic reconstruction for the Permian of south Devon

DEPOSITIONAL FEATURES: SANDSTONES, MUDSTONES AND CONGLOMERATES

Sandstone forms a surprisingly small proportion of a sequence which has the name New Red Sandstone: but it must be accepted that this title was conferred initially in the English Midlands where sandstone building stones were the most obvious representatives of the unit. In Devon the great preponderance of breccia, due to the much more extensive Palaeozoic uplands, puts sandstone in third place, as a lithology, with mudstone in second place. Conglomerates are confined to the one level at the base of the Triassic, and are dealt with in this section because they are part of the Triassic riverine sedimentary regime rather than the Permo-Carboniferous fanglomerate complex.

The sandstones, in fact, have a dual origin, aeolian and fluvial; and the fluvial sandstones were deposited in two environments, alluvial fan fringes and river channels. The latter two intergraded, but in the coast section they appear as distinct facies. The prominent sandstone headlands east of Exmouth are discrete river-channel sands which occur as lenses in a floodplain complex of mudstone and minor sandstone, the whole known as the Exmouth Formation. West of the River Exe both aeolian and alluvial-fan sands interfinger, and correlation across the Exe Estuary is made difficult by this facies change.

Aeolian sands

Red sands with regular, evenly-curved cross-bedding are beautifully displayed at Dawlish, largely protected from carvers of initials by the railway that runs along the sea wall. These sands are the remnants of desert dunes, formed from sand grains blown across the fan surfaces by a wind from the southeast, and the complex truncation planes in these beds show where dunes were partly eroded and then overlaid by others. The largest single cross-bed unit, 20 m thick, is in the cliff behind Coryton Cove (SX960760), showing that these dunes attained heights of 20 m or more at times.

The aeolian sands can be traced northwards to the Clyst valley east of Exeter, where they occur in quarries and road cuttings near the Sandy Gate motorway interchange (SX 965913). The Clyst Sands, of which the latter locality is the type section, correlate with the Dawlish Sands to the south but they appear to commence lower in the succession, at the level of the Crediton Breccias or possibly the Knowle Sandstones. There is a clear case of lateral facies change at the eastern end of the Crediton Valley Cuvette, where the fan breccias interfinger with aeolian sands which have been blown into the cuvette from the eastward, in an up-fan direction.

The interplay between aeolian and fan sedimentation is, however, best seen between Dawlish and Langstone Rock. Along the Dawlish sea wall the basal aeolian sand beds alternate with breccia horizons and pass up into uninterrupted sands; but some of these sands are water-deposited, lacking the characteristic aeolian bedding and including some mudstone layers. Higher still there are breccia-filled channels in the sands showing that, at this point, the down-fan water flow was overwhelming the up-fan wind-blown sand. There was a battle for supremacy, and for a time the aeolian sands built up sufficiently to deflect the down-fan flow of the sheet-floods, but finally the latter conquered the field. This may have been due, as suggested by Smith *et al.* (1974), to a climatic change at the end of the Lower Permian which increased rainfall over the British region, concurrent with the expansion of the Zechstein Sea.

Other dur   are found, higher in the section, in the channel-sand complex of Straight Point (Exmouth Formation) and in the Otter Sandstones above the Budleigh Salterton Pebble Beds. At Roundham Head, Paignton, a sand lens some 6 m thick within the Tor Bay Breccias also shows aeolian cross-bedding but is notable for its composition—a good proportion of rounded slate fragments—compared to the quartz grains that make up the majority of the Dawlish aeolian sands. In common with all other such sands, the high degree of rounding and the surface textures testify to wind action, and, apart from the thin haematitic coating that gives the red colour to the rock, these surface textures are the oldest found in sedimentary rocks that have not been altered diagenetically (Krinsley *et al.*, 1976).

The wind directions that produced the fields of sand dunes are interesting, as they blew from an easterly direction, unlike the prevailing westerly winds of today. Moreover, they blew from that quarter only, in a regime typical of the tropical trade-wind belts, where northeasterly winds blow throughout the year. Under these conditions a particular dune form, the *barchan* or half-moon dune, is built up. Studies of the attitude of aeolian cross-bedding in the Dawlish Sands and elsewhere (*Figure 7.6*) demonstrated the presence of this type of dune, formed by a wind from the south-southeast (Laming, 1966) during the Permian, which could be related to early palaeomagnetic results (Laming, 1954, 1958; Shotton, 1956). It is interesting to note that the wind direction indicated by the sand lens in the Tor Bay Breccias is different—from the northeast—suggesting that a significant change in climatic geography occurred at the beginning of the Permian.

Fluvial sands

In volume, fluvial sand probably exceeds aeolian sand by a wide margin, but shows rather fewer interesting features. Many of the sand beds are associated with the distal portions of alluvial fans, where, after a flood, water which had sunk into the permeable gravel of the fan flowed out of the toe, bringing sand with it to produce miniature fans and braided stream deposits. Water supplied in this way could, at times, support a less inconstant river flow in the centre of the basin: evidently this occurred in the Upper Permian, where the Exmouth Formation consists of channel-sand bodies at various levels in a mudstone-and-minor-sandstone flood-plain or ephemeral-lake complex.

The Exmouth sand bodies form prominent headlands east of the mouth of the River Exe, where marine erosion has worn back the cliffs of interbedded mudstone to form broad sandy beaches. Straight Point shows the most interesting features, but this is used as a military gunnery range; the sandstones are seen nearly as well at Orcombe and Rodney Points (SY 020796). Near Orcombe Point some poorly preserved plant fragments have been found, in a rubbly bed at the base of the sandstone, and malachite mineralisation was reported (Carus-Wilson, 1913); native copper has been found nearby (Harrison, 1975).

The cross-bedding of these sandstones is moderately coarse, concave upwards, planar and trough forms being present; these compare with cross-bedding in fluvial sands of present day and of the past, such as the Millstone Grit, and their origin is attributed to a low-gradient, meandering river complex which migrated periodically across a featureless (and probably unvegetated) muddy flood plain. Aiding this interpretation is the presence of mudstone lenses in the sand bodies and scoured channel bases where they cut initially into flood-plain muds.

An interesting fluvial channel can be seen near Kenton, in a vertical road-side exposure 100 m north of the church. Here a semicircular channel has been cut into aeolian sands, leaving vertical walls more than a metre high on each side, and the infill is sand obviously derived from nearby sand dunes but

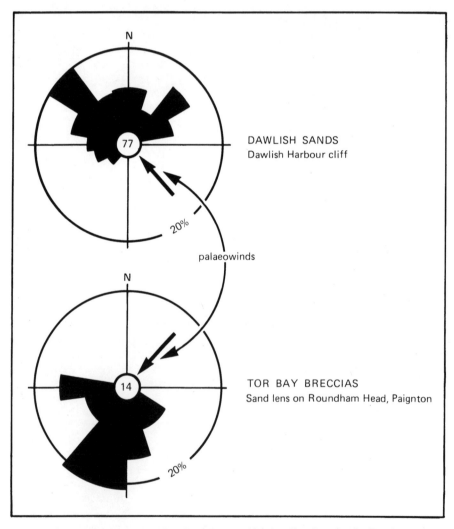

Figure 7.6. Rose diagrams of aeolian cross-stratification directions for the Permian of south Devon

incorporating a few fragments and mud pellets. The conflict of transport direction and sedimentation/erosion between wind and flood can be fully visualised in this one exposure.

In the Triassic succession, above the Budleigh Salterton Pebble Beds, the Otter Sandstones have been studied by Henson (1971) who found it to be composed of a number of sand bodies, comparable to those near Exmouth but with very little mudstone in between. Above the base the sands are aeolian, but the rest are fluvial. Derivation of the latter appears to be from the west and southwest, so the general basin-ward flow pattern is maintained, despite the powerful intervention of the Pebble Beds river.

Mudstones

The mudstones and other fine-grained rocks form, in fact, the majority of the Upper Permian and Triassic succession, approximately three-quarters of the 1,100 m thickness. Much of the Exmouth Formation and nearly all of the Littleham Mudstones in the Permian, and the Mercia Mudstone Group in the Triassic, are fine-grained, part of a very widespread depositional regime that in the Triassic extended across most of lowland Britain. A vast volume is involved, and it represents a major transfer of material from the upland source areas to the desert sedimentary basins. In Devon, the levelling of the Variscan uplands must have been close to complete at the end of the Triassic, partly due to erosion and partly to the infilling of the basin to the brim: but 'shoreline' facies have not been preserved anywhere nearer than the Mendips.

The mudstone formations, commonly known as 'marls' but lacking a sufficient carbonate content to justify the term, are composed of a variety of similar rock types from siltstone through mudstone to clay, with some interbedded thin fine-grained sandstone. Most of the mudstones have a blocky-weathering habit, with the block corners spalling off at the surface to form sub-spherical lumps: this is typical of much of the 'Keuper' Marl in many parts of Britain, and it is present in the Permian Littleham Mudstones just as in the Mercia Mudstones of 'Keuper' age. The clay mineralogy of the Littleham Mudstones indicates that they were formed on a flood plain on the margins of a larger desert basin (Henson, 1973). Sedimentary structures in the Mercia Mudstones include oscillation ripple-marks and rhythmic laminations which Klein (1962) regarded as being of lacustrine origin, though sedimentary structures that characterised lake environments in other 'Keuper' outcrops were not seen. The assemblage of sedimentary features in the Devon 'Keuper' is consistent with them having been formed in temporary lakes on a flood plain, fed by impersistent rivers that must have originated by groundwater from the breccia fans that were, in turn, recharged by flash floods at rare intervals. Thus the mudstones are products of the total semi-arid sedimentary regime and could have occurred, therefore, at any time during New Red Sandstone time. As the source areas were reduced by erosion and the basin areas enlarged by deposition, however, the proportion of mudstone must have increased greatly compared to coarser-grained sediments and are thus typically found in the latter parts of the sedimentary cycle.

The mudstone sequences are a dark red colour with streaks of pale green, the streaks seeming to be associated with more permeable horizons and tiny fissures. Interstitial haematite is present in the red rocks, with up to 9.1 per cent Fe_2O_3, and the pale green spots originated by reduction and removal of the haematite during diagenesis or, as demonstrated by Durrance *et al.* (1978), to inhibition of oxidation in sediments that may have been buried under about 1,000 m of overburden at the time. Organic fragments in them are almost certainly the cause of these spots, while thinly-spread plant debris caused the pale layers; permeability variations undoubtedly influenced the distribution. Nodules in the formation contain some interesting minerals: uranium and vanadium minerals were identified in grey nodules (Durrance and George, 1976), while other nodules, related in many cases to beds of sandstone directly above them, contain native copper, niccolite (nickel arsenide) and traces of native silver, cobalt arsenide and coffinite—a uranium-vanadium silico-phosphate (Harrison, 1975). The native copper was found as oval discs up to 16 cm long and 4 mm thick, with curled edges, assaying 99.6 per cent copper.

Conglomerates

This rock type is treated separately from the other coarse-grained rocks, the breccias, being significantly different in character, origin and depositional environment. Moreover, the conglomerate formations constitute the commencement of the Triassic upwards-fining cycle, albeit a much finer-grained one overall than the Carboniferous-Permian cycle beneath.

The main representative of the basal Triassic conglomerates is the Budleigh Salterton Pebble Beds (the formation name retains this traditional form despite attempts to convert the lithological term to 'Conglomerates'), seen very well in the cliffs west of Budleigh Salterton. Some 30 m of mixed conglomerate beds and sand lenses are well displayed, above a beach composed of more-polished versions of the pebbles from the cliff. These 'Budleigh pebbles' include distinctive types of purple quartzite which turn up on beaches in many parts of South West England, suggesting that the formation was once very much more widely spread. The source of these pebbles is not certain, though long ago Vicary and Salter (1864) had matched the quartzite to Ordovician rocks in Normandy on the basis of the fossils that can be found in some of the pebbles (notably *Orthis budleighensis*). Although a northerly direction of flow is shown by pebble imbrication, it is possible to doubt that a river capable of transporting cobbles up to 40 cm in size would be able to move them as much as 250 km, from one side of a desert basin to the other. A less distant source could have been a Palaeozoic ridge in the vicinity of Start Point, which might once have existed as an eastward continuation of the Start Point-Eddystone-Dodman ridge in which Ordovician quartzites similar to those known in Cornwall may have been present.

A number of other resistant pebble types are present in the conglomerate, notably quartzite and contact metamorphic rocks of the type that appear in the Langstone Breccias—implying that the Dartmoor upland was still supplying material of good size at this time. These types, in fact, are more common further north in the laterally-equivalent Uffculme Conglomerates, fed apparently from the west rather than from the south. Intriguing also are the Carboniferous Limestone pebbles of south Wales affinities in the Milverton Conglomerates, which laterally interfinger with the Uffculme Conglomerates in west Somerset and which appear to have been derived from the vicinity of Exmoor—there may once have been an outlier of Carboniferous Limestone there. This and other questions on the pebble beds await answers from more systematic research, but it seems evident that all three formations mentioned were laid down at the same time. A climatic control seems an inescapable conclusion.

The base of the Pebble Beds in the cliffs at Budleigh Salterton is hard to locate because of slippage of material from above, but where seen there is nothing remarkable to note: it seems to be a simple erosional junction as would be expected from a large river cutting a new channel across a flood plain. The top of the formation, however, shows an interesting desert pavement where the topmost pebbles of the river deposit evidently were subjected to sand abrasion before being covered by the basal aeolian sands of the Otter Sandstones. At beach level, most of the dreikanters in this bed have been removed by collectors, but where still seen they show a coating of iron and manganese oxides that probably represents a 'desert varnish'—in fact, the only place in the sequence where such a feature is well shown. Evidently this surface was left uncovered for a long time, and the sand dunes that finally overran it came from a relatively distant source to the east or southeast. Wind-faceted pebbles also occur on Woodbury Common.

PALAEOCLIMATE

Clues to the climatic regime under which the New Red Sandstone was deposited are many, but an entirely consistent picture does not emerge. On the one hand is evidence, particularly derived from present-day comparisons, that alluvial fan sedimentation formed a major component of the depositional regime and that these were due to a low-rainfall semi-arid climate; while throughout most of the sequence there are seen the results of river and flood action so violent that they must have matched in intensity the disastrous 1952 Lynmouth flood in north Devon.

First it must be said that there were variations in the climate during New Red Sandstone times, and quite probably between mountain and plain areas also. The cessation of widespread aeolian deposition and the incoming of coarse conglomerates are two events for which a climatic cause has already been invoked, but there may have been others less easily distinguished. Indeed, was the onset of aeolian sedimentation also due to climatic causes—a reduction of rainfall, a change in wind direction? Or was it due rather to the development of large sand flats on the margins of alluvial fans, vulnerable to wind deflation, so that the dunes had an adequate source area to begin their growth? There certainly appears to have been a change of dominant wind direction from northeasterly in the late Carboniferous (Tor Bay Breccias) to southeasterly at the end of the Lower Permian (Dawlish Sands) (*Figure 7.6*), and since both wind directions were from basin to upland the idea of topographic deflection cannot reasonably be considered. Accumulation and subsequent preservation of the sands themselves may have been due in part to the adverse gradient of the fans and nearby mountain slopes causing some slackening of the velocity of the sand grains, as in present-day situations; but the wind directions, firmly preserved as aeolian cross-bedding, must have been due to regional atmospheric pressure distributions in the tropical latitudes in which Britain then lay. These varied presumably because of geographical changes—did a Carboniferous marine gulf persist to the south of the area late into the Stephanian?

Other variations must undoubtedly have occurred across the region due to topography. Convectional air movements would have been the sole source of rainfall, and orographic amplification of these would result, as today, in the mountain areas receiving more rain than the plains. Combined with the sparse vegetation cover, which aided rapid run-off, with steep valleys funnelling the rain clouds and the resultant torrents, one can visualise how the powerful flows of water needed to move large boulders could have been generated. Moreover, although stream energy would have been used up in transporting sedimentary material, the addition of large quantities would significantly increase the bulk of the floods and their density, both factors that could help buoy up the boulders. Perhaps the bed slopes of the mountain valleys were sufficient that the downward drag of the sedimentary material was sufficient in itself to accelerate the flood, giving rise to autosuspension phenomena—and the resulting sedimentary deposits might therefore be regarded as 'subaerial turbidites'.

Once onto the alluvial fan surface, however, the permeability of the underlying fan gravels must have been sufficient to drain away much of the water in the flood, at least in mid-fan and down-fan areas. This would have produced the reverse of the normal situation, with the water dropping out of the suspension rather than the sediment doing so. The water entered the groundwater reservoir of the fans, however, and, as in present-day fans, it would provide a moderate source of flow for fan-toe streams, depositing sands and muds beyond the fringing gravel slopes.

Two major elements are lacking from the palaeoclimatic picture, however. One, vegetation, may merely be a function of preservation, since the strongly oxidising red-bed environment removed all but very minor evidence of plant life. The other is evidence of aridity: desiccation cracks, salt crystals and salt deposits are unusual, rare or completely absent. The Somerset salt field (Whittaker, 1975) shows that the prerequisites for salt deposition existed in the region, but conditions were not right for salt to form in Devon. Crystal pseudomorphs occur in the Mercia Mudstones, as do gypsum veins; desiccation cracks have been described from the Livermead and Ness Formations (Laming, 1966) but Klein (1962) found none in the mudstones. The presence of aeolian sands is not evidence of aridity, only of dryness sufficient for surface deflation that may have been periodic.

A semi-arid climate, one that is seasonally or occasionally wet with long dry periods, is therefore indicated. Winds that blow consistently from the same quarter are strong evidence of a location within the Trade Wind belt, generally between 15° and 30° north of the equator at the present day, and on the basis of palaeowind directions from as far north as Lossiemouth, Scotland (Shotton, 1956) it would seem that the whole of Britain lay in this belt. It was argued (Laming, 1958) that the Trade Wind belt was, for the most part, geometrically determined and therefore unlikely to have been at any other latitude in the geological past, so the position of Britain relative to the equator must have been radically altered since the Permian. Palaeomagnetic studies supported this conclusion.

The question of vegetation is not simply one of preservation, however, since other red-bed sequences of similar age have yielded many more fossil remains than the Devon New Red Sandstone, where they are very rare. Moreover, spore analyses (Warrington, 1971) have produced only derived Carboniferous forms from the majority of the succession; datable miospores were obtained from the Mercia Mudstones but other beds that were of a suitable lithology, including one in which actual plant fossils were found, yielded none.

The major difference with the present-day desert fringe (savannah) areas, with which the New Red Sandstone palaeoclimate may be compared, is the presence of grasses. Generally recognised to have evolved in the Cretaceous, grasses have brought to otherwise sparsely vegetated areas both a coherent soil-binding rootmat and a supply of organic material—the one having geomorphic effects and the other geochemical. Both go some way towards explaining why red beds are far less widespread in Cretaceous and later strata, the reduced supply of loose surface material and the greater amount of organic matter producing strata with less bulk and more carbon to inhibit the development of red iron oxides.

The savannah areas of the present day lie in the climatic transition between equatorial rain forest, generally reaching 8° latitude north and south of the Equator in the African continent, and the subtropical sandy or stony desert which lies generally between 15° and 30° latitude. Other continents show the same transition but boundaries are varied due to oceanic influences. Between 8° and 15°, particularly in the great Sahel zone from Senegal to the Sudan, the tropical savannah belt is characterised by an alteration of wet and dry seasons, with a high variability in the rainfall. Vegetation consists of tall grasses and drought-resistant trees, and rivers show radical seasonal changes from dry beds to bank-full floods. Closer to the desert fringe only grasses are present and rainfall is even more variable. Cultivation at the present day disguises much of the natural condition of these zones, but the highly variable rainfall produces periodic droughts that may last for years, causing famine and great suffering.

It is in this zone that the New Red Sandstone of Devon is visualised as having been formed. Lacking grasses in the Permian and Triassic, rainfall could have been considerable at times yet given rise to little or no vegetative cover capable of withstanding the intervening dry years. Violent flooding could have occurred when the tropical low-pressure belt moved anomalously far north, which might have been as infrequent as once a century or more; in between, smaller rain storms probably acted merely to redistribute the fan sediments in local terms or to cause impermanent rivers to deposit sand patches. Wind action would have had plenty of opportunity to build up dunes, which might have been unaffected by all but the most violent of rainstorms. Even today, the southern cultivated fringes of the Sahara Desert experience problems from advancing sand dunes, and the process operated in the past without any interference from man.

Northern Nigeria forms an interesting climatic analogue for the Permian climate, and has geological features that suggest parallels with New Red Sandstone conditions. Red weathering of bedrock and spheroidal boulder formation are two directly comparable features, though of course these can be developed elsewhere; more to the point is the area's relationship to the sandy desert to the north and rain forest to the south, and its vegetation restricted mainly to types evolved since the Triassic. North of this area there are bare arid regions, many of which still appear to support some human habitation, provoking the question as to whether the bareness is due to natural causes or to overgrazing. Ephemeral lakes are present at this latitude, the largest being

Lake Chad which defies the efforts of topographers to plot its actual extent. It is fed by rivers from uplands to the south which periodically cause it to expand considerably, but in no sense does it match the type and extent of the 'Keuper' lakes or flood plains of the Triassic. For a physiographic analogue of these one must look more to the deserts of Australia or, bearing in mind the fundamental vegetational developments since the Triassic, to areas of present-day savannah or even tropical fringes.

PALAEOWINDS AND PALAEOMAGNETISM

Evidence of the position of Devon during the Permian and Triassic has come from consideration of the wind patterns and remanent magnetism of igneous and sedimentary rocks in the sequence. One for the first palaeomagnetic determinations made on sedimentary rocks anywhere was on red mudstones from Sidmouth (Clegg et al., 1954), which gave a result indicating that the geomagnetic pole in the Triassic was 34° to the east of true north today. The implication was that Britain had rotated anitclockwise to its present orientation since that time, and their figures indicated a palaeolatitude of about 16° north, which indicated northward drift of some 35° (4,000 km). Such findings were less easily accepted two or three decades ago, so it was fortunate for the progress of this new branch of geophysics that independent confirmation came from the south-southeasterly wind direction deduced from the Dawlish Sands. Laming (1954) compared the present-day wind regime of Egypt and the Sahara with that of Permian Britain, and interesting parallels showed up if the British Permian geographic north lay some 45° eastward of the present direction. The coincidence of this deduction with the palaeomagnetic results was striking (Laming, 1954, 1958).

Palaeomagnetic studies have now been made on a wider variety of sedimentary and volcanic rocks from the New Red Sandstone in Devon (Cornwell, 1957; Zijderveld, 1967) which confirm the general picture: rotation of about 30° and a palaeolatitude of about 10° north. The geomagnetic equator was indicated by Zijderveld to lie across southern Spain: despite possible relative movement of the Iberian Peninsula to the rest of Europe since then, it is possible that tropical rainforest covered Spain in the Permian. Geomagnetic pole positions for the Exeter lavas and the Mercia Mudstones lay in the Sea of Okhotsk and Vladivostok areas respectively, suggesting that drift of some 1,000 km may have taken place from earliest Permian to late Triassic, some 80 Ma; the rate of 12 mm per year is not rapid. Movement probably occurred oblique to the Equator.

STRUCTURE

The tectonic environment of the New Red Sandstone is nowhere comparable to the Variscan and other fold belts that underlie it in Britain, though in south Devon there is rather more structural disturbance than elsewhere: possibly the only overturned New Red Sandstone strata in Britain occur near Kingskerswell, along a line of steep dips associated with a partly-buried basement ridge of Devonian rocks extending eastwards from the Kerswell Arch (SX 882677).

To dispel ideas that there was a post-Permian orogeny, it must be said that the structural effects seen in the red beds are entirely consistent with passive response to basement block-faulting, the basement being the Devonian and associated rocks that formed the original mountainous upland. The steep dips are northwards, suggesting that a soft cover of New Red Sandstone strata were tilted and 'peeled-off' an upthrown block to the south. Moreover, north–south normal faults which cut the Oddicombe Breccias escarpment north of Barton (Watcombe Heights) can be related to a section of the basement ridge, the Barton inlier of Devonian limestone (*Figure 7.7*). This indicates that the inlier moved differentially to other sections of the ridge, and the cover responded passively.

Generally, however, the pattern of dips and faulting is one of gentle angles (5°–20°) and normal faults mainly parallel to the strike and with small throws. This structural regime characterises the coast section, particularly from Watcombe northwards, and then eastwards into the Jurassic strata of the Dorset coast. Structures mapped in the Upper Cretaceous of the Haldon Hills (Durrance and Hamblin, 1969) have not been directly traced into the New Red Sandstone beneath, but they are of basically similar style; larger-throw faults in the red beds have likewise not been traced into the Cretaceous.

The age of the movements is indicated by this last observation, since the structural style is unchanged as far up the succession as Wealden strata in the Weymouth district. In that area there are the pre-Albian tectonic movements, folding and faulting that produced the Poxwell Anticline in Jurassic and Lower Cretaceous rocks. If the line of that fold is projected westwards across Lyme Bay, it meets the shore in the vicinity of Torquay, so it is entirely possible that the age of the block faulting is also pre-Albian. There is no evidence of Alpine fold structures as are found in the Isle of Wight.

Tertiary wrench faulting affected the area, with the Sticklepath Fault cutting through the end of the Crediton Valley Cuvette outcrop and through the Newton Abbot-Tor Bay area. Offshore geophysical surveys by the author in 1967 and 1969 located at least three parallel fault traces in the bay and off Berry Head, as well as north–south faults of lesser extent. The trace of the fault is also evident behind the Palm Court Hotel where, before its construction, red sandstone beds were exposed beside the Devonian limestone of the adjacent Rock Walk (Pengelly, 1862), and a fault plane seen by the author in a temporary exposure at Torbay Hospital in 1967 showed 30 cm of clay gouge with embedded limestone fragments. Its orientation matched the fault trace mapped in the vicinity (*Figure 7.7*).

Faults on the shores of Tor Bay, and others mapped geophysically underwater, have a more usual northerly orientation, some with an easterly downthrow. The best known is the Crystal Cave Fault, which can be traced northward from Broadsands to the corner of Saltern Cove along the face of the cliff; for part of the distance it is faced on the seaward side by red sandstone with thin beds of very angular limestone breccia, obviously very close to source. Along the whole length of the fault is a massive development of calcite crystals that fill a shatter zone up to 10 m thick.

Contemporaneous Movements

A distinct but not well exposed angular unconformity is present between the Watcombe Formation and the Oddicombe Breccias north of Torquay.

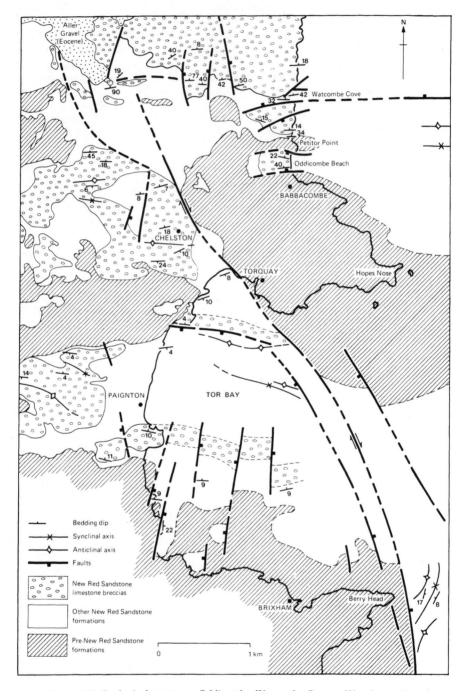

Figure 7.7. Geological structure: Oddicombe–Watcombe–Barton–Kingskerswell

The only clear exposure is in Whitsand Cove (SX 927676, *Figure 7.7*) north of Watcombe, where slate-fragment sandstones and fine breccias of the Watcombe Formation are truncated by the overlying coarse grained Oddicombe limestone breccias. An angular discordance of 9° indicates that a tilt of this amount affected the Watcombe Formation at the time, accompanied by an inrush of large amounts of limestone detritus. In Watcombe Cove itself the unconformity can be detected among fallen blocks on the north side, and in Petitor Cove beneath the cliffs of Oddicombe Breccias on the north side (*Figure 7.2*), where Watcombe sandstones dip at 20° steeper than the overlying breccias.

The implication of these observations is that Watcombe strata were deposited north and west of the present-day Torquay Promontory, and probably above it also, and at the time little or no limestone detritus was being supplied. Tectonic movements, most likely basement block-fault uplift similar to the pre-Albian activity later on, resulted in the red beds becoming tilted northwards at angles of at least 20°, and to the west the effect was to initiate erosion of a limestone massif—or at least to divert its erosion products into the Watcombe area for the first time. The material cannot have come from the Torquay Promontory itself as the rounding of the limestone fragments (*Figure 7.3*) indicates that they had undergone some 16 km of transport before being deposited. The limestone mass west of Kingskerswell is the most likely source, assuming some transport down canyons before reaching the cuvette fans. The date of these movements may be put at late Stephanian on the basis of the probable age of the Oddicombe Breccias.

Another unconformity is present in the Broadsands outlier (SX 896581), very local in occurrence and situated at the bottom of the Tor Bay Breccias there. It is possibly connected with sandstone dykes found along the shore at Shoalstone Beach (SX 934567), and other locations east of Brixham, where hard red sandstones with calcite veins and cementation occur in fissures in Devonian limestone. Their orientation is mainly east–northeast–west–southwest (Richter, 1966) with some approximately north–south. There are, in fact, two generations of dykes, the darker red ones cutting subparallel paler red ones. A similar colour contrast distinguishes the local unconformity, though the nature of the darker red constituent and its relationship to the unconformity is not known.

LOCALITIES

Tor Bay Area

7.1. *Shoalstone Beach, Brixham* (SX 934567): red sandstone dykes in Devonian limestone can be seen on the foreshore east of the swimming pool. There are two sets, both near-vertical and trending roughly east–west.

7.2. *Broadsands-Saltern Cove, Goodrington* (SX 897581): red sandstones of the Tor Bay Breccias are faulted against Devonian rocks along a cliff face trending north–south, noted for its extensive calcite mineralisation in the Crystal Cave Fault zone. Approach from the Broadsands car park or Waterside campsite, at low tide only.

7.3. *Waterside (Oyster) Cove, Goodrington* (SX 895587): the best exposure of the basal unconformity is found in a small cove east of Goodrington church reached via Oyster Bend and a railway footbridge. Coarse Tor

Bay Breccias overlie purple Devonian slate and sandstone; note particularly the zone of red weathering of these beneath the unconformity. The shape of the original land surface can be seen on the underside of the breccias where they overhang. Fossil burrows occur in the breccias just north of here.

7.4. *Roundham Head, Paignton* (SX 896598): a well-marked lens of aeolian sand occurs within the Tor Bay Breccias forming the headland, best seen on the south side, beyond the end of the promenade (low tide only).

7.5. *Livermead Head, Torquay* (SX 905627): sandy breccias of the Livermead Formation include encrustations and replacements by chalcedonic silica (beekite) on limestone fragments, like little yellowish warts. Access from Cliff Park Road and steps to the beach, low tide only. The use of hammers at this locality is discouraged.

North of Torquay

7.6. *Oddicombe Beach and Petitor Point, Babbacombe* (SX 927662, *Figure 7.3*): the type section of the Oddicombe Breccias, 350 m thick, and faulted against the Devonian limestone of Petitor Point at the north end of the cove. Here, at the 'Gentlemen's Bathing Place', cemented grey limestone talus deposits host cavities filled with red sandstone and calcite nodules; these are mountain-side features that were subsequently buried by red beds. The point itself is a re-excavated mountain of limestone. Access from the beach (reached by cliff railway) or from Petitor Road and cliff paths, which lead also to locality 7.7.

7.7. *Petitor Cove, Babbacombe* (SX 927664): shattered Devonian limestone surface overlain unconformably by sandstone of the Watcombe Formation is clearly seen in the northern side of the cove. Massive Oddicombe Breccias form a large cliff on the north side, but the second unconformity at the base, over the Watcombe strata, is obscured and visible only as a dip discordance.

Teignmouth-Dawlish Area

7.8. *The Ness, Shaldon* (SX 939718): fine-grained breccias of the Ness Formation include large boulders of porphyry and other rocks, most easily seen strewn on the foreshore. Sand dykes and features such as rain prints and desiccation cracks can also be found, especially at the south end of the beach. Access is from the Ness car park through the Smugglers Tunnel.

7.9. *Smugglers Lane, Holcombe* (SX 957747): typical Teignmouth Breccias are seen in this small cove, with several well marked sand dyke structures and some porphyry boulders. Walk down the lane from Dawlish Road (car parking very difficult in the lane) or approach along the sea wall from Teignmouth.

7.10. *Coryton Cove, Dawlish* (SX 961761): type section of the quartzite-porphyry-aureole-feldspar Coryton Breccias, with sand dykes at one level; the transition to the Dawlish Sands is well displayed. The latter show a single aeolian cross-bedded unit 20 m thick in the cliff behind the cove. Approach from Beach Road over the railway footbridge, or by cliff paths.

7.11. *Dawlish Sea Wall and Langstone Rock* (SX 967770-978779): the complete succession of Dawlish Sands is shown in the cliffs from the station to Langstone Rock, easily observed from the sea wall. Aeolian and fluvial sands with some interbedded breccia beds are best seen at the two footbridges, reached from A379 opposite Westcliff Road or beside the Rockstone Hotel. Nearer the headland are channel-fill breccia lenses; the sequence grades up into the Langstone Breccias (type section) mainly quartzite and aureole fragments, seen also across the water at the Beacon, Exmouth (SY 002807). The headland can be approached from Dawlish Warren car park also.

East of Exmouth

7.12. *Orcombe and Rodney Points, Exmouth* (SY 019796): the Exmouth Formation is composed of thick channel-sandstone lenses, which form the headlands, and interbedded mudstone-thin sandstone sequences which form the majority of the succession. Both facies can be seen at this locality and eastward to Straight Point (a gunnery range); several small faults cut the sand lenses and rare plant fossils and malachite have been found. Approach from Marine Drive or cliff paths.

7.13. *West Cliff, Budleigh Salterton* (SY 060816): the Budleigh Salterton Pebble Beds are well displayed in the cliffs, with Littleham Mudstones to the westward. The Pebble Beds are overlain by the Otter Sandstones, aeolian and fluvial. Approach from the lane beside the Rolle.

7.14. *Chit Rocks, Sidmouth* (SY 120869): Otter Sandstones, faulted against Mercia Mudstones, form a small headland. Easy access from West Peak Hill Road.

Inland Localities

7.15. *Kenton*, road cutting 100 m north of the Church (SX 958835): semi-circular fluvial channel cut in aeolian sands on east side of a narrow busy main road, close to a closed-off cross lane (beware of fast traffic).

7.16. *Dunchideock, School Wood Quarry* (SX 875873): amygdaloidal basalt with a variety of minerals in the amygdales; brecciated flow-tops with sedimentary infills.

7.17. *Kennford*, road cuttings on A38 and M5 (SX 915867): excellent exposures of the type section of Kennford Breccias, porphyry-quartzite-feldspar fragments; best seen in the access roads north of the village but also forming the huge Matford Cut on the M5 to the northeast (pedestrians are not allowed on the motorway).

7.18. *Sandy Gate, Exeter* (SX 965913): an old quarry in aeolian Clyst Sands is located above the westbound carriageway, accessible to the agile. On the opposite side, a large face of aeolian sand shows a mudstone layer (marked by vegetation) offset by small faults. The Bishops Court sand quarry, to the north, is the type section of the Clyst Sands.

7.19. *Heavitree Quarries, Exeter* (SX 953918): Murchison's type section of the Heavitree Breccias, similar to the Kennford Breccias; these pass up into Clyst Sands at the top of Quarry Lane.

7.20. *Newton St Cyres* (SX 873978): type section of St Cyres Breccias in road cuttings in the village: fine-grained sandy breccias with feldspar and granitoid fragments as well as quartzite; sand dykes can be seen in places.

7.21. *Sandford, north of Crediton* (SS 829026): a rare exposure of the Cadbury Breccias lies just inside a field gate on the right of the lane about 50 m north of the last house. Fragments almost entirely of Culm quartzite, a few vein quartz and Pilton Shale fragments also.

7.22. *Posbury Quarry* (SX 815978): volcanic neck and lava-bomb agglomerates, overlain by red sandstone.

7.23. *Sampford Peverell Quarries* (ST 035148): showing Sampford Peverell Breccias derived from nearby distinctive Carboniferous grey limestone outcropping to the northwest. Approach the quarries through lanes north of the Grand Western Canal, or see outcrops in the village away from the main road.

7.24. *White Ball Hill, Devon-Somerset Border* (ST 090180): White Ball Sands overlying Uffculme Conglomerates in road-side pits beside the A38. The sands show a range of red and pale green-buff colours.

7.25. *Ramsey Quarry, Langford Budville* (ST 107222): lateral transition from Uffculme Conglomerates to the Milverton Conglomerates (containing Carboniferous limestone pebbles of south Wales affinities). Approach by footpath from the common north of the village.

Chapter Eight

The Marine Rocks of the Mesozoic

The Mesozoic Era is characterised world-wide by a major marine transgression which culminated in the late Cretaceous. At maximum sea-level height, there was a considerable depth of water over the greater part of northwest Europe, with only small isolated land masses remaining in northwest Scotland, Wales, Brittany, Scandinavia, south Germany and the Massif Central of France. Such land masses were probably of low relief, thereby providing little, if any, detrital material into the surrounding shelf seas. During Mesozoic times the North Atlantic Ocean began to open up to the west of the British Isles, and the effects of this developing ocean can also be detected in the Mesozoic succession of South West England.

This great marine transgression was not, however, a simple single event. It proceeded in fits and starts, and on a few occasions it distinctly looked as though it was in reverse. The rate of change also varied greatly, and this has resulted in an equally varied and interesting succession of rocks. In Devon only part of the record is preserved, although in the offshore areas surrounding the county it is a little more complete. Large areas of the county (together with Cornwall) remained as an island throughout the greater part of Mesozoic, being finally drowned in Turonian times—the first of the two Upper Cretaceous sea-level maxima. During the earlier part of the Mesozoic the depositional areas within and around Devon were never far from a coastline, although nowhere are shoreline deposits preserved.

It is notoriously difficult to estimate transgression rates, or even to calculate the depth at which a sedimentary stratum was laid down, but geologists still try to produce models of these depth changes in the hope that they will convey some impression of the history of an area. This present account is no exception, and in *Figure 8.1* the stratigraphic succession of the marine Mesozoic rocks of Devon is outlined together with a tentative transgression-regression model. Shown alongside are the principal tectonic events that have also influenced the depositional history of the area. The model is based on the work of Hancock and Kauffman (1979) and Hart and Bailey (1979), and discussions with other stratigraphers and palaeontologists. The sequence of events shown created a stratigraphic succession from the Rhaetian (Triassic) to the Danian (Palaeocene) embracing nearly all of the Mesozoic Era and passing over into the Cenozoic.

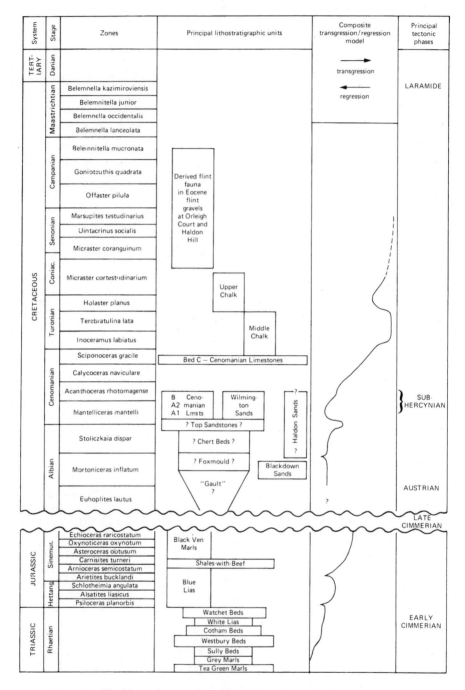

Figure 8.1. The Mesozoic succession of South West England with sea-level changes and tectonic events

TECTONIC HISTORY

The Variscan Orogeny, and its associated granitic intrusions, led to the formation of a major topographic ridge with a tendency for uplift, running through Devon, Cornwall and the Isles of Scilly, westwards at least as far as Haig Fras. There is no evidence that this ridge was present in southeast Devon, but even in the west the ridge was almost buried in Permian and Triassic times by a thick sequence of red-beds which built up in post-orogenic, intracratonic basins. In the Triassic the tectonic style changed from a compressional to a tensional regime (W. H. Ziegler, 1975; P. A. Ziegler, 1975), and a series of narrow rifts and graben with intervening horsts developed over the whole of northwest Europe; subsiding areas included the Bristol Channel area and the Celtic Sea (north and south). This taphrogenic rifting has been related, by many workers, to the development of the North Atlantic Ocean rift zone (P. A. Ziegler, 1975).

By the late Triassic the mountainous topography of northwest Europe which followed the Variscan Orogeny was already very much reduced, and the initiation of the early Cimmerian earth movements in the Rhaetian (probably also associated with a general eustatic rise of sea level) began the first of the major Mesozoic marine transgressions. Because of the reduced topography this marine incursion quickly drowned most of northwest Europe, apart from those areas like Devon and Cornwall (Cornubia) that remained islands for a considerable part of the Mesozoic. In southern England the Celtic Sea Basins, the South West Approaches Basin, the Bristol Channel Basin, and (in particular) the Wessex Basin all began to subside. In general this was a simple subsidence, despite the complexity of the Variscan fold belt that formed the basement, and in the basinal areas deposition was as near continuous as it can ever be. However, fluctuations in the rate of sediment input and minor tectonic events produced slight interruptions of deposition and folding at the margins of the basins.

The Jurassic was dominated by the further development of the rift systems begun in the Triassic. This process was only slightly upset by the mid-Cimmerian earth movements, but the effects of the late Cimmerian movements were far more widespread. The late Cimmerian produced minor uplift and erosion in Portlandian times, but also marked a change in the stress patterns over Europe: the previous extensional phase gave way to a more compressional phase as the Alpine movements began to dominate the structural evolution of Europe. Thus in the Cretaceous minor earth movements in the Aptian-Albian (the Austrian phase) created many structures described in the literature as pre-Albian (Drummond, 1970), and recognised in the Torquay district as well as farther east around Weymouth (Laming, 1965). However, the subsidence (transgression) resumed in the mid and late Albian, became important in the early and mid-Cenomanian, and reached a peak early in the Turonian. After an important relative regression in the late Turonian (Hancock and Kauffman, 1979; Hart and Bailey, 1979) the transgression continued until another maximum was reached in Campanian times. This was followed by an apparent regression in the Maastrichtian–Danian. Minor movements in the Cenomanian, possibly related to the Sub-Hercynian tectonic phase, have been described from southeast Devon. Over large areas of Europe the Sub-Hercynian movements of Cenomanian to Maastrichtian times created inversions in many marginal troughs, such as the Weald, English Channel, and the North Holland-South North Sea Basin (Hancock, 1976). These became accentuated in the Maastrichtian–Danian movements of the Laramide phase which, coupled with a major regression of sea level, effectively terminated marine sedimentation over large areas of northwest Europe, and produced the isolated basins of deposition that characterise the remainder of the Palaeogene.

Throughout the Mesozoic the North Atlantic Ocean was forming to the west of the British Isles, but it is only in the late Cretaceous that open marine conditions can be recognised over the northwest Europe and North Atlantic Ocean area. It is no coincidence that the establishment of this environment marked the onset of chalk deposition over the greater part of northwest Europe, and geochemical investigations by Pomerol and Aubry (1977) have attempted to relate the characteristic black flints of the chalk with volcanic processes occurring at the Mid-Atlantic Ridge.

STRATIGRAPHY AND CORRELATIONS

Just as the changes in environment from the arid desert landscapes of the Triassic to the open marine seaways of the Upper Cretaceous produced very different lithologies, so they also produced a dramatic change in the fossils found enclosed in those sedimentary rocks. The stratigrapher has therefore to change his allegiances to fit the environment, and use plants (Warrington, 1976) and vertebrates (Walker, 1969) in the Triassic, ammonites in the Jurassic and Lower Cretaceous (Dean et al., 1961) and planktonic foraminifera and calcareous nannoplankton in the Upper Cretaceous (van Hinte, 1976).

Fortunately, difficulties of correlation are few. In the Lower Jurassic the ammonite zones were actually first described from the Lower Lias succession of the Devon–Dorset coastline, and conditions during the deposition of the chalk in the Upper Cretaceous were so uniform that a zonation based on benthonic foraminifera can be applied over the whole of Europe. Indeed, it is possible to use a zonation based on planktonic foraminifera that has worldwide application (despite statements to the contrary by Rawson et al., 1978). Only the extension of the use of the Lower Cretaceous ammonite zones, which were established in the succession of the Gault Clay in southeast England, to the Greensand facies of South West England, is still a problem.

THE 'RHAETIC'

The 'Rhaetic', as the rocks between the distinctive Triassic and distinctive Jurassic are known, has had an involved stratigraphic history (Pearson, 1970). The 'Rhaetic' is an informal lithostratigraphic term that only has local application within the United Kingdom. It is an unfortunate term, being too close in sound and form to the Rhaetian the standard chronostratigraphic subdivision of the Upper Triassic, as defined in the (marine) Alpine region using the distribution of *Choristoceras* spp., and other ceratite ammonoids. This ceratite fauna is completely absent from the United Kingdom succession, and even the distribution of an important bivalve fauna (such as *Rhaetavicula contorta* (Portlock)) is controlled by the onset of marine conditions. The usage of the terms 'Rhaetic' and Rhaetian was discussed fully by Pearson (1970).

The 'Rhaetic' is best seen on the Somerset coast (*Figure 8.2*) in the magnificent cliff sections between Watchet (ST 070435 to ST 033436) and Lilstock (ST 172452), while on the south coast of Devon parts of the succession are exposed between Axmouth and Pinhay Bay (SY 320908). These successions were described in detail by Richardson (1906, 1911), while Audley-Charles (1970a and b) reviewed their stratigraphy and conditions of deposition. In 1960 Hallam provided a detailed account of the White Lias (Langport Beds) of Pinhay Bay, Charton Bay (SY 300900) and Tolcis Quarry (ST 281010) near Axminster, the only inland site still available for study.

The Tea Green Marls are the lowest unit and comprise some 15 m of green-grey calcareous shale and mudstone at Culverhole (SY 274894). The dominant clay mineral is illite, with subordinate chlorite and an illite-montomorillonite interstratification (Cosgrove, 1975). These pass upwards into the Grey Marls that are best seen on the Somerset coast in St Audrie's Bay (ST 105430). The upper part of the Grey Marls (Sully Beds) at Lilstock have yielded specimens of *Rhaetavicula contorta*, that already indicate the presence of near-marine conditions, and probably points to little, if any, stratigraphic break between the Grey Marls and the 'Rhaetic'. The environment at this time is thought to have been characterised by low relief, with deposition occurring in bodies of shallow water cut off probably by low barriers from the open sea.

Richardson (1906, 1911) described some 6 m of 'Rhaetic' at Culverhole, the base of the local succession being taken at the base of a thin black gritty shale. This shale, in fact, appears to be a local development at the base of the Westbury Beds, and at times when the foreshore is clear of modern beach deposits it can be seen piped into the upper surface of the underlying Grey Marls. The remainder of the Westbury Beds is made up of black laminated shales with abundant bivalves, including *Rhaetavicula contorta*, *Chlamys valoniensis* (Defrance) and *Protocardia rhaetica* (Escher v.d. Linth). At Lilstock, poorly developed thin argillaceous limestone beds (commonly crowded with fossils) are also present within the Westbury Beds succession. The overlying Cotham Beds are best seen at Lilstock, where 1.5 m of pale, slightly pyritous, greenish-grey limestone alternate with poorly fossiliferous shales. At the top of this unit there is a thin (0.2 m) algal limestone that expands northeastwards into the famous 'landscape marble' of the Bristol area.

The overlying White Lias (Langport Beds) is 8 m thick on the Devon coast, in Pinhay Bay, and forms an interesting succession of fine-grained calcilutites. Hallam (1960) described this succession in detail, documenting shallow-water depositional features such as intraformational conglomerates, corrosion surfaces, mud-cracks, algal laminae and slump beds. These limestones also have a very high strontium content, probably indicating that the original sediment was a fine-grained aragonitic mud. Fossils are uncommon, although *Liostrea hisingeri* (Nilsson) can usually be found throughout the succession. The uppermost bed of the White Lias (the 'Sun Bed') is only 5 cm thick on the coast but still displays beautifully preserved U-shaped burrows (*Diplocraterion* sp.) and occasional desiccation cracks. These are best seen adjacent to the fault in the centre of Pinhay Bay. There is some evidence to suggest that the top of the White Lias has been eroded, and this hiatus may be represented by the Watchet Beds of the Bristol Channel area.

The beginning of the White Lias marks a distinct change in the clay min-

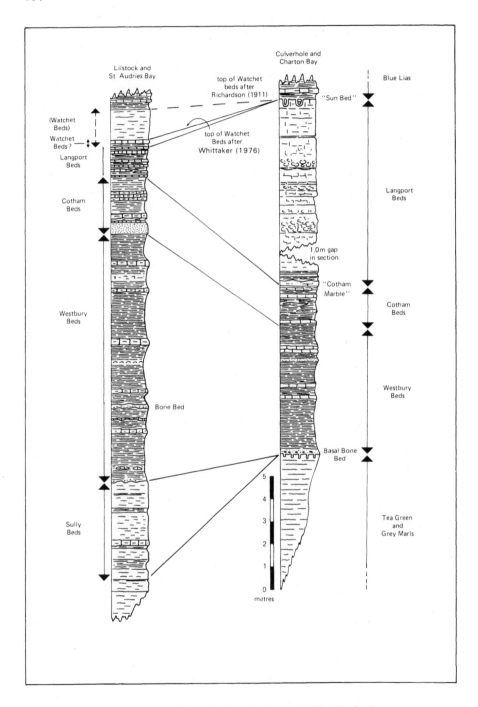

Figure 8.2. The 'Rhaetic' succession in South West England

eral assemblage (Cosgrove, 1975) with the appearance of an illite-chlorite-kaolinite association. This may indicate the extension of marine erosion towards the upland areas of South West England.

THE JURASSIC

Only the lowermost Jurassic is present in Devon and the immediately adjacent areas of Somerset and Dorset, but almost all the available sections are internationally famous. The fundamental work of Lang (1914 to 1936) described these sections, especially those between Axmouth and Charmouth, in bed-by-bed detail, and even today his work is a basis for all further research. The classic ammonite study of Dean *et al.* (1961) was established on the same sections.

As shown in *Figures 8.3 and 8.4*, the lithological units represented in the Pinhay Bay to Charmouth section are, in ascending order, the Blue Lias (29 m), the Shales-with-Beef (21.5 m) and the Black Ven Marls (43 m).

The Blue Lias not only comprises the *Planorbis, Liassicus, Angulata, Bucklandi* and *Semicostatum* (pars) Zones, but has in its lowermost 4 m strata described as the Pre-planorbis Beds. These are lithologically identical with the remainder of the Blue Lias, but contain a limited fauna that is dominated by *Liostrea hisingeri*. They are inferred to be Jurassic, even though they precede the appearance of the first Jurassic ammonite, *Psiloceras planorbis* (Sowerby). In Dorset and Somerset *Psiloceras planorbis* appears at precisely the same level (Whittaker, 1976), an almost identical lithological succession being recognised across the whole area. *Psiloceras planorbis* is more abundant in the Bristol Channel sequence than in the Pinhay Bay succession, although no physical barriers to migration are thought to have existed between the two areas.

The Blue Lias is dominated by limestone-shale cycles, which give a characteristically banded appearance to weathered cliff sections. These cycles have been described in detail by Hallam (1964). The carbonate part of each cycle is usually between 5 cm and 30 cm thick, and many are argillaceous calcilutites. Some are nodular in appearance, while others have almost planar upper and lower surfaces. The alternating shales are generally somewhat thicker.

Even the casual observer must wonder about the cause of these remarkably uniform, and laterally persistent, cycles whilst amongst geologists this has been a contentious issue for several decades. The debate hinges on how much these cycles reflect primary sedimentological differences, or depend on the rhythmic segregation of calcium carbonate during diagenesis. The strongest evidence for a primary origin is the presence of intense bioturbation, with lighter coloured rock, richer in $CaCO_3$, penetrating downwards into darker rock less rich in $CaCO_3$, and vice versa. Within the overall cyclical succession there are also thin horizons of finely laminated bituminous shales. These are paper thin, and contain few fossils apart from occasional fish debris. They seem to indicate near-anaerobic bottom conditions, and these, again, appear to be primary in origin.

However, the situation cannot be as simple as that. A number of the limestones are definitely concretionary in appearance, and seem to indicate a migration of $CaCO_3$ onto some 'nucleating horizon'. It is interesting to note that the limestones are approximately 85 per cent $CaCO_3$, while the shales are

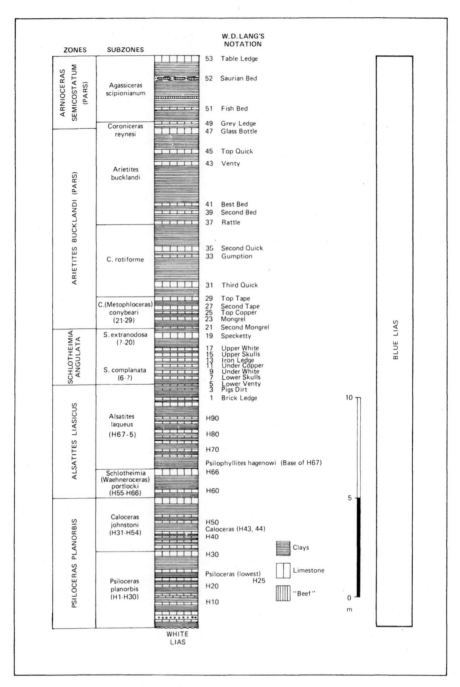

Figure 8.3. The Lower Jurassic succession of the Devon–Dorset coast: Blue Lias

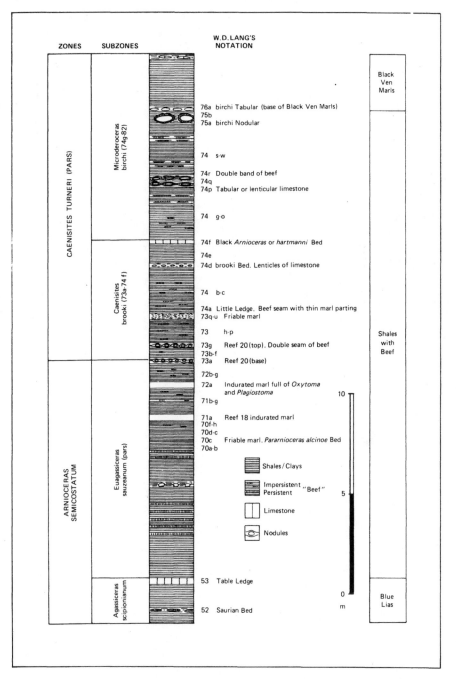

Figure 8.4. The Lower Jurassic succession of the Devon–Dorset coast: Blue Lias–Black Ven Marls

generally 35 per cent $CaCO_3$, and although this will probably have been altered during diagenesis there must have been at least some initial difference in composition to begin the migration. It is also evident that the thickness of the alternating sedimentary units remain constant regardless of the thickness of the total succession. In general, therefore, the thicker the sedimentary sequence, the more limestone it contains, but the lack of limestone where the successions is thinner is not due to wholesale loss of strata at non-sequences, as the very accurate palaeontological record indicates a full succession. Probably there were some initial sedimentological differences that have been accentuated by later diagenesis.

The Pinhay Bay to Lyme Regis section yields vast numbers of fossils, including abundant *Plagiostoma giganteum* Sowerby, *Gryphaea arcuata* Lamarck *Calcirhynchia calcaria* Buckman, and large specimens of *Cenoceras* sp., a Lower Jurassic nautiloid. The ammonite fauna is especially rich, although some of the zonal species can be a little difficult to find. At certain levels one has almost solid ammonite pavements (such as in the *Bucklandi* Zone (Bed 29 of Lang) where very large specimens of *Coroniceras* (*Metophioceras*) *conybeari* are abundant. At many levels there is a distinctive mottling of the rocks produced by bioturbation of the sediments. Throughout the Blue Lias fossil material from ichthyosaurs, plesiosaurs, and fish is common.

The overlying Shales-with-Beef are seen between Seven Rock Point (SY 328909) and the Cobb (SY 339916) where they display a typical 'paper' shale lithology, interspersed with thin, laterally impersistent, beds of fibrous calcite (the 'beef'). The latter probably formed during early diagenesis. Fossils are generally crushed or badly preserved, except in the limestone or calcareous nodules. The Black Ven Marls are better seen at Charmouth.

The environment of deposition for the whole of the Lower Jurassic succession was probably a shallow sea, as witnessed by the infauna and trace fossils, but higher in the succession there is less carbonate which probably indicates increasing depth of deposition. Also, the benthonic fauna of the Shales-with-Beef and the Black Ven Marls is certainly less abundant and less diverse than that of the Blue Lias and it would appear that with the passage of time bottom conditions become less hospitable.

THE JURASSIC-CRETACEOUS HIATUS

While the full Jurassic succession can be seen along the whole of the Dorset coast, it is very difficult to assess where the margins of deposition lay, and in Devon there is no direct evidence of further marine deposition until the Upper Greensand (mid Cretaceous). However, palaeogeographic maps of the Jurassic produced by W. H. Ziegler (1975) and P. A. Ziegler (1975) suggested Jurassic deposition in southeast and northeast Devon, although the greater part of Devon (as part of Cornubia) remained an island. Throughout the Jurassic, deposition may have continued with a little or no interruption in the South West Approaches, Bristol Channel and Celtic Sea Basins.

The major transgressive pulse that began in the Rhaetian continued throughout the Jurassic, but minor tectonic movements in the Upper Jurassic probably reduced the area of deposition, an effect which was emphasised by a phase of regression (or basin infilling). Thus Cosgrove (1975) noted that kaolinite disappears from the clay mineral association in the Portlandian, leaving only illite and montmorillonite, possibly indicating that the sea had retreated from the Dartmoor area. This set of changes essentially reduced the upper-most Jurassic depositional areas to a series of shallow lagoons and swamps (in the 'Purbeck'), with a limited areal distribution, and even affected the South West Approaches Basin where Middle Jurassic strata are overlain by Lower Cretaceous and/or Upper Cretaceous rocks. However, above the Cinder Bed (Rawson et al., 1978), which marks the local base of the Creta-ceous System, kaolinite re-appears and remains well established in the overly-ing Wealden and later Cretaceous strata.

The late Cimmerian movements at the end of the Jurassic provided the slight angular unconformity between the underlying Jurassic and the overly-ing Cretaceous strata.

THE CRETACEOUS

The Cretaceous rocks of southeast Devon contribute to the scenic beauty of the area: forming wooded hills of the Upper Greensand around Axminster and Sidmouth and the magnificent chalk cliffs of Whitecliff (SY 232892) near Seaton, and Beer Head (SY 220880). These rocks form the most westerly Cretaceous strata exposed on land in southern England, and almost the whole succession shows signs of having been deposited in near-shore or shallow marine conditions. The only exception to this are rocks of Turonian age, which contain a planktonic microfauna suggestive of slightly deeper water conditions (Hart and Weaver, 1977).

The succession is readily divisible into four units: 'Gault', Upper Green-sand, Cenomanian Limestones and Sandstones, and the Chalk (*Figure 8.5*). Each unit has its own particular problems, either in terms of palaeoecology, sedimentary petrology or stratigraphy.

The 'Gault' and Upper Greensand are an almost totally clastic succession. Together they occupy a large tract of country east of Exeter (*Figure 8.5*), but they also crop out on the Haldon Hills west of the River Exe. The Cenoma-nian Limestones and Sandstones form a thin, but distinctive, lithological unit between the Upper Greensand and the overlying chalk. The chalk outcrop is limited to isolated patches on hilltops in the Axminster–Seaton area, even though it was once the most geographically extensive stratigraphic unit. Its present distribution has a very important effect on the Upper Greensand, as noted by Tresise (1960): where the Upper Greensand is capped by Chalk, calcareous faunas are often recorded but where the Chalk cover has been removed by erosion, the Upper Greensand is decalcified to a considerable depth, producing the so-called 'Blackdown facies'.

The Gault and Upper Greensand

In Devon the 'Gault' is a very silty, slightly glauconitic clay. This has been dated as Middle Albian (*Loricatus* Zone) by Hancock (1969), and Rawson et al. (1978). The 'Gault' grades upwards, almost imperceptibly, into the Upper Greensand, which in its normal development (around Beer and Seaton) can be divided into a lower (Foxmould) unit and an upper (Chert Beds) unit. The uppermost Upper Greensand is often separated off as the Top Sandstones (Jukes-Brown and Hill, 1900; Tresise, 1960; Smith, 1957; Rawson et al., 1978).

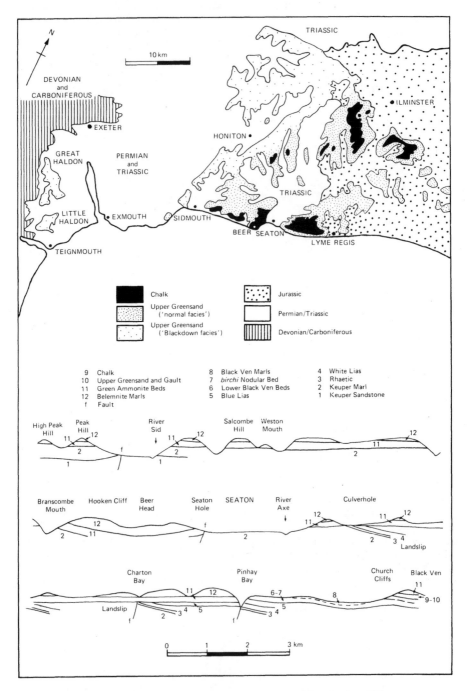

Figure 8.5. Geological map of southeast Devon with coastal profiles

In the Beer-Seaton area the Foxmould comprises some 26 m of grey-green glauconitic sands that contain large courses of calcareous concretions or 'cowstones'. The latter can be very fosiliferous, being crowded with *Exogyra* spp., and serpulids (*Rotularia concava* (Sowerby)). Between the Foxmould and the overlying Chert Beds is a thin, nodular, glauconitic limestone crowded with *Exogyra* spp., and other bivalves. Associated with this limestone are two or three thin, highly glauconitic, sandstones, that usually weather dark blue-green.

The Chert Beds are a series of coarse calcarenites with an average thickness of about 20 m. At several levels these sandstones are coarsely nodular, while other levels are very coarse-grained, containing large subangular fragments. The chert itself (Tresise, 1960, 1961; Hart, 1973) occurs either as isolated nodules or as lenticular bands which are generally parallel to the bedding. Some of the chert bands are remarkably continuous and are almost parallel-sided. The majority of the cherts are yellow-brown in colour although some grey chert is known. Each nodule (or lens) usually has an external skin of more porous stone, which appears to be either less chertified, or full of included sedimentary grains, some of which have been leached out by weathering. In thin section the cherts are seen to be formed of masses of fibrous chalcedonic silica. Several workers on the South West England Cretaceous disagreed with Tresise's (1960, 1961) view that these cherts were formed as soft, gel-like masses on the sea floor, because of the presence of cross-stratification, quartz grains, and fossils within the cherts. In section the needles of chalcedonic silica can be seen to have grown, layer upon layer, on small included fragments in a regular and well oriented way. Even planktonic foraminifera can be found, surrounded by enveloping masses of silica fibres. The presence of well preserved trace fossils within the cherts would also seem to suggest that they could not have formed in gel-like masses on the sea floor, but are a product of silica redistribution during diagenesis. The Top Sandstones are usually chert-free, coarse calcareous sandstones. They are mainly free of glauconite and appear very similar to the Eggardon Grit of Dorset (Kennedy, 1970; Drummond, 1970; Carter and Hart, 1977). The immediately overlying cobble conglomerate, described in detail by Ali (1976), is very limited in its distribution, being best developed in the area of Beer Harbour (SY 230891).

As indicated by Tresise (1960) the 'Blackdown facies' can be seen in the area between Blackdown Hill in the west, Yarcombe (ST 245084) in the east and Peak Hill (SY 105860) in the south. In the Blackdown Hills there are some 30 m of coarse, whitish-brown sands. In the upper levels of the succession are layers of siliceous concretions that contain the famous silicified Blackdown fauna. Bivalves and gastropods predominate, but there are also ammonites and echinoids.

Palaeontologically the Upper Greensand (*Figure 8.5*) is dominated by a bivalve fauna that is long-ranging and almost certainly facies controlled. At best this can only be used to indicate either an Albian or Cenomanian 'aspect' (Hamblin and Wood, 1976). Ammonite evidence is sparse and limited to a handful of specimens. Moreover, the majority of the known fauna also comes from two very restricted horizons: the 'Ammonite Bed' of the Purbeck Coast, and the top of the Chert Beds at Shapwick Quarry (SY 312918) near Lyme Regis (Hamblin and Wood, 1976). This probably indicates levels of preferen-

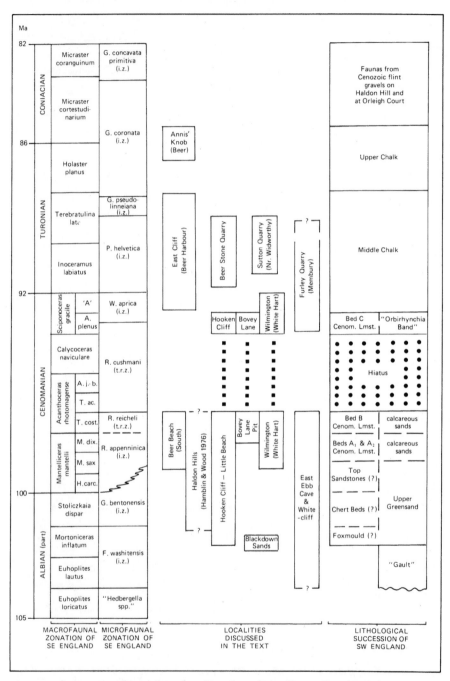

Figure 8.6. The Cretaceous succession of southeast Devon (i.z.—interval zone; t.r.z.—taxon range; ●●●●—hiatus: ■■■■—linked sequence)

tial preservation, and indeed many of the specimens from the Purbeck Coast are in the form of relatively resistant, phosphate casts.

Hancock (1969) regarded the Blackdown Sands as being *Varicosum* and/or *Orbignyi* Subzone age, and gave a reasonable faunal list as evidence. He also regarded the Foxmould on the coast as being *Variscosum* and *Auritus* Subzone age. The overlying Chert Beds have usually been ascribed to the *Stoliczkaia dispar* Zone (Hancock, 1969; Kennedy, 1970; Drummond, 1970; Hamblin and Wood, 1976; Rawson *et al.*, 1978). The lowest levels of the Haldon Hills succession were regarded as *Auritus* Subzone age by Hamblin and Wood (1976), following the discovery of *Stomohamites* sp. and they thus equated the immediately overlying molluscan fauna (of 'Albian aspect') with the Blackdown fauna and the Foxmould. The coral bed on Haldon, which lies stratigraphically higher, contains a coral and bivalve fauna reminiscent of the type Cenomanian, but ammonites ascribed to this level appear to indicate the *Stoliczkaia dispar* Zone, and the highest dateable horizon in the Haldon Hills has yielded three Cenomanian ammonites. However, the important Albian ammonite fauna from Shapwick Quarry which is found at the boundary between the Chert Beds and the Top Sandstones, is definitely of *Stoliczkaia dispar* Zone age.

The acceptance of a *Stoliczkaia dispar* Zone age for the Chert Beds nevertheless ignores some of the micropalaeontological data. As indicated by Carter and Hart (1977) *Orbitolina* faunas are found at many localities in South West England. At one locality, Wolborough (SX 855700) near Newton Abbot, *Orbitolina* are so abundant as to dominate the total fauna. Both this Wolborough *Orbitolina* fauna and another recovered from cherts collected in Bullers Hill Quarry (SX 882848) have been statistically compared to a fauna of *Orbitolina concava concava* (Lamarck) from the type Lower Cenomanian of the Sarthe, France. The Bullers Hill Quarry fauna compares favourably with the Sarthe fauna and appears to be of early Cenomanian age, which agrees with the ammonites described by Wood (1971). The Wolborough fauna appears to be very slightly older, although it is impossible to be very precise. At the present time it can only be regarded as earliest Cenomanian or latest Albian in age. Detailed studies on the internal structures of the fauna may eventually allow more accuracy. Hamblin and Wood (1976) have, however, stated categorically that the *Orbitolina* fauna (at some levels) has to be Albian, and must not be compared to Cenomanian faunas from elsewhere. Unfortunately the only Lower Cretaceous *Orbitolina* faunas from the upper part of the succession in the North Atlantic area come from Flemish Cap (Aptian) and Portugal (Albian). If the palaeogeography of W. H. Ziegler (1975) and the geological data from the English Channel (Curry *et al.*, 1970, 1971; Andreieff *et al.*, 1975) is accepted there appears to have been little chance of the *Orbitolina* fauna migrating into the United Kingdom from the south during the greater part of the Albian, especially as this genus always occupies near-shore clastic environments. Unfortunately the matter cannot be resolved until further work is completed.

On the Devon coast there is some evidence to suggest that the Upper Greensand was slightly folded or warped (Smith, 1957, 1961a; Hart, 1971; Carter and Hart, 1977) before the deposition of the overlying Cenomanian Limestones. Folding has also been detected on Haldon Hill by Durrance and Hamblin (1969).

Cenomanian Limestones and Sandstones

The folding of the Upper Greensand on the Devon coast is associated with a fracturing of the Top Sandstones, into which early Cenomanian chalky fissure deposits have been let down (Ali, 1975, 1976). These fissures are best seen at the south end of Beer Harbour, across the anticlinal structure first described by Smith (1957). The thickness and distribution of the Cenomanian Limestones and the Cenomanian Sandstones are controlled by these structures (*Figure 8.7*) and have a trend which is about north-northeast–south-southwest. They have been described by Smith (1957, 1961a, 1961b, 1965), Kennedy (1970), Drummond (1970), and Carter and Hart (1977). As shown in *Figure 8.8* the Cenomanian Limestones on the coast represent a reduced correlative of the Cenomanian Sandstones of the Bovey Lane and Wilmington exposures. The sequence has been described in detail by Jukes-Browne and Hill (1903).

The lowest unit (Bed A) of the Cenomanian Limestone succession varies in thickness from zero at Small Point to a maximum development of 5–5.5 m at the east end of Hooken Cliff (SY 220880). In all sections where the full succession is present Bed A can be separated into a lower (Bed A1) and an upper (Bed A2) portion. Bed A1 is a coarse calcareous grit, with abundant grains and small pebbles of quartz. Also included within the lowest levels of Bed A1 are large, glauconitised, cobbles of calcareous sandstone (probably from the underlying Top Sandstones), together with a very rich fauna of *Ceriopora ramulosa* (Michelin) a coral-like bryozoan. Bed A2, which is much more widespread than A1, is a shelly limestone, with no large cobbles, and

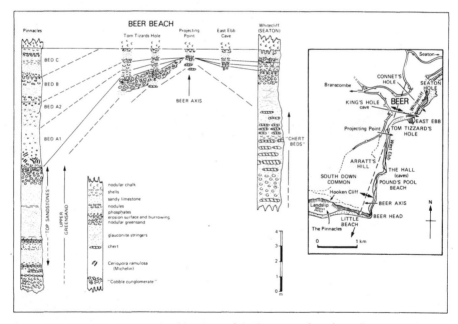

Figure 8.7. The Cenomanian Limestone of the Beer area of southeast Devon (map after Smith, 1957)

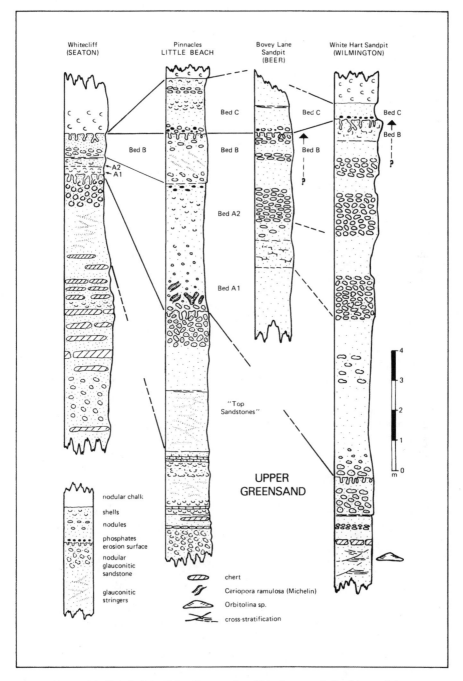

Figure 8.8. Correlation of the Cenomanian Limestones and Sandstones between Hooken Cliff, Bovey Lane and Wilmington, southeast Devon

relatively little sand content. The ammonite faunas of Beds A1 and A2 (Kennedy, 1970) are clearly reworked, as many of the fossils are included within definite pebbles. *Mantelliceras* spp. is quite common, as are the echinoids *Catopygus* and *Sternotaxis*.

The overlying Bed B is a hard white sandy limestone, that in many places is distinctly glauconitic. Much of the glauconite is concentrated into lenses some 10–20 cm in length. These have been variously ascribed to bioturbation or sediment winnowing by bottom currents. Ammonites are very rare in Bed B, but echinoids (particularly *Sternotaxis* spp.) are common throughout. The upper bedding plane of Bed B is a marked erosion surface that shows a thin layer of brownish phosphatised pebbles, from which a good fauna (brachiopods, echinoids and ammonites) can usually be extracted. Kennedy (1970) recorded a mixture of upper Lower Cenomanian and lower Middle Cenomanian ammonites from Bed B, the latter date agreeing more closely with the suggestions of Carter and Hart (1977). Recently, specimens of *Rotalipora reicheli* Mornod have been identified from the calcareous sands at Bovey Lane and Wilmington that correlate with Bed B of the limestone succession (*Figure 8.8*). This species is associated with the Lower-Middle Cenomanian boundary in southeast England and northern France.

The surface between Beds B and C is heavily phosphatised, probably indicating exposure on the sea floor for a considerable period of time. In many places it is difficult to say from which bed some fossils really come, as they are rigidly cemented to the boundary. Kennedy (1970) was able to identify (at some localities) a zone of phosphatised and glauconitised pebbles 30 cm thick in the top of Bed B.

Bed C (the *Orbirhynchia* Band of Smith, 1961a) is very locally developed. A complex pebble bed may be present at the base in which ammonites can be found in several preservation modes. Some are chocolate-brown phosphatised abraded casts, while others are green-coated, and only slightly phosphatised; some are completely unphosphatised. Kennedy (1970) studied these three faunas, and gave a *Calycoceras naviculare* Zone dating for the most phosphatised ammonite fauna. The unphosphatised fauna included *Actinocamax plenus* (de Blainville), *Orbirhynchia wiesti* (Quenstedt), and ammonites indicative of the *Actinocamax plenus* Zone (now the lower portion of the *Sciponoceras gracile* Zone).

In the Hooken Cliff section Bed C is 1.75 m thick, and composed of very hard nodular slightly glauconitic, sandy limestone. Many coarse grains or quartz and phosphate are present. Detailed microfaunal analysis (Hart, 1975; Carter and Hart, 1977) has shown that three principal subdivisions can be recognised. Bed C expands in thickness northwards, and can be seen again in the Bovey Lane Sandpit, Beer (SY 216900) where an incomplete section yields 1.5 m of soft, glauconitic calcareous sands. A temporary section in the Beer Stone Quarries showed over 4 m of glauconitic and sandy chalk, forming the greatest thickness of Bed C yet recorded. Northwards towards Wilmington, Bed C is reduced to a thin (0.5 m) calcareous greensand with a phosphatic conglomerate at its base. The same feature, though less well exposed, was recorded from Membury (Kennedy, 1970; Hart, 1975; Carter and Hart, 1977).

In a detailed palaeo-environmental study of these limestones Ali (1975, 1976) described features characteristic of the near-shore depositional environment, including the blocky, calcareous sandstone of the Top Sandstones, which were interpreted as a Lower Cenomanian beach deposit. Nearly all the limestones (A1, A2, B, and C) show signs of current activity, reworked pebbles and boulders, and damaged and abraded fossils—all features of a near-shore, shallow marine situation.

Smith (1957, 1961a, 1961b, 1965) was able to correlate this set of limestones along the Devon coast from Dunscombe (SY 164880) in the west to Pinhay in the east, and clearly demonstrated how the thickness of every bed, and the limited distribution of Bed C, was controlled by the underlying structures. Inland, however, the limestones rapidly give way to calcareous sandstones, with courses of calcareous stone or nodules. At Bovey Lane Sandpit (Carter and Hart, 1977), Bed C is a soft glauconitic calcareous sand, that contains abundant *Orbirhynchia* spp. Beds A and B cannot effectively be separated, as both are lithologically similar. Further north, at Wilmington (SY 209998), the famous Wilmington Sands (Kennedy, 1970; Carter and Hart, 1977) attain a total thickness of 13.5 m. Some years ago a pit in the base of the quarry exposed the base of the sand succession: at the bottom was a coarse cross-stratified sandstone (with *Orbitolina* sp.) which has a rather broken, irregular upper surface. This was overlain by a bouldery basal conglomerate, above which is the sand succession exposed at present. The calcareous sandstone succession (especially in the upper rubbly layers, locally known as 'grizzle') is highly fossiliferous and an exceptional echinoid (*Sternotaxis* and *Catopygus*), bivalve (*Inoceramus* and *Chlamys*), gastropod (*Pleurotomaria*), and ammonite (*Mantelliceras, Schloenbachia,* and *Turrilites*) fauna can be collected with little difficulty. The ammonite fauna may be compared with that from the Cenomanian Limestones on the coast, and indicates an early Cenomanian age (*Mantelliceras saxbii–Mantelliceras dixoni* Zone), with some lower Middle Cenomanian (*Turrilites costatus* Zone) elements.

West and northwest of the Scilly Isles are marginal calcareous sandstones and dolomites of presumed Cenomanian age.

The Chalk

The overlying Chalk also reflects the pre-Cenomanian Limestone depositional controls. In areas where the Cenomanian Limestone succession is reduced, the lower part of the Turonian appears very nodular, with evidence of penecontemporaneous erosion and the formation of incipient hardgrounds (Kennedy and Garrison, 1975). This is particularly well developed at the eastern end of Beer Harbour, although part of the sea cliffs there are now obscured by the concrete sea defences. Where the Cenomanian Limestone succession is better developed (such as at Hooken Cliff) the Lower Turonian is much thicker and in the vicinity of Beer Village has developed the peculiar facies known locally as the Beer Stone. Between the Chalk and the underlying Cenomanian Limestones there is also a rich rolled phosphatised fauna preserved in what is known as the Neocardioceras Pebble Bed (Kennedy, 1970; Rawson *et al.*, 1978). It contains a rich fauna of re-worked ammonites, representing the upper part of the *Sciponceras gracile* Zone of the uppermost Cenomanian, and includes Lower Turonian ammonites preserved in nodules or pebbles.

In southeast Devon the whole of the Middle Chalk and the lowest two zones of the Upper Chalk are represented, although exposures of any value are rather limited. Undoubtedly the most accessible section is that at Beer

Harbour, where one can follow an almost uninterrupted sequence of chalk from the top of the Cenomanian Limestones to the top of the local succession on Annis' Knob (*Figure 8.9*).

In southeast Devon only the Cretaceous succession above the Cenomanian Limestones is entirely within the chalk facies, but in southeast England the Cenomanian is also represented by chalk (Kennedy, 1970; Drummond, 1970; Carter and Hart, 1977; Rawson *et al.*, 1978). Chalk is a soft, pure, white limestone that characterises several continents during the Upper Cretaceous; its properties and formation have been discussed by Hancock (1976).

Chalk is generally composed of particles of two very distinct size ranges. The coarser grain size fraction (10–100 micrometres) is dominated by the foraminifera and ostracoda. The remainder is composed of 'calcispheres', and macrofaunal debris, notably echinoid plates and spines, bivalve prisms and bryozoa. The finer fraction (0.5–4 micrometres) is composed almost entirely of the rays, plates, and laths of Haptophyceae (algae). When living, each alga is formed of between seven and twenty coccoliths, of variable and distinctive shape. These are shed continually during life and at death, thereby providing a carbonate rich 'snow' to the sea floor. As these algae were producing this carbonate in enormous quantities within the surface waters of the Cretaceous shelf seas, the changes now seen in the lithological features of the chalk must have been caused by local bottom conditions and the processes involved in diagenesis.

The presence of nodular chalks has been discussed by Kennedy and Garrison (1975), and Hancock (1976) provided a discussion on the formation of flints. Flints are demonstrably secondary in origin, never being found in intraformational chalk conglomeratic beds, or with a chalk epifauna. They must, however, have been completely formed and lithified prior to erosion and incorporation in the Palaeocene Bull Head Bed at the base of the Thanet Sands in southeast England. Many flints occur parallel to bedding, while others cross the bedding at varying angles. Higher in the succession flints are cavernous, containing the highly problematical 'flint meal'. Hart and Bailey (1979) have also discussed the distribution of nodular chalks and flints, with respect to their proposed water depth model for the Albian-Santonian interval.

The lowest Turonian (Zone of *Inoceramus labiatus*) is normally without flints but displays several (at least six) rhythmically bedded, nodular chalk beds and hardgrounds. The lowermost levels of the succession are very hard and 'gritty', with grains of quartz and glauconite. The nodular chalks are often stained yellow-brown, indicating relative degrees of phosphatisation. This probably represents prolonged periods of exposure on the sea floor and/or periods of erosion and reworking of the partly lithified sediment. The total thickness for this zone is just under 10 m, while even as close as Pinhay Bay, to the east, there is double that thickness. Fossils are rare, apart from *Inoceramus* sp., and Hart and Weaver (1977) failed to obtain any microfauna from the lowest 2 m of the succession. However, the upper part of the zone does contain a very diagnostic Lower Turonian microfauna identical with that of other successions in southeast England and northern France. The 'Beer Stone' is a locally developed hard chalk, 5 m thick, best seen in the Beer Stone Quarries between Beer and Branscombe (SY 208881). The peculiar texture is provided by the finely comminuted *Inoceramus* debris. Near Brans-

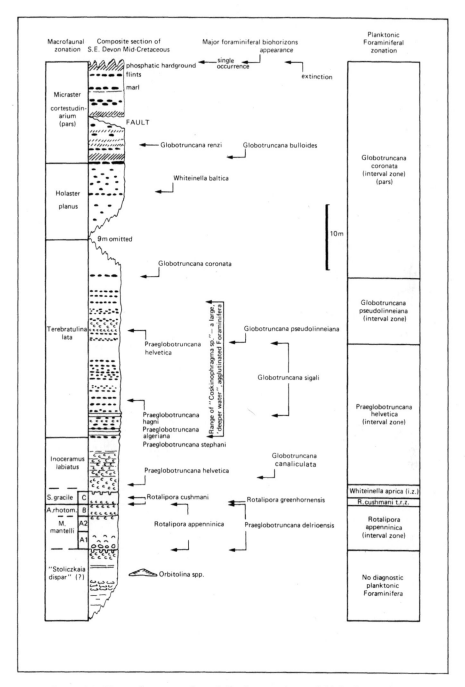

Figure 8.9. Upper Cretaceous foraminiferal succession and biohorizons of Beer, southeast Devon

combe Mouth (at the western end of Hooken Cliff) the overlying *Terebratulina lata* Zone rests directly on the Cenomanian Limestones, the *Inoceramus labiatus* Zone having been overstepped (Smith, 1961a). Further west, in Berry Cliff (SY 190880), there is a fuller succession of the *Inoceramus labiatus* Zone again, but with a distinctive echinoid fauna.

The *Terebratulina lata* Zone consists of massive, marly chalk with abundant courses of black flints (Hart and Weaver, 1977). Traditionally the base of the *Terebratulina lata* Zone has been drawn at the first prominent course of flints, above which are two distinctive hardgrounds. Higher in the cliff, and very useful for visual correlation, are the 'two foot band' and the 'four foot band': beds of exceptionally marly cream-coloured flint-free chalk. The *Terebratulina lata* succession is much thicker at Hooken Cliff, although access makes the Beer Harbour succession much more attractive to observers. The zonal fossil (*Terebratulina lata* (Etheridge)) is relatively abundant, especially in the middle of the zone, and although most specimens are very small (about 1 cm diameter), they can usually be found with little difficulty. The microfauna described by Hart and Weaver (1977) is exceptionally interesting as it includes several species of southern aspect, all regularly used for international correlation. The distribution of *Praeglobotruncana helvetica* (Bolli) is perhaps the most important, but the occurrence of *Globotruncana coronata* (Bolli) and *Globotruncana sigali* Reichel is also significant (*Figure 8.9*).

The presence of *Bdelloidina* sp. (*Coskinophragma* sp. of Hart and Weaver, 1977) has been taken to indicate deeper water (Hart and Bailey, 1979), and can be found all over southern England and northern France at this level. Hart and Bailey take this to be the first of the Upper Cretaceous sea-level maxima, and as such is regarded as the time of likely submergence of Cornubia. Detrital grains (Smith, 1961b; Hancock, 1969) became very rare in the Turonian, and (as suggested by Smith) it would appear that the source of the sediment no longer existed at that time. In a full account of this Turonian transgression Hart and Bailey gave no figures for the depth of the sea, although clearly something in excess of 300 m would have been expected in the centre of the Wessex Basin.

The placing of the *Terebratulina lata-Sternotaxis planus* Zone boundary in the Beer Harbour succession is very difficult, as there is a gap in the exposures at this level. The thickness of the *Sternotaxis planus* Zone is therefore estimated at between 15 m and 20 m. The chalk at this level is again very nodular, and hard. Flints are less common than in the *Terebratulina lata* Zone. The exposed section resumes in the bluff known as Annis' Knob, just above the cliff path to Seaton. This has been described by Jukes-Browne and Hill (1904), Kennedy and Garrison (1975), and Bailey (1975). The boundary between the *Sternotaxis planus* Zone and the overlying *Micraster cortestudinarium* Zone (the Turonian-Coniacian boundary of Rawson *et al.*, 1978) was placed by Jukes-Browne and Hill (1904) at the base of a prominent black tabular flint, 0.13 m thick. The chalk below this tabular flint is very hard and nodular, with a fauna that includes *Orbirhynchia cuvieri* (d'Orbigny). Above the tabular flint the irregularly weathered cliff is characterised by the presence of very hard nodular chalks that, in places, contain yellow-brown phosphatic horizons. These have been described as incipient hardgrounds by Kennedy and Garrison (1975). Echinoids are particularly common in these hardgrounds (Bailey, 1975), and many appear to be the zonal fossil. Nodular chalk, with flints and incipient hardgrounds in places, can be seen to the top of the cliff. A further succession of softer, less nodular chalk can be found immediately up hill and round the corner from Annis' Knob, and is clearly brought down by a minor fault. The planktonic Foraminiferida are again very useful for correlation at this level, and the fauna includes the Coniacian marker *Globotruncana renzi* Gandolfi.

Chalk deposition clearly occurred over a wide range of Devon, as a Cretaceous fauna is well known from Eocene flint gravels on Haldon Hill and from Orleigh Court in north Devon (Jukes-Browne and Hill, 1904; Rogers and Simpson, 1937; Hancock, 1969; Rawson *et al.*, 1978). The Haldon flint gravel fauna preserves a record from (probably) the *Sternotaxis planus* Zone to low in the *Belemnitella mucronata* Zone of the Campanian, while at Orleigh Court there is a similar stratigraphic range represented. It seems no coincidence that both flint gravel faunas extend to the mid-Campanian, as that was probably the second of the Upper Cretaceous sea-level maxima. From offshore areas it is well known that the succession continues up into the Maastrichtian–Danian (Curry *et al.*, 1970, 1971; Warrington and Owens, 1977; Rawson *et al.*, 1978). The Campanian and Maastrictian are much softer chalks, and yield good, well preserved microfaunas. The upper levels of the Maastrichtian contain *Racemiguembelina fructicosa* (Egger), just as in the classic successions of Denmark. These chalks are overlain immediately, and apparently conformably, by the soft cream biocalarenites of the Danian. Hancock (1969) and Rawson *et al.* (1978) commented on the lack of Lower to Middle Turonian fossils in the flint gravel faunas, even though most authors seem to agree on a high sea level at that time. The suggestion by Rawson *et al.* (1978) that there may also be a Lower to Middle Turonian hiatus in the offshore areas, does not seem to be supported by the presence of several sites at which Lower to Middle Turonian chalks have been described by Warrington and Owens (1977). The faunas listed by P. J. Bigg in that account, from this level, are almost identical with those recorded by Hart and Weaver (1977) from Beer.

End of Marine Deposition

The regression at the end of the Maastrichtian–Danian was created by the combined effects of the Laramide earth movements, coupled with a major eustatic change (Hancock and Kauffman, 1979). These two events affected the whole of northwest Europe, and reduced the depositional areas of the post-Danian Palaeogene to small isolated basins.

LOCALITIES

8.1. *St Audrie's Bay, Somerset* (ST 105430): 'Rhaetic'-Lias. For access park at the above grid reference, which lies at the end of the private road leading through The Belt. After following the zig-zag path to the beach one can follow the full succession westwards (towards Watchet) through the Keuper Marls, the Tea Green Marls, the 'Rhaetic' and into the Blue Lias. The appearance of the first *Psiloceras planorbis* can usually be located on the headland where the cliff turns westwards.

8.2. *Lilstock, Somerset* (ST 172452): Grey Marls–'Rhaetic'. For access park at the above grid reference, immediately north of the village. Follow the road northwards towards the old harbour, and at the coast turn eastwards. After walking around a bend in the coastline the best section is visible in the cliffs forming a northwest facing slope, immediately before a small headland (which can get cut off at high tide). The succession includes the whole succession from the Grey Marls to the top of the 'Rhaetic'.

8.3. *Watchet to Blue Anchor Somerset* (ST 070435 to ST 033436): this coastal section includes beautifully preserved exposures of the whole Triassic to Lower Jurassic transition. Many parts of the succession can be better seen at the above localities, although the presence of evaporites in the Upper Triassic is a feature of this section. These are best seen just east of the western end of the section, at Blue Anchor.

8.4. *Pinhay Bay, Devon* (SY 320908): Lower Jurassic. While the whole cliff succession from Axmouth (SY 254900) to Lyme Regis is worthy of investigation it is a very long haul, and should not be attempted on a rising tide. Only a few access points are available through the landslip, and in high summer many of these are difficult to find in the thick vegetation. Pinhay Bay is best approached from the Cobb, in Lyme Regis, the walk via Seven Rock Point giving a continuous section through the Lower Jurassic.

8.5. *Hooken Cliff, Devon* (SY 220880): Cretaceous. The section from Branscombe Mouth to Beer Head contains a magnificent section from the Red Marls of the Triassic through the whole Cretaceous succession; Foxmould to Upper Chalk. The section can be approached either from the beach at Branscombe Mouth (SY 208882) or from the cliff path from Beer (SY 224880). In the latter case there is a long climb down the landslip, which, although not dangerous, should only be tackled with care.

8.6. *Whitecliff, Devon* (SY 232892): Cretaceous. This gives a very clear view of the Foxmould, Chert Beds and Cenomanian Limestone succession, and can be approached from Seaton Hole in the north or from Beer Harbour in the south. In either case as low a tide as possible is required, and care should be taken to avoid being trapped by the rising tide.

8.7. *Shapwick Quarry, Devon* (SY 312918): Chert Beds—Middle Chalk. This quarry is just north of the Lyme Regis–Sidford road (A3052) about 2.8 km west of Lyme Regis. The old quarry is within the Middle Chalk, but near the entrance there are workings in the upper levels of the Chert Beds. A reduced succession of the Cenomanian Limestones is also visible.

8.8. *Beer Harbour, Devon* (SY 230891): Cenomanian Limestone—Upper Chalk. On arriving at the base of the slipway proceed south to the first small promontory, on which can be seen a reduced succession of the Cenomanian Limestones, resting on the broken surface of the Upper Greensand. This succession then continues all along this side of the bay down to Beer Head. At the opposite end of the bay, near the new jetty the lower levels of the *Inoceramus labiatus* Zone can be examined. Following the cliff succession back towards the centre of the bay, the first line of flints, the base of the *Terebratulina lata* Zone, is easily found. Where the beach huts begin follow the sloping path up the cliff past soft chalks of the *Terebratulina lata* Zone. Turning sharply onto the cliff path to Seaton one arrives at Annis' Knob, with the prominent tabular flint taken to mark the Turonian–Coniacian boundary. It should be noted that East Devon District Council have imposed a ban on hammers and sample collection in the Beer Harbour area, as past indiscriminate geological activity has ruined many of the best sections.

8.9. *Bovey Lane Sandpit, Devon* (SY 216900): Cenomanian Sandstones. This overgrown sandpit is now almost impossible to investigate as the land owners have fenced in the face. Beds A, B and C of the Cenomanian Sandstones succession were at one time easily accessible in this quarry.

8.10. *Beer Stone Quarries, Devon* (SY 215895): Middle Chalk. These working quarries give a very good section through the Turonian, including the famous Beer Stone, which was quarried out in a series of caves that are still visible.

8.11. *Wilmington, White Hart Sandpit, Devon* (SY 209998): Cenomanian Sands. This famous quarry is still worked; access is immediately opposite the White Hart Public House at the western end of the village. The section changes rapidly as the sands are worked, but there is usually a good section of the fossiliferous sands. Much of the fauna, however, is easily collected from the loose material on the quarry floor.

8.12. *Membury, Devon* (ST 275043): Middle Chalk. This old chalk pit is still available at the side of the lane from Membury to Furley, although the section is badly degraded. The interesting basement beds described by Kennedy (1970) and Carter and Hart (1977) are no longer visible (Hart, 1975) without extensive excavation. The Middle Chalk can still be collected, although fossils, apart from *Inoceramus*, are very rare.

8.13. *Bullers Hill Quarry, Devon* (SX 882847): Upper Greensand. At the side of the Forestry track, just past the Forestry Headquarters, the old quarry can be found showing some 5.8 m of grey–white tourmaline rich, quartz sands. Thin grey cherts are visible in the section, but brown, fossiliferous cherts can be found on the quarry floor.

8.14. *Babcombe Copse, Devon* (SX 869767): Upper Greensand. This old quarry, now filling with water, is best approached from a lay-by on the western side of the main Exeter–Torbay road (A380). The succession described by Hamblin and Wood (1976) is now fast disappearing and the section may soon be useless.

Chapter Nine

The Tertiary Sedimentary Rocks

Two broad divisions of the Tertiary sedimentary rocks can be recognised in Devon: flint bearing gravels, of Palaeocene and Eocene age; and kaolinitic clays, lignites and sands of the Eocene-Oligocene Bovey Formation. The flint gravels include residual varieties which were derived from solution of the underlying Chalk *in situ*, and attest to the former extent of that formation over much of Devon. Younger gravels, which were probably deposited by rivers flowing over extensive plains stretching away from the foothills of Dartmoor eastwards to Dorset, are also included. In contrast, deposition of the Bovey Formation, although also largely fluvial (but with important lacustrine components), was in the main closely confined to partly fault-controlled basins which formed along the Sticklepath Fault (*Figure 9.1*).

The two main onshore basins, the Petrockstowe Basin in North Devon and the Bovey Basin between Bovey Tracey and Newton Abbot in the south, have long been famous for the valuable deposits of ball clay which they contain. The name 'ball clay' originates from an early method of extraction in which the clays were cut into 'balls' about 0.25 m (9 inches) square and weighing 13 to 33 kg. In modern usage, a ball clay is a fine-grained, highly kaolinitic sedimentary clay, the higher grades of which fire to a white or near-white colour in an oxidising atmosphere (Highley, 1975). Lignite has also been extracted from the Bovey deposit, and the basin represents the only major reserve of lignite in Great Britain.

Recently, a third major Tertiary basin has been discovered in the offshore area east of Lundy Island (Fletcher, 1975). It lies on the northwest extension of the Sticklepath Fault and contains Bovey Formation strata comparable to those at Petrockstowe and Bovey.

PALAEOCENE AND EOCENE FLINT GRAVELS

Deposits of flint-bearing gravels of probable Palaeocene and Eocene age are found in four main areas in Devon (*Figure 9.1*). The gravels are of two types: a residual facies derived from the *in situ* solution of chalk, and a younger fluvial facies. Both facies are present on the Haldon Hills (Hamblin, 1973a); the gravels bordering the Bovey Basin are mainly fluvial, with minor occurrences of residual gravels in solution pipes in Devonian limestones near Kingsteignton (Brunsden *et al.*, 1976). The Orleigh Court deposit, near Bideford in north Devon, considered to be Pliocene by Rogers and Simpson (1937), is now thought to be at least in part a Palaeocene residual gravel.

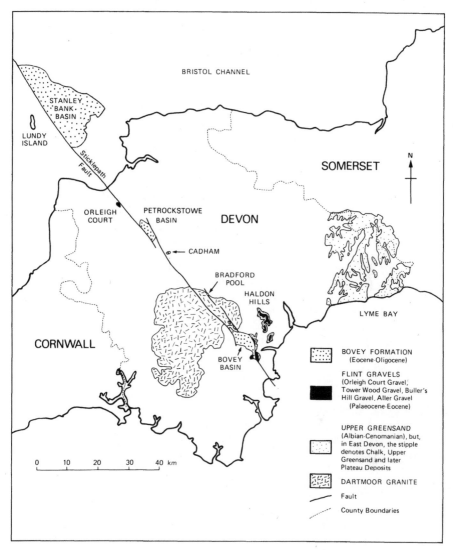

Figure 9.1. Distribution of Tertiary deposits in Devon

Isaac (1979) indicated that both residual and fluvial Tertiary gravel facies are present around Sidmouth, and possibly in other parts of east Devon, but the fluvial Mutters Moor Gravel is now thought to be Pleistocene in age (K. Isaac, personal communication). An interesting feature of the east Devon deposits is the presence of a well-developed lateritic weathering profile which includes the local development of silcrete.

Haldon Hills

On the Haldon Hills, Hamblin (1973a) was able to distinguish and map two divisions of the Haldon Gravels which are of Palaeocene-Eocene age

(*Figure 9.2*). The oldest of the two units was named the Tower Wood Gravel; it rests on the Albian-Cenomanian Upper Greensand and is overlain by the Buller's Hill Gravel which contains lenticular bodies of kaolinitic clay.

Tower Wood Gravel. The type locality of the Tower Wood Gravel Member is Tower Wood Quarry (SX 876856) at the northern end of Great Haldon. The Tower Wood Gravel is an unbedded deposit up to 8 m thick of large, closely packed, unabraded pale grey flints in a matrix of white, locally brown-stained, clay and a little sand. Most of the flints have been peripherally shattered *in situ* by Pleistocene frost action so that the deposit comprises large, generally horizontally aligned, cores of flint in a matrix of flint chips and clay. The junction with the underlying Upper Greensand is marked by a sandy flint-free bed up to 12.5 cm thick containing well rounded quartz and schorl pebbles. The matrix clay of the gravel, studied by Hamblin (1973b), typically consists of crystallographically well-ordered 'china clay' kaolinite with a little disordered 'ball clay' kaolinite and illite, although in some samples the clay was similar to that in the Buller's Hill Gravel and believed to have been washed down through the strata from that source.

The unabraded nature of the flints and the purity of the deposit led Hamblin (1973a) to conclude that the Tower Wood Gravel was formed by the *in situ* solution of chalk before deposition of the Buller's Hill Gravel. The basal bed of sand with pebbles is thus the residuum after solution of flint-free Lower Chalk, deposited as a basal gravel during the westward transgression of the Chalk sea. Hamblin (1973a, 1973b) suggested that the matrix of the Tower Wood Gravel was derived not from solution of chalk, but from a hydro-thermal kaolinite deposit, probably in the area of the Dartmoor granite; solution of the chalk and introduction of the matrix probably took place almost simultaneously, although it is uncertain whether the clay was washed in steadily as the chalk dissolved, or whether it formed a body of sediment on top of the chalk and was let down as solution proceeded. The latter is probably more likely in view of the unabraded nature of the flints. A Dartmoor source is supported by the presence of a granite-derived heavy mineral suite in the Tower Wood Gravel matrix.

Buller's Hill Gravel and Included Clay Bodies. The Buller's Hill Gravel Member, named from Buller's Hill Quarry (SX 882847) in the northern part of Great Haldon, comprises up to 10 m of abraded flint gravel with subordinate clay bodies and sand beds. It rests on the Tower Wood Gravel, and Hamblin (1973a) stated that it has not been seen to cut through it to rest on the Upper Greensand. The gravel itself is pale brown or pale grey in colour, comprising abraded and chatter-marked grey flints up to 30 cm across, with smaller, better rounded and less frost-shattered pebbles of vein quartz, schorl, quartzite, altered Carboniferous shale and chert; the matrix is sand and clay. The matrix contains a granitic suite of heavy minerals (including abundant tourmaline); the clay comprises approximately equal amounts of ordered and disordered kaolinite, the proportion of illite being higher than that in the Tower Wood Gravel (Hamblin, 1973b). Sand in beds associated with the gravel is similar to that of the matrix.

Horizontal lenticular bodies (up to 11 m long, 8 m wide and 2 m thick) of grey to white clay occur within the gravel. They consisting mainly of disordered 'ball clay' kaolinite with a variable illite and silt content. Hamblin

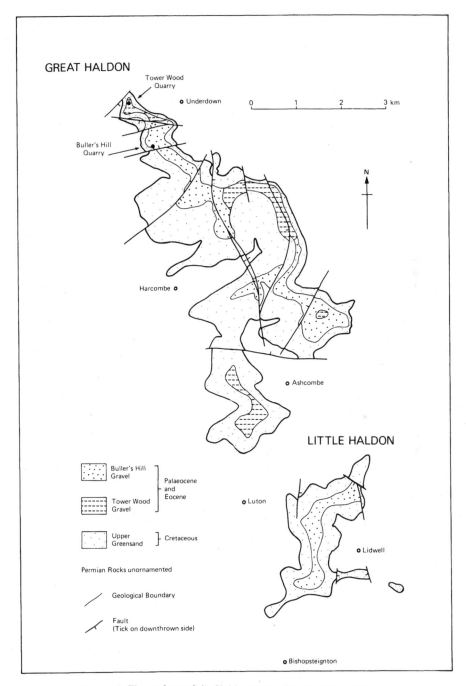

Figure 9.2. The geology of the Haldon Hills (after Hamblin, 1973a)

(1973a) considered that the clay, originally a continuous sheet overlying the Buller's Hill Gravel, had been incorporated into the Buller's Hill Gravel by cryoturbation and solifluction processes occurring during the Pleistocene. As supporting evidence for this hypothesis, may be cited the markedly different clay mineralogies of the clay bodies and the clay in the matrix of the Buller's Hill Gravel; the lack of sedimentary structures connecting the clay with the surrounding gravel; and the lack of a granitic heavy mineral suite in the clay bodies comparable with that in the gravel. Alternatively, the clay could be the same age as the gravel, representing the sediment infill of abandoned channels of streams draining down from Carboniferous rocks.

Although the direct evidence of original fluvial sedimentary structures is lacking in the Buller's Hill Gravel (periglacial processes have destroyed any such evidence), Hamblin (1973a) considered that the gravel had a fluvial origin because of the kaolinitic matrix of the gravel (which is common in fluvial deposits but rare in marine, and the presence of the well-ordered form of kaolinite as in the Tower Wood Gravel argues against later conversion of other clays to kaolinite); and the generally low degree of abrasion of the flints, which indicates fluvial rather than marine transport. Also, the pebble content indicates derivation from higher ground to the west whereas, if the deposit were a beach shingle, marine transgression from the east would have added exotic pebbles from the east.

Hamblin thought that the Buller's Hill Gravel was deposited on a wide flat plain by sheet-flooding in a savanna climate, and Edwards (1973) suggested that both the Buller's Hill Gravel and Aller Gravel were deposited by braided streams flowing over an extensive flood-plain extending east and south of the Dartmoor upland.

Contemporaneous fossils have not been recorded from the Haldon Gravels, but the Chalk fossils contained in the flints indicate that the earliest possible age for all the units is the Senonian stage' of the Upper Chalk. The Eocene units of the Haldon Gravels cannot be younger than the formation of the Bovey Basin, since from that time they would have occupied an upland area adjacent to the subsiding basin; the lower part of the Bovey Formation is of Eocene age, so the initiation of the basin and therefore cessation of Buller's Hill Gravel deposition took place in the Eocene. By mineralogical comparison with deposits farther east, the Tower Wood Gravel was dated by Hamblin (1973a) as pre-Bagshot Beds and possibly pre-Reading Beds, and thus might in part be of Palaeocene age. The clay bodies in the Buller's Hill Gravel, regarded by Hamblin as post-dating that gravel, must also pre-date the formation of the Bovey Basin, and would be Eocene but post-Bagshot Beds. A tentative correlation with some part of the Bracklesham and Barton Beds was suggested by Hamblin.

The Bovey Basin

Tower Wood Gravel. Gravels correlated with the Tower Wood Gravel occur on the fringe of the Bovey Basin near Kingsteignton where, in cuttings for the Kingsteignton–Newton Abbot by-pass, Brunsden *et al.* (1976) mapped large gravel, sand and clay-filled solution pipes extending to depths in excess of 30 m in Middle Devonian limestone. In the solution pipes, Upper Greensand rests on and in cavities within the Devonian limestone and is generally overlain by Aller Gravel. In places, a gravel of unabraded but mainly broken flint nodules occurs between the Upper Greensand and the Aller Gravel. This gravel can be correlated with the Tower Wood Gravel of Haldon, and may have been derived from solution of the chalk cover in a similar way.

Aller Gravel. Fluvial flint gravel, named the Aller Gravel by Edwards (1973), outcrops along the eastern side of the main Bovey Basin and around the Decoy Basin south of Newton Abbot (*Figure 9.3*). A small exposure of Aller Gravel overlain by Staplehill Gravel on the fault-bounded southern side of the Ringslade Clay Pit (SX 846726) has also recently been noted (B. L. Jones and M. R. Harvey, personal communication). The gravel is Eocene, and normally rests on the Upper Greensand. Both Aller Gravel and Upper Greensand dip beneath the Bovey Formation in the Decoy and main basins.

The Aller Gravel comprises up to 25 m of grey to brown flint gravel and sand with subordinate lenticles of white to pale grey silt and silty clay. The gravel consists dominantly of chatter-marked flints, only moderately abraded; other components include vein quartz, tourmaline and quartz-tourmaline rock, Upper Greensand chert, Lower Carboniferous chert, Upper Carboniferous sandstone, dolerite and metadolerite, hornfels and other fine-grained aureole rocks. Rapid lateral and vertical changes in grain size occur, and the gravel contains lenticular channel-fills which rest on other channel-fills with curved to straight erosional junctions. The sediment in the channels is generally poorly sorted and commonly shows large scale cross-bedding. Some channel-fills contain fining-upward sequences from gravel at the base to silty clay at the top. Fine-grained deposits are uncommon and occur mainly as small lenses of silty clay, representing deposition in abandoned depressions or channels. Intraformational clay clasts also occur within beds of sand or gravel. Edwards (1973) considered that the sedimentary features of the Aller Gravel correspond most nearly to those described from modern braided streams which have numerous channels containing coarse sand and gravel but relatively rare clay and silt.

Relationship of Buller's Hill and Aller Gravels. Many authors (such as De la Beche, 1839; Clayden, 1906) favoured a correlation between the gravels of Bovey and of Haldon. Reid (in Ussher, 1913), however, considered that the flint gravel around the Bovey Basin was a marginal facies of the Bovey Beds (now Bovey Formation) and to a large extent derived from the Haldon deposit.

Edwards (1973, 1976) favoured a direct correlation between the Aller Gravel and the Buller's Hill Gravel, and considered that both gravel deposits were part of the same sheet, which originally extended over the area of the Bovey Basin and Haldon Hills in the interval immediately preceding the subsidence of the Bovey Basin. But, in view of the considerable thickness of the Eocene Bovey Formation in the axial part of the Bovey Basin, he did not exclude the possibility that deposition of the Aller Gravel and its correlatives was continuing at the time of initiation of the Bovey Basin, and that the focus of deposition of the gravels shifted to a geographically restricted area along the axis of the newly formed basin, while deposition had ceased on the Haldon Hills and in marginal areas of the Bovey Basin. In the central trough area of the basin there might be a continuous transition between a thick basal gravel sequence and Bovey Formation lithologies. Nevertheless such gravel, if it exists, would be concealed deep in the axial part of the Bovey Basin and

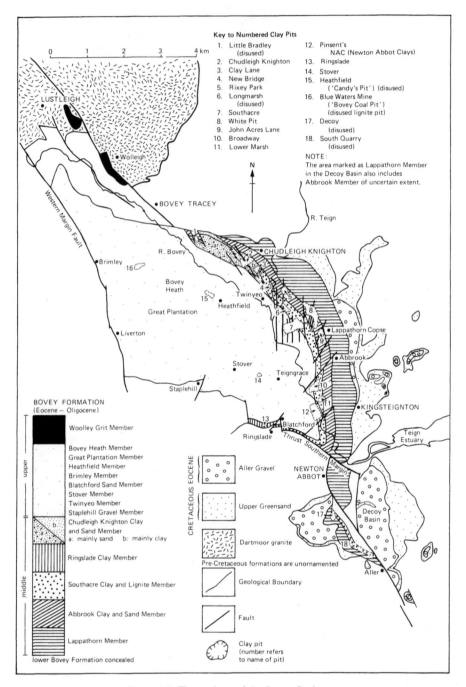

Figure 9.3. The geology of the Bovey Basin

would not affect the arguments for direct correlation of the Aller Gravel and Buller's Hill Gravel.

Alternatively, Hamblin (1974) concluded that the Aller Gravel was formed during the earliest stages of downfolding of the Bovey Basin, after the formation of the Haldon Gravels on an easterly-dipping plain, but still early enough for the Aller Gravel itself to be downfolded in the Bovey Basin. Also, that if the Haldon clay bodies represent a sedimentary unit which once overlay the Buller's Hill Gravel, then the Aller Gravel must be later, and there can be no overlap in the formation of the two gravel units.

Orleigh Court

The north Devon outlier of flint gravel at Orleigh Court (SS 493223), 1 km southwest of Littleham, was first noted by Vancouver (1808), but first described by Rogers and Simpson (1937). Recent work has been reported by Edmonds et al. (1979). The poorly exposed deposit covers an area some 300 m by 400 m and consists of nodular flints in a matrix of buff and brownish-red fine to coarse variably silty and clayey sand. The maximum thickness of the deposit is uncertain; Rogers and Simpson (1937) recorded 7.6 m of gravel resting on yellow clay at the Rookery, near the centre of the deposit. They described two divisions of the deposit: a 'lower' buff-coloured gravel with small and relatively few flints; and an 'upper' gravel which ranges in colour from brownish-grey, through yellowish-brown, to black and red, and is crowded with flint nodules, indurated ferruginous grit also being present. Mineralogically, the 'lower' gravel is characterised by a tourmaline-zircon-andalusite suite, the 'upper' gravel by a zircon-rutile-tourmaline suite. Recent work suggests that the 'upper' gravel of Rogers and Simpson is the original deposit and the 'lower' gravel a later downslope weathering product or 'head' derived from it. Numerous fossils of Upper Chalk age, all occurring as flint casts and most completely unabraded, are present in the 'upper' gravel.

Rogers and Simpson (1937) suggested that the deposit was laid down on the 400-foot (120 m) erosional Pliocene platform, either in a small basin or as part of a more extensive deposit. The unabraded nature of the flints suggests, however, that the Orleigh Court Gravel is a residual deposit of probable Palaeocene age, comparable to the Tower Wood Gravel of south Devon, and similarly derived from the *in situ* solution of chalk. The sand and clay matrix was either washed in steadily as the chalk dissolved, or was present on top of the chalk and was incorporated into it as solution proceeded.

East Devon

The Geological Survey maps of Wellington (Sheet 311) and Sidmouth (Sheet 326/340) show extensive deposits of 'Clay with flints and chert (in part Eocene)' mantling the Chalk and Upper Greensand tablelands of east Devon. This designation includes a wide variety of deposits, the complexities of which remain largely unravelled.

Two groups of deposit were distinguished by Woodward and Ussher (1911): clays and gravels with quartz, quartzite and other materials as well as chert and flint; and clay with flint and chert. The deposits were thought to be of Eocene age. It was not found possible to determine the exact relations of the two sets of deposits (Ussher, 1906). The 'clay with flint and chert' accumulation varies with the rocks that it rests upon: it is a brown or red-brown clay with unworn and broken flints where it rests on chalk, and varies from a yellowish or brown clay with fragments of flints and of chert, to a more or less loamy clay with chert fragments, where it rests on Upper Greensand.

In the Sidmouth area, Isaac (1979) recognised two stratigraphic units above the Upper Greensand. The lowest unit (the Peak Hill Gravel) is a residual deposit produced by the lateritic weathering of chalk. It consists of 2–7 m of reddish-brown clay and slightly sandy clay with unabraded flints, which vary from being tightly packed to having an open fabric in which the flints rarely touch. Silcrete is locally developed. The succeeding unit, the Mutters Moor Gravel, is a mixture, up to 20 m thick, of fluvial gravel and a regolith accumulation, and is essentially composed of reworked residual gravel. The flints are slightly abraded, and irregular beds of sand and clay are present; the clay matrix commonly contains a high proportion of sand. At the base of the unit is a well developed lateritic weathering profile with a zone of silcrete development, 1.5 m thick at the top. The lateritic profile and silcrete was thought to have formed in Tertiary times under conditions similar to those developed in semi-arid and arid regions of Australia, USA and parts of Africa. The Mutters Moor Gravel is now thought to be Pleistocene in age (K. Isaac, personal communication).

DEPOSITS OF THE BOVEY BASIN

The Bovey Basin, in common with the Petrockstowe Basin and the offshore Stanley Bank Basin in the Bristol Channel, lies on the line of the northwest–southeast trending Sticklepath wrench fault, to which its development and that of the other basins is intimately related. As shown in *Figure 9.3*, the Bovey Basin is filled with a succession of Eocene to Oligocene kaolinitic clay, silty clay, lignite and sand belonging to the Bovey Formation (Edwards, 1976). The depth of the basin has been determined as 1245 m using geophysical (gravity) surveying (Fasham, 1971). Vincent (1974) commented that, in Fasham's study, the depth-density data determined from the Petrockstowe Basin were used, on the assumption that the lithologies in the two basins were comparable. However, deep boreholes drilled in the central part of the Bovey Basin have since shown that substantial thicknesses of lignite are present (in one case 50 per cent of a total thickness of over 300 m of Bovey Formation is lignite), while in the Petrockstowe Basin lignite is of minor importance. Such thicknesses of low density material are bound to affect the gravity anomaly and reduce the estimate of the overall depth of the basin. Vincent considered that the maximum depth of the Bovey Basin is likely to be in the order of 1,100 m.

The Bovey Basin is divisible into two parts: the deposits of the main part of the basin have an approximately rhomboid-shaped outcrop, extending for some 11 km between Bovey Tracey in the northwest and Newton Abbot in the southeast, the smaller subcircular part of the basin lies south of Newton Abbot and is called the Decoy Basin. In the main basin, the Bovey Formation is underlain along its eastern outcrop by the Eocene Aller Gravel and by Cretaceous Upper Greensand; elsewhere it is faulted against or rests upon Devonian and Carboniferous rocks. In the Decoy Basin, the Bovey Formation overlies or is faulted against Permian breccias, Upper Greensand or Aller Gravel.

The Bovey Basin is the principal source of the ball clay supplies of the United Kingdom, the great majority of the mainly opencast workings extending in a line for some 5 km along the eastern outcrop of the main basin between Chudleigh Knighton and Newton Abbot. Ball clay is no longer extracted from the Decoy Basin. In the past, lignite has been extracted from the Bovey Basin and, although major extraction has ceased, the Bovey Basin remains the only major reserve of lignite in Great Britain.

Stratigraphy of the Bovey Formation

Although a two-part stratigraphic division of the Bovey Formation along the eastern crop of the main basin and in the Decoy Basin has been recognised for many years, Edwards (1970, 1976) divided the bulk of the known strata into three members. In ascending order, these are the Abbrook Clay and Sand Member, the Southacre Clay and Lignite Member, and the Blatchford Sand Member. Subordinate clay facies were also recognised on the northern and southern margins of the main basin at Chudleigh Knighton and Ringslade, and the Staplehill Gravels on the southern margin of the main basin and the Woolley Grit near Bovey Tracey were also included in the Bovey Formation. He recognised that perhaps 900 m of sedimentary strata lay concealed beneath the known stratigraphy.

With an extension of the deep drilling programme carried out by the ball clay companies in previously unknown parts of the Bovey Basin, and the ability to 'fingerprint' individual clay seams by sophisticated chemical, mineralogical and ceramic tests, the complex stratigraphy, particularly of the post-Southacre Member part of the sequence, has now been largely unravelled. Particularly, the results of exploration by Watts, Blake, Bearne and Company PLC, were set out by Vincent (1974), and the account here is largely built on his elaboration of the stratigraphy of Edwards (1976). Nevertheless, many gaps remain in unexplored parts of the basin and in areas where publication is not possible for commercial reasons. Combining these studies, the scheme for the stratigraphy of the Bovey Formation shown in *Table 9.1* is suggested. The term lower Bovey Formation, an informal subdivision of the formation, is proposed for the estimated 750–900 m of sediment which probably lies concealed beneath the Lappathorn Member in the central part of the Bovey Basin. However, there is no evidence of the nature of these strata, except by analogy with the trough deposits of the Petrockstowe Basin, which are silt, sand, gravel and extremely silty clay with some lenticular developments of laminated brown silty clay (Freshney, 1970). Also, the middle Bovey Formation is here used to embrace the consistent tripartite division into Lappathorn Member, Abbrook Clay and Sand Member, and Southacre Clay and Lignite Member, recognised along the eastern outcrop of the main basin. The upper two units are the main ball-clay producing horizons. The structure and stratigraphic relationships in the eastern part of the Bovey Basin are shown in *Figure 9.4*.

While Edwards (1970, 1976) grouped all the grey and white sand, gravelly sand, and sandy clay above the Southacre Clay and Lignite Member into the Blatchford Sand Member, Vincent (1974) distinguished seven members in the post-Southacre sequence which characteristically lie with marked discordance upon each other. The majority of these consist of sand and clay, but two (the Twinyeo Member and Brimley Member) are dominated by lignite. It is pro-

Table 9.1 LITHOSTRATIGRAPHIC SUBDIVISIONS OF THE BOVEY FORMATION

upper Bovey Formation
    Woolley Grit Member
    Bovey Heath Member
    Great Plantation Member
    Heathfield Member
    Brimley Member
    Blatchford Sand Member
    Stover Member
    Twinyeo Member
middle Bovey Formation
    Chudleigh Knighton Clay and Sand Member
    Ringslade Clay Member
    Southacre Clay and Lignite Member
    Abbrook Clay and Sand Member
    Lappathorn Member
    Staplehill Gravel Member
lower Bovey Formation
    Not exposed at surface: possibly silts, sands, gravels and extremely sandy clays.

The division of the Bovey Formation into upper, middle and lower subunits is informal.

posed here to include these seven post-Southacre Member units in the upper Bovey Formation. The stratigraphic positions of the Woolley Grit and Staplehill Gravel are uncertain; the Staplehill Gravel is tentatively included in the middle Bovey Formation, and the Woolley Grit in the upper Bovey Formation.

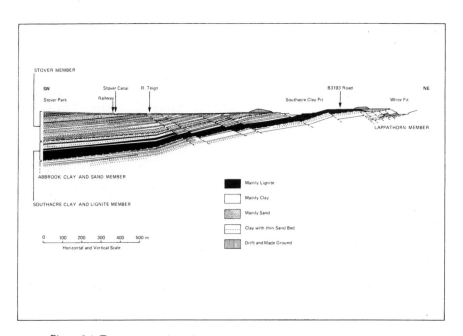

Figure 9.4. Transverse section of the eastern part of the Bovey Basin from Stover Park to White Pit (after Vincent, 1974)

The Lappathorn Member. Vincent (1974) distinguished this member from beds previously included in the basal part of the Abbrook Clay and Sand by Edwards (1976) and called by him the 'pink mottled facies', but not formally defined as a separate member. The member, named from Lappathorn Copse (SX 865752) near Sandygate, comprises red and pink mottled silty clay and sand. Being of no commercial significance, it is rarely penetrated by boreholes so that details of the stratigraphy are uncertain, nor is its thickness precisely known, although it is probably around 80 to 100 m in the Lappathorn area. The member rests in places on the Aller Gravel, elsewhere on the Upper Greensand, or, near Chudleigh Knighton, on Carboniferous rocks.

The Staplehill Gravel Member. The poorly-exposed Staplehill Gravel comprises an uncertain thickness (possibly around 10 m) of pale grey gravel and sand with clasts of quartz, pale grey Carboniferous chert, dark grey igneous rock, sandstone, and rare flint and Greensand chert. An exposure at Ringslade Clay pit, 3 km east-southeast of the main outcrop, shows about 1 m of Aller Gravel resting on Devonian slate and overlain by pale grey gravel with quartz, grey Carboniferous chert and dark grey igneous rocks, probably the Staplehill Gravel. If, as seems likely, these beds dip northwards beneath the Ringslade Clay, the Staplehill Gravel might equate with part of the Lappathorn Member which similarly rests on Aller Gravel near Sandygate.

The Abbrook Clay and Sand Member. This member, named from Abbrook Farm (SX 866745) near Kingsteignton, as re-defined by Vincent (1974), comprises 40–52 m of generally non-carbonaceous high-silica grey clay and sand; sideritic concretions and spherulites are developed at some horizons. Carbonaceous clay is rare, one thin band occurring in the middle of the sequence, and a group of brown clays with occasional thin lignite bands is developed in the upper part of the member in the southern part of the main basin. In a pit at Rixey Park (SX 852763) the member contains two thin beds of dark carbonaceous clay showing well developed brecciation with angular clasts of dark and pale clay in a matrix of dark clay. Farther north in the main basin sand becomes more dominant, and carbonaceous clay is absent. The lateral changes in the member are shown in *Figure 9.5.*

The Southacre Clay and Lignite Member; Chudleigh Knighton Clay and Sand Member; Ringslade Clay Member. The Southacre Clay and Lignite Member in its type area at Southacre (SX 855754) comprises lignite and carbonaceous brown and black clay, with subordinate grey to greyish-brown silty clay and sand. It is about 30 m thick in the Denistone area, and about 60 m thick in the type area. The relationship of the Southacre lignite to the Chudleigh Knighton and Clay Lane sequence was not clear to Edwards (1970), but Vincent (1974) resolved this problem, showing that the Southacre lignite horizons pass laterally via a brown clay sequence in the Clay Lane area into the Chudleigh Knighton Clay and Sand (*Figure 9.5*). The Chudleigh Knighton clays, in turn, pass laterally into a sand sequence in the Little Bradley area. Edwards (1976) showed that when traced towards Ringslade, the Southacre clays and lignites of the Teignbridge area pass into a sequence of grey and brown clay with one lignite horizon (the 'Ringslade Clay' of

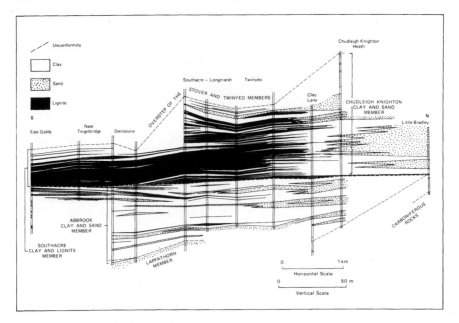

Figure 9.5. Longitudinal section of the Bovey Basin from Newton Abbot to Chudleigh Knighton (after Vincent, 1974)

Edwards). At the margins of the basin, the Ringslade Clay contains in places grey clay with blue, brown, grey and red mottles due to included fragments of Devonian slate ('Ringslade Mottled Clay').

Near the top of the Southacre sequence is the 'Parks Seam', a pale grey clay, silty in places with laminated and burrowed portions; it is characterised by the presence of well-ordered kaolinite, low alkalis and abundant siderite, and forms a major marker band. In places the seam contains well developed vertical rootlets; elsewhere it contains intraformational breccias comprising clasts of grey laminated clay in a grey clay matrix. Brecciation is also seen in the Chudleigh Knighton Clay and Sand, in which intraformational breccias of angular to subrounded grey clay clasts in a matrix of similar clay have been recorded (Edwards, 1976). Rootlet horizons are also present in the Ringslade Clay.

The lignites in the middle part of the Southacre sequence overstep the Abbrook Member to the south to lie directly on the carbonaceous clays of that member in the East Golds area (*Figure 9.5*). The Southacre Member is overlain by the Blatchford and Stover Members in the south, while to the north the Twinyeo Member lenses in between the Stover and Southacre Members.

The Twinyeo Member. This member, named after Twinyeo Farm (SX 846761) by Vincent (1974), comprises at least 90 m of lignite with brown clay and rarer grey to greyish-fawn clay. It overlies the Southacre Member with slight angular unconformity, but is overstepped with considerable discordance by the Stover Member which lies directly on the Southacre Member along most of the eastern crop of the main basin.

The Stover Member. This member was named by Vincent (1974) from Stover (SX 844741); it comprises dominantly grey sand and silty sand, gravelly at some horizons, and with subordinate beds of clayey silt, sandy clay and rare lignite bands. The 'Stover Clay' of Edwards (1976) is now included in this member, and shows distinctive brecciation and disturbed-bedding structures. The Stover Member is overstepped in the extreme south of the main basin by the Blatchford Sand Member and is overlain by the Brimley Member before lensing out between the Brimley and Twinyeo Members.

The Blatchford Sand Member. Edwards (1970, 1976) used the term Blatchford Sand Member to describe all the post-Southacre sand and clay strata. Vincent (1974), however, restricted its use to the southern area of the main basin, where it is almost exclusively sand. It is named from Blatchford Farm (SX 851729). The member is over 90 m thick in the Blatchford and West Golds area, and comprises grey sand, very coarse to gravelly in parts of the sequence, consisting predominantly of angular to subangular quartz grains and with lesser amounts of tourmaline and quartz-tourmaline grains. The washed sand is indistinguishable from a washed sand formed as a by-product of china clay production, and the Blatchford Sand is almost certainly of granitic origin.

The Blatchford Sand Member, according to Vincent, oversteps the Stover Member to rest on the Southacre Member in the extreme south of the main basin), but elsewhere the Stover Member intervenes. Vincent considered that, after deposition of the Blatchford and Stover Members, the focus of deposition moved to the north and west, and that there is a large unconformity between the earlier and later groups of members.

The Brimley Member. The Brimley Member of Vincent (1974) is named from Brimley (SX 799771) and comprises 73–104 m of lignite and brown clay, becoming dominantly sand in the topmost part. It includes the beds in the Bovey Coal Pit (Blue Waters Mine) which thus belongs to a higher stratigraphic sequence than the Southacre lignites with which they had previously been correlated. To the south, the Brimley Member rests on a northward-feathering sequence belonging to the Stover Member, but farther north it rests directly on Twinyeo Member lignite; it is overlain by the Heathfield and Great Plantation Members. Between Great Plantation and Heathfield, the Brimley Member is mainly sand in the west, passing eastwards through clays to a dominantly lignitic facies.

The Heathfield Member. This member overlies the Brimley and Twinyeo Members and is generally overlapped to the west by the Great Plantation Member and Bovey Heath Member. It has a small outcrop area at Heathfield (SX 835760), from where it is named. Clays were formerly extracted from the Heathfield Member at the Heathfield Pit ('Candy's Pit', SX 832761). The member is about 40 m thick and comprises in its upper part brown clay with occasional lignite, becoming sandy towards the base.

The Great Plantation Member. Vincent (1974) named this member after its thickest development in boreholes in the Great Plantation (SX 820757) area, it having no definite proven outcrop owing to being extensively and discordantly overlain by the Bovey Heath Member. It comprises over 107 m of mainly sideritic clayey silt, bluish-grey silty clay, and sand. Sandrock (siderite-cemented sand) occurs at some horizons. The workings at Halford (SX 810746) are thought to have been in this member.

The Bovey Heath Member. This was named from Bovey Heath (SX 824765), and was considered by Vincent (1974) to be the youngest member of the Bovey Formation. It is present over much of the northwest of the main basin, and lies almost horizontally over the Great Plantation and Heathfield Members. The thickest section is about 49 m of dominantly grey and yellow, muddy gravelly sand.

The Woolley Grit Member. This member comprises up to 24 m of grey hard siliceous coarse sandstone and conglomeratic sandstone outcropping in two small partly fault-bounded outliers northwest of Bovey Tracey (*Figure 9.3*). The first recorded occurrence is at Woolley (Wolleigh on newer maps) near Wolleigh House (SX 802797), but a second, more extensive outcrop of conglo-meratic grit was discovered in Higher Knowle Wood (SX 792810) near Lust-leigh by Blyth and Shearman (1962). From this area they described cross-bedding and channel structures and noted the presence of poorly preserved coniferous wood fragments indicating an age range from early Cretaceous to late Tertiary, and considered that the Woolley Grit should be regarded as a fluvial facies of the Bovey Formation rather than a lateral equivalent of the Cretaceous Haldon Greensand.

The association of similar chalcedonic conglomerate and Bovey Formation silt and clay at Sandy Park north of Chagford (Edmonds *et al.*, 1968) supports the inclusion of the Woolley Grit in the Bovey Formation. Certain of the gravelly quartz and quartz-tourmaline bearing sands from the upper Bovey Formation would resemble the Woolley Grit if cemented, and the member is tentatively included in the upper Bovey Formation.

Age of the Bovey Formation

In the absence of animal fossils and suitable minerals for radiometric dat-ing, the age of the Bovey Formation is based on the evidence of the plant macrofossils and, more recently, on pollen. Although Chandler (1957, 1964) emphasised the difficulties of dating Lower Tertiary floras, as tropical vegeta-tion of Lower Tertiary type persisted almost unchanged from lowest Eocene into the Oligocene, material from the Bovey Coal Pit, Heathfield Pit, and 'a pit at Kingsteignton' is of Oligocene (possibly Middle Oligocene) age. She noted, however, that while the beds in the Bovey Coal Pit were probably Oligocene, the beds below must be older, so that the possibility of Eocene strata being present in the lower part of the Bovey Formation could not be excluded. Again, the highest beds might range up into the Miocene.

G. C. Wilkinson and M. C. Boulter (personal communication) have studied the pollen from a borehole east of Heathfield (at SX 839758), and recorded *Anacolosidites* and *Pompeckjoidaepollenites* from its base at 290 m, which sug-gests that Eocene strata are present at greater depth with the beds above being Middle to Lower Oligocene.

Depositional Environments

Sedimentation of the Bovey Formation is considered to have taken place largely on river flood-plains and, to a lesser degree, in short-lived lakes. There is evidence throughout much of the succession for shallow water deposition in the form of breccia horizons (some of which possibly formed by desiccation of clay on flood plains), rootlet horizons and channel sands. Sedimentation and subsidence of the basin must have kept pace during much of Bovey Formation time. Records of rootlet beds from the Southacre Member, the Ringslade Clay and the Stover Member, indicate that some vegetation grew *in situ*, but most of the lignite in the Bovey Basin was derived from dense *Sequoia* forest growing on the uplands north and west of the basin. Chandler (1964) found that *Sequoia* was always much broken and battered and commonly represented by rather small fragments obviously transported from a distance.

The Southacre Member lignites and clays were probably deposited on flood plains in which backswamps were widely developed. During periods of flood, masses of *Sequoia* fragments were transported into the basin and deposited with clays in the overbank area. Repeated exposure of the overbank clays caused desiccation which produced brecciated horizons such as those in the Parks Seam and other silty clays of the Southacre sequence; these also contain rootlets and burrows. From Southacre towards Chudleigh Knighton and Little Bradley, the Southacre Member passes into an area where deposition in channels was probably more important; in the Chudleigh Knighton clay pits, several sand bodies occur filling channels cut in underlying clay beds. Also at Chudleigh Knighton are distinctive intraformational breccias consisting of angular to subrounded clay clasts set in a matrix of similar clay, interpreted as the product of streams which eroded the exposed banks of mud flats and later deposited the muddy suspended load with the more consolidated mud clasts.

In the Abbrook sequence few sedimentary structures are seen; although sand-filled channels have been recorded (Edwards, 1976) and deposition probably also took place on a flood plain with more actively meandering streams. Intraformational breccias are present in the carbonaceous seams of the Abbrook Member, and probably indicate subaerial exposure during deposition of the sequence. Probably the lignite and brown clay sequences of the Twinyeo and Brimley Members were also deposited on flood plains in a way similar to that of the Southacre lignite-brown clay sequence.

Thick sequences of granitic sand and gravelly sand with no discernible sedimentary structures are present in many of the upper Bovey Formation members; the very poorly sorted sediments were possibly deposited in conditions of high energy turbid flow by streams on the lower parts of alluvial fans. Marked discordances were probably caused by rapidly shifting foci of subsidence between the upper Bovey members.

Edwards (1976) suggested that there were two main phases of sedimentation in the Bovey Basin. The first phase was probably entirely fault controlled, and up to 700 m of sediment was deposited in a central trough area, but during the second phase faulting (except for the western margin fault) and subsidence became less active (or the rate of sedimentation increased) so that the sediments spread out over a wider area to overlie older rock with unfaulted contacts. Sediments in the Decoy Basin were deposited only during

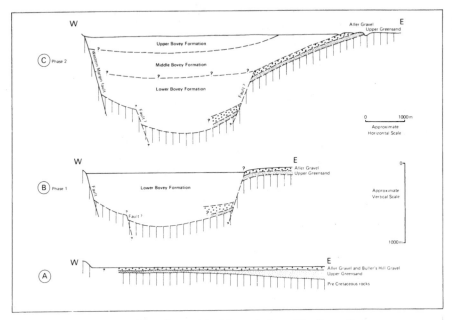

Figure 9.6. Tentative model of the evolution of the Bovey Basin

the second phase. It is thus possible to correlate the first phase with the lower Bovey Formation and the second with the middle and upper Bovey Formation. This hypothetical sequence of events is summarised in *Figure 9.6.*

Derivation of the Sediments

Most early workers (Reid, in Ussher, 1913; Groves, 1931) considered that the Bovey deposit was entirely derived by decomposition of the Dartmoor granite, the feldspar breaking down to give kaolinitic clays and the quartz grains contributing to the sands. However, it is now considered that the Bovey sediments were derived both from a granite source and from a weathering mantle developed on a variety of other rocks. Bristow (1968) suggested that much of the Bovey Formation in the Petrockstowe Basin has been derived from a weathering mantle developed on Carboniferous rocks during the early Tertiary. Vincent (1974) also showed that while the coarse material in the Abbrook Member of the Bovey Basin is of granitic origin, the disordered nature and fine grain size of the kaolinite argue against derivation from a granite source, possibly because the granite was only partially unroofed. Nevertheless, in the Southacre Member, the kaolinite of the clays is well ordered and probably derived from hydrothermal kaolinite developed on granite. Interestingly, in the southern part of the Southacre Member outcrop, mixtures of kaolinite of both types occur.

The Twinyeo Member clays of the lignitic facies are typically composed of medium-order kaolinites showing little similarity to the two preceding members. The kaolinites may be mixtures of well-ordered kaolinite derived from erosion of hydrothermally altered granite and poorly-ordered particles derived from a weathering mantle on country rock, or the kaolinite may have been derived from the weathering of feldspar in the granite. Vincent (1974) stated that 'the seam in the middle of the Stover Member is the last known occurrence of a comparatively well-ordered kaolinite in the sedimentology of the Bovey Basin, and it must be assumed that, at this time, the last china clay deposits were being removed from the southeastern part of Dartmoor'.

The later members all contain abundant sand of granitic origin, and the kaolinite appears to have been derived from the weathering of granitic feldspar. The Great Plantation Member contains abundant siderite, and Vincent (1974) suggested that the source of the carbonate was Devonian limestone to the southwest of the Bovey Basin from which carbonate-rich waters were derived before entering an environment with near-neutral pH values.

Structure of the Bovey Basin

The Bovey Basin lies on the line of the major northwest–southeast wrench fault system, the Sticklepath Fault, which trends across the county and upon which a dextral shift of some 1.3 km is apparent at the margin of the Dartmoor granite near Lustleigh. A splay fault from the Sticklepath system oriented north-northwest–south-southeast forms the western margin of the main Bovey Basin. Fasham (1971) suggested that this fault has a vertical throw of some 700 m, dipping easterly at 63°. There is no evidence of a wrench component at its southern end. The fault probably operated continuously during Bovey Formation sedimentation.

South of Newton Abbot, a fault with the same trend as the Sticklepath Fault separates the outcrop of the Bovey Formation to the west from that of the Aller Gravel to the east. The Decoy Basin and main Bovey Basin are also almost separated by an east–west trending ridge of Devonian slate in the Newton Abbot area, probably mantled by a thin cover of Lappathorn Member strata. Along the eastern and northern side of the main Bovey Basin, the Bovey Formation (Lappathorn Member) lies with unfaulted contact on the Carboniferous, Upper Greensand and Aller Gravel. Here the presence of Upper Greensand and Eocene flint gravel sequences comparable with those on the Haldon Hills demonstrate the degree of downwarping that has taken place during the formation of the Bovey Basin. Around the Decoy Basin, Upper Greensand and Aller Gravel dip inwards beneath a central area of Bovey Formation.

Between Ringslade and Knowles Hill (Newton Abbot), on the southern margin of the main basin, up to 40 m of Devonian slate rest on the Bovey Formation. Bristow and Hughes (1971) suggested that the slates have been thrust northward over the Tertiary beds, the thrust being a rejuvenation of thrusts of Variscan date in the Bickington and Holne area. However, it remains possible that local low-angle reverse faulting or rotational slip could account for the observed relationships.

The Bovey strata are affected by a number of small faults. Vincent (1974) noted that most of these are oblique-slip normal faults. He also noted that in that part of the main basin northwest of a line from Staplehill to Rixey Park, faulting is generally parallel to the trend of the Sticklepath Fault while, southeast of that line, the direction of faulting is more variable, being between northeast–southwest in the south and north–south near Preston Manor. In particular, a fault with the Sticklepath trend occurring in the northern end of Preston Manor 'White Pit' appears to be truncated by north–south trending faults, suggesting that these were later.

DEPOSITS OF THE PETROCKSTOWE BASIN

The structural and sedimentary early Tertiary basin at Petrockstowe, north Devon, lies in a topographic depression trending northwest–southeast a few kilometres to the south of Great Torrington. It is both flanked and cut internally by elements of the Stickelpath Fault Zone. Natural exposures of the beds are extremely rare, most of the available information coming from open pits, mines and boreholes.

Structural Framework

The deposition of the strata and their preservation is due mainly to their involvement in the Sticklepath Fault Zone (*Figure 9.7*). This zone, which is up to 800 m wide, is made up of several individual faults which vary in importance along their length. It has a total dextral movement of about 3 km in this area, as shown by the displacement of the New Red Sandstone in the Crediton Valley Trough which lies to the southeast of the Petrockstowe Basin. Interestingly, the transfer of dextral wrench movement from one fault of the zone to another adjacent fault in an *en échelon* fashion causes a downwarp in the intervening ground. In the relatively brittle Upper Carboniferous rocks, this downwarp is partly accommodated by a series of cross-faults trending between northeast–southwest and east-northeast–west-southwest. Although the more plastic Tertiary deposits in this downwarp generally deform by gentle folding, some of the Carboniferous faults break through the Tertiary beds.

The southeastern margin of the basin is formed by a stratigraphic junction, but here the contact is irregular, with a salient of Carboniferous outcrop jutting out into the basin in a northwesterly direction at Wooladon Moor. This spur, which continues as a plunging ridge under the Tertiary strata, is possibly defined on each flank by faults, and forms a threshold which separates the deeper part of the basin to the north from a small relatively shallow sub-basin to the south.

A section constructed in a northwest–southeast direction across the basin displays a broad asymmetrical synclinal structure with the deeper part (about 780 m) biased towards the northwestern end (*Figure 9.8*). This asymmetry is also indicated by the position of the residual gravity anomaly, the stratigraphy, and the sedimentology of the deposits.

Sections drawn across the basin in its northwestern part (*Figure 9.9*) show a deep fault trough flanked by a shelf on its northeastern flank. Folding is present in the southeastern part of the shelf, probably caused by the movement on the faults in this area, one of which can be seen to cut the steep limb of a fold. However, this is exceptional, as in most cases the faults do not rupture the Tertiary strata, but produce monoclinal rolls. Monoclinal structures of this kind used to be seen commonly in the old ball clay mines, particularly those driven southwestward towards the large internal northwest–southwest fault bounding the shelf. Here the clay seams roll over from a dip of around 10° to dips in excess of 45° and then back again to 10° or less.

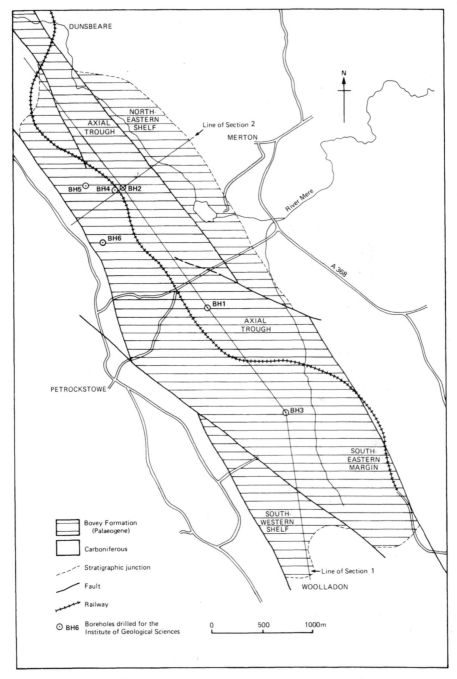

Figure 9.7. The Petrockstowe Basin with positions of the sections shown in Figures 9.8 (section 1) and 9.9 (section 2) (after Freshney *et al.*, 1979)

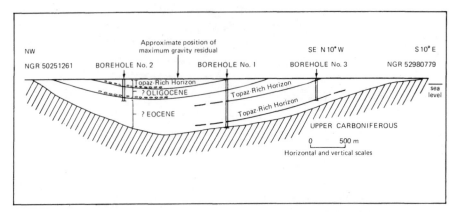

Figure 9.8. Longitudinal section of the Petrockstowe Basin (after Freshney *et al.*, 1979)

Each of the monoclines probably formed in response to the presence of a small subsidiary basement fault occurring approximately parallel to the shelf margin fault. The miners found that in tunnels approaching this fault the seams steepened to 60° or more, and mining had to cease.

In the axial trough of the basin, the presence of a broad fold can be deduced from the correlation of strata in boreholes. It trends northwest–southeast and its axis passes west of Petrockstowe Station (SS 517105). Its eastern limb dips 15–20° southwest in the north and 10° west in the area around the IGS Borehole No. 1. The western limb dips 8–10° east. In the southeast of the basin, drilling and quarry operations have shown the presence of many faults and a monoclinal structure.

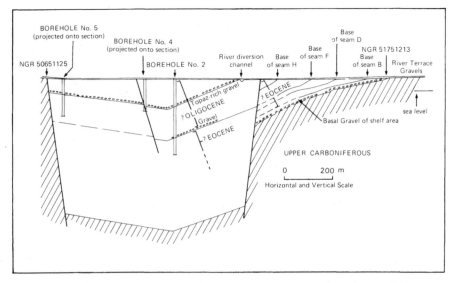

Figure 9.9. Transverse section of the northwest part of the Petrockstowe Basin (after Freshney *et al.*, 1979)

Lithologies

Clay. By far the commonest sedimentary material present in this basin is clay, although it is usually very sandy and forms part of a continuous sequence which passes into fine clayey sands. Abrupt contacts between different lithologies are thus infrequent, except for certain horizons which are the erosive bases of depositional cycles. Within these cycles, clay, which is dominantly kaolinite, may comprise anything between 8 per cent and 55 per cent of the material, while quartz in the form of fine sand and silt varies between 8 per cent and 70 per cent, and clay micas between 10 per cent and 38 per cent. The kaolinite is crystallographically disordered and is very fine grained. Some smectites are occasionally present in small quantities, particularly in the more carbonaceous clays.

Most of the clays show little internal structure, although in places a crude irregular banding is seen, especially in the carbonaceous brown clays where there can be signs of burrowing. Also, the seat-earth types of clay commonly show brecciation, besides containing rootlets. However, polygonal desiccation cracks filled with either sand or sandy clay are common, and in places the polygons may be over 1 m across with the cracks over 1 m deep. In the northeastern part of the axial trough there occurs a finely laminated reddish brown clay. The laminae, which are in the order of 1–10 mm thick, show a distinct grading from very sandy clay just above a sharp, slightly erosive base to a less sandy top. These clays are also characterised by the presence of whole-leaf fossils.

Sand and silt. The sands and silts usually contain a rather high proportion of clay. Grading analyses show the grain size distribution in most of the sands to be divided into two main populations; the coarser, with a grain size greater than 0.085 mm, is relatively well sorted and was probably transported by saltation, while the finer, with grain sizes less than 0.085 mm, is poorly sorted and was probably transported in suspension.

The sand and silt grade particles are almost wholly quartz and are subrounded to sub-angular with some almost spherical transperent grains. Like the clays, most of the sands exposed in the clay pits show little sign of internal structure. In one of the few exposures of sand in the axial area of the basin, and in some of the borehole cores, however, there is some lamination and cross-lamination. Lamination is also to be seen in sands which contain brown organic matter. These sands show irregular laminae picked out in pale fawn and deep reddish-brown colours, the laminae ranging in thickness from 1 mm to 20 mm. Small scour channels are commonly present, as are abundant micro-faults.

Other sands and silts showing good lamination are those associated with the finely laminated clay mentioned above. These show the same features as the clay and there is usually a gradation from one to the other. As might be expected the erosional features at the base of each sand lamina are more strongly developed than at the base of the sandy clay laminae.

Clay pellets are common within the sands, especially near the erosional base of a sand unit. In some cases larger disrupted clay masses are present in the sands, but this is often associated with large scale disruption and brecciation of unconsolidated strata, perhaps involved in contemporaneous tectonic movement.

Gravel. The gravels consist almost entirely of sub-rounded pebbles of vein quartz, with some flints and rotten green Carboniferous sandstone up to 10 cm in size. Flint and sandstone pebbles are particularly common in the basal gravels of the deposit. Pure gravel is very rare, most being a very sandy gravel or a gravelly sand. In the one exposure of gravel in a stream diversion channel (at SS 514117) rough bedding, wedge bedding and channelling can be seen.

Lignite. Compared with the Bovey Basin, there is little lignite at Petrockstowe. That which is found occurs both as fragmented lignite or as a highly lignitic clay with the vegetable matter in a finely divided form making up to 80 per cent of the sediment. Beneath each lignitic horizon there is usually a rootlet bed indicating the former presence of an *in situ* vegetation cover.

Stratigraphy

Figure 9.10 shows the generalised stratigraphy constructed from Borehole No. 1, near the deepest part of the basin. A rather argillaceous section, with abundant brown clay and lignite, occurs down to a depth of about 135 m, below which the beds are greyer and more sandy down to 316 m. A band of gravel and sand occurs immediately below this and, in turn, overlies a sequence of sandy clay and clayey sand (which is deeply red-stained between 564 m and 601 m). The lowest part of the succession consists of lignitic laminated sand underlain by sand and gravel which are coarse at the base, at a depth of 660.5 m.

A correlation based on topaz-rich sand bands and pollen analysis by C. Turner (personal communication) is shown in *Figure 9.8*. The pollen analysis suggested that the youngest strata (Oligocene) occur in the northern half of the basin, a few hundred metres to the northwest of Borehole No 1.

The total thickness of the strata obtained from Boreholes No 1 and No 2 is in the order of 760 m, the age of the topmost 260 m being Lower-Upper Oligocene and the strata below 260 m being Eocene (C. Turner, personal communication). The dating of the basal deposits is very doubtful owing to their unsatisfactory yield of pollen and spores, but it is possible that they are as old as early Eocene.

Northeastern Marginal Shelf Area. The succession in the northeastern marginal shelf area is between 70 m and 122 m thick, and consists of nine fining-upward cycles (A to I in *Figure 9.10*) of sandy clays with abundant seams of brown lignitic, almost sand-free, clays. Coarse sands and gravels only occur in the basal few metres. According to Turner the succession correlates with beds between 110 m and 385 m depth in Borehole No 1.

Wooladon Shelf Area. This small shelf area in the southwestern part of the basin is similar in many respects to the northeastern marginal shelf, with a succession that is made up of several fining-upward cycles containing relatively sand-free brown lignitic clays. It is only 65 m thick, and its correlation with other areas is doubtful. However, strong overlap of strata occurs from the centre of the basin towards the southern margin, which makes it improbable that the base of the deposit at Wooladon is the same age as the base in Borehole No 1.

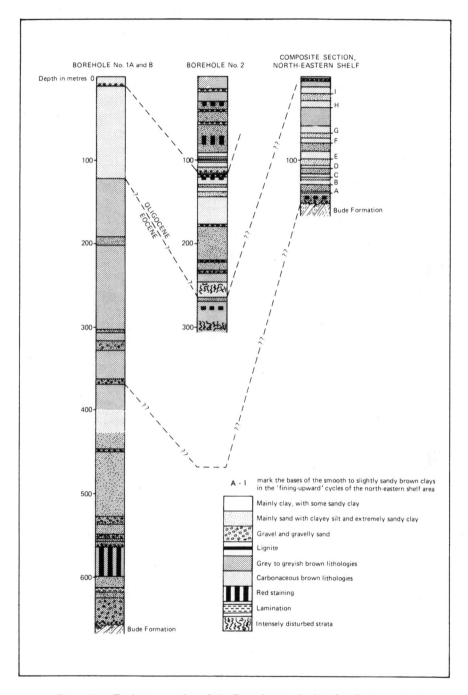

Figure 9.10. Tertiary succession of the Petrockstowe Basin (after Freshney *et al.*, 1979)

Table 9.2 SUCCESSION OF THE BOVEY FORMATION IN THE SOUTHEASTERN
PART OF THE PETROCKSTOWE BASIN, NORTH DEVON

Lithology	Approximate thickness (m)
Mainly clayey sand and sand clay with thin beds of silty clay; red and yellow staining common; rare carbonaceous bands	30
Productive ball clay sequence: up to 12 fining-upward cycles; sand generally clayey and fine grained, locally with lignitic fragments; rare bands of quartz gravel; ferruginous staining in some cycles; clay usually pale grey, rarely carbonaceous	120
Sand and very silty clay, usually with red and yellow ferruginous staining; spherulitic siderite common; poorly defined contact with underlying very weathered Bude Formation sandstone and shale	30

The base of the Tertiary succession here is exposed in the southwestern part of Wooladon Pit (SS 529078). Stained sandy clay rests on a rather impersistent fine quartz gravel; in turn, these rest on mudstone and sandstone which have been intensively weathered during the early Tertiary, so that the sandstone now appears as soft green sand and the mudstone as soft bluish-grey, rather kaolinitic, clay (Bristow, 1968). At various times during the excavation of the pit, channel structures have been uncovered. One of these, which was infilled by a mixed sequence of sand and clay, was so large that only one side was exposed. In another, the channel exposed was 2 m wide and 1.5 m deep, and filled with fine-grained sand only.

Southeastern Area. The stratigraphy in this area is less well defined than in the northeast. There are fewer obvious marker bands, and the presence of folding, faulting and sand-filled channels which trend northwest–southeast makes for difficult interpretation. Recently, however, intensive drilling by ECC Ball Clays Ltd has enabled the succession shown in *Table 9.2* to be established (B. L. Jones, personal communication).

This succession is comparable with that in the northeastern area, especially in that the productive sequence has a similar number of cycles, but this southeastern succession is far sandier and contains almost no lignite or lignitic clays.

Sedimentation and Conditions of Deposition

One of the most striking features of the sequence, which has relevance to the depositional environment, is its cyclical nature. The cycles fine upwards and are typical of alluvial sedimentation. Examples from various parts of the basin are shown in *Figure 9.11*. The cycles in the axial part of the basin commence with a sand which is usually coarse and in places gravelly. The basal strata show a strong erosional contact which can be as steep at 40° or more with respect to bedding. Above, the sand fines upwards, passing into a sandy clay which usually contains a rootlet network in its topmost part: this is a typical seat-earth. Another feature of this seat-earth horizon is the development of various iron compounds, the nature of which probably depends on

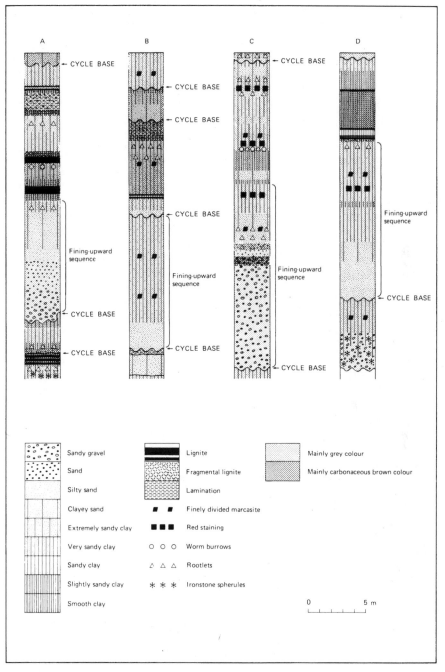

Figure 9.11. Sedimentary cycles in the Tertiary succession of the Bovey Basin (after Freshney *et al.*, 1979)

the water-table level and the effect of the organic content of the seat-earth on the overlying sediment at the time of deposition. In saturated swampy conditions with abundant plant matter, the iron compound which formed was marcasite, usually in a finely divided form. In conditions of low water table, with plenty of aeration of the soil, the iron formed as red and yellow oxides. In some cases it appears that marcasite was formed first in swampy conditions and then subsequent uplift of the strata caused oxidation to give red and yellow staining.

The cycles in the northeastern shelf area differ from those in the axial area by showing a general reduction in mean grain size. The base of each cycle, although still erosive, also presents less evidence of channelling, and the basal sediment is a fine silty sand. Additionally, the clays above the seat-earth are in many cases almost sand-free and commonly very lignitic, particularly just above the seat-earth. It is these clays which form the chief commercial ball clays in the basin.

The reduction in mean grain size from the axial area to the marginal parts of the basin, accompanied by a gradual decrease in grain size northwestwards along the axis, and the presence of fining-upward cycles, all point to alluvial sedimentation from a river flowing along the basin towards the northwest. Also, the cycles with coarse sand and gravel over a highly erosional base are restricted to the axial area, and the cycles of finer mean grain size to the margins; this strongly suggests that a river system was present, broadly differentiated into an axial zone in which mainly channel deposits, along with some overbank material, were laid down, and flanking areas in which the deposits were almost exclusively of an overbank and backswamp nature.

Interpretation of the sediment grain sizes points to a distinct differentiation into lag gravels, point-bar sands and other sands having a high clay content, which probably represent levee and crevasse splay deposits. The sandy clay lithologies have a dominant suspension population, and can probably be classified as flood-plain deposits with the finer grained, more lignitic, clays originating in backswamps. The axial area of the basin thus contains sediments deposited within channels and a considerable quantity of coarser overbank material, but the northeastern marginal shelf consists almost exclusively of overbank deposits. During some periods the overbank area must have been fairly dry, with a vegetation cover and oxides forming the soil profile, while in others it was very swampy with much development of marcasite.

From grain-size analysis, the finely laminated clay and silty sand appear to form a very distinctive group of pelagic character, and the graded nature of the laminae and the unabraded leaf remains point to deposition in a lacustrine environment, or certainly in ponds of water existing longer than the usual overbank flood. It is notable that these sediments are restricted to the northwestern axial area, which is also the area of the basin with the maximum thickness of strata and largest residual gravity anomaly. Thus, although fault movement and attendant subsidence of the basin was usually just enough to maintain either fluvial sedimentation or non-deposition, there were occasions when this subsidence was sufficient in the northwest to allow the development of temporary lakes.

Most of the cyclicity in the deposits is probably due to the normal action of a meandering river, but some of the broader cyclicity is almost certainly caused by the tectonic control of subsidence and sedimentation.

Provenance of the Sediments

The fining of grain size northwestwards along the axis of the basin suggests that the river flowed from the southeast. The heavy minerals present in the sands are for the most part probably reworked from rocks ranging between Carboniferous and Cretaceous in age. They consist of well-rounded zircon, some zoned tourmaline, heavily abraded staurolite, apatite, brookite, etc. Some of the well rounded zoned zircons may well have originated directly from exposed areas of the Cornubian batholith, but more likely they are reworked from the New Red Sandstone. There are, however, a number of distinct horizons where relatively unabraded topaz crystals occur with large unabraded fragments of reddish-brown to green zoned tourmaline. In places this material occurs in such abundance that it is almost certainly a direct derivation from aplitic or pegmatitic bodies associated with the Dartmoor granite. The most prominent horizons are in the basal gravel of Petrockstowe and at about 300 m depth in Borehole No 1. The gravel, being almost all vein quartz, comes from the ubiquitous veins which occur in the Carboniferous rocks. The fine sand and silt are derived from the fine-grained silty sandstone of the Carboniferous succession, and the kaolinite and illitic clay minerals from the illite and chlorite assemblages of the Carboniferous shale. Almost all the sediments in the basin are derived from the reworking of Carboniferous sandstone and shale, with perhaps some New Red Sandstone and Mesozoic rocks. During the Tertiary, these older rocks were deeply weathered in subtropical to tropical conditions, as shown by the deep weathering profile seen in boreholes which have penetrated below the bottom of the basin (Bristow, 1968).

The absence of flint in most of the gravels in the Petrockstowe Basin suggests that the Chalk cover of the area had been largely eroded away before sedimentation began there. It seems that even the sheets of residual and fluvial flint gravels, of which the sediments at Orleigh Court are a remnant, must also have been mostly reworked and dissipated.

The flora preserved in the Petrockstowe Basin, with such plants as palm trees and magnolia, suggests a subtropical climate. Beds of the same age in Hampshire and in the London area contain a fauna characteristic of tropical to subtropical seas. A weathering mantle in which kaolinite is produced from chlorite and illite is also highly suggestive of subtropical or tropical conditions.

OTHER TERTIARY DEPOSITS

Cadham Farm

The Palaeogene deposit at Cadham Farm (SS 584030) is very poorly exposed, but temporarily pits show that at its eastern end it consists mainly of sand and a basal gravel giving a total thickness of 12 m, with red and purple-stained plastic clays of unknown thickness at its western end. The deposit lies between faults adjacent to the Sticklepath Fault and, although no faults can clearly be seen to cut it, it is probable that its deposition and preservation are connected with movement on that zone. At one locality (SS 589019) near Jacobstowe, red-stained clays can be seen in the bed of a stream with the layers of red staining pulled into a vertical position and dislocated by a series of small faults trending northwest–southeast. The deposit is probably an isolated remnant formed by the same river which carried sediment to the Petrockstowe Basin.

Sandy Park

Near Sandy Park in the Chagford area, old excavations, reputedly for tin (Edmonds *et al.*, 1968), show traces of pale clay and clayey sand at Bradford Pool (SX 700910) and at an excavation at Great Tree Farm (SX 705906). At an excavation at Parford Wood (SX 712902) there are loose blocks of quartz tourmaline conglomerate and quartz grit with a chalcedonic cement, the rock being very similar to the Woolley Grit. The deposits are probably fluvial and associated with rivers flowing towards the Bovey Basin. The tin almost certainly occurred as a Tertiary alluvial deposit.

Stanley Bank Basin

This basin of Palaeogene deposits lies offshore between Lundy Island and Morte Point, and was first detected by a seismic reflection survey during the mapping of the continental shelf by the IGS (Fletcher, 1975). The survey showed a basin some 32 km long northwest–southeast, and 17 km wide, truncated along its southwestern margin by the Sticklepath Fault. Seismic refraction work by Brooks and James (1975) indicated a depth of 340 m. The basin has been drilled by the IGS and found to consist of grey clay (some showing red mottling), lignitic clay, silt and lignite. One lignite bed is 5 m thick. The deposit thus has a continental aspect which is similar to the strata of the Petrockstowe Basin. Its age is Middle Oligocene, similar to that of the upper parts of the Petrockstowe sequence, and it may be that both these deposits formed from the same river, which fed into the sea off the north Devon coast.

ENVIRONMENTAL EVOLUTION

Dating of the Bovey Formation relies exclusively on the evidence of plants and their pollen. The earlier studies of the floras of the Bovey Basin (in particular by Chandler, 1957) used plants to indicate an Oligocene or Middle Oligocene age for parts of the middle and upper Bovey Formation, although Chandler did not exclude the possibility of Eocene or even Miocene strata being present. More recent palynological work on the Petrockstowe, Bovey and Stanley Bank Basins by C. Turner, M. C. Boulter and G. C. Wilkinson (personal communication) indicates that the deposits range in age from Upper (possibly Middle) Eocene to Upper Oligocene. Thus the Bovey Formation is equivalent in age to much of the succession found in the Hampshire Basin.

It is possible that Devon was an area of non-deposition in the Maastrichtian, at the end of the Cretaceous Period. Solution of chalk to form residual gravel might even have begun in late Cretaceous times, and it probably continued through the Palaeocene. The source of the kaolinitic clay matrix of the residual gravels remains a matter of debate. These probably formed on a plain of extremely low relief, in a climate of seasonal rainfall; this would lead to a low or rapidly fluctuating water table, aiding solution of chalk and removal of material in solution.

Tensional forces associated with the opening of the northern part of the North Atlantic Ocean in late Palaeocene-early Eocene times activated block faults and uplift, which particularly affected the Dartmoor massif. Consequently, the residual gravels were eroded by streams flowing off upland areas to form fluvial deposits such as the Buller's Hill Gravel and Aller Gravel. Possibly some fluvial gravel was deposited over areas of residual gravel, beneath which chalk continued to dissolve to produce further residual gravel. The streams were braided and the sediments were deposited under conditions of seasonal rainfall in a semi-arid climate. The aridity is suggested by the occurrence of silcrete horizons in the gravels of east Devon.

In the early Eocene, compressional forces reactivated major wrench faults in southern England (such as the Sticklepath Fault) causing the development of basinal downwarps which accommodated great thicknesses of fluvio-lacustrine strata.

The clay minerals and heavy minerals in the Petrockstowe and Bovey Basins indicate the gradual unroofing of the Dartmoor granite. The lower part of the Petrockstowe deposit contains horizons in which topaz is the predominant heavy mineral, being most abundant in the basal sand and gravel. The topaz probably was derived from pegmatites and aplites in the roof area of the granite. In the Bovey Basin, the disordered nature and fine grain size of the kaolinite in the Abbrook Member argue against derivation of this member, and possibly lower members, from a granitic source; possibly again because the granite was only partially unroofed. In the Southacre Member, however, the kaolinite is well ordered and was probably derived from *in situ* kaolins developed on the granite. Thus, much of the roof material of the Dartmoor granite was removed by the later Eocene. In the upper Bovey Formation, the sources of kaolinite were more variable, reflecting diminishing supplies of *in situ* kaolin and possibly derivation of kaolinite from both granitic and non-granitic sources.

Although there had been continuing northward continental drift of Britain from its tropical latitude during the Permian, Devon still lay within a subtropical to tropical climatic belt in early to early-middle Eocene times; and it was under these conditions, with attendant heavier rainfall, that the Bovey Formation was deposited. This is in contrast to the more arid conditions prevailing during deposition of the fluvial flint gravels. This change in climate was conducive to the formation of deep lateritic weathering profiles, erosion of which contributed the detritus which filled the Petrockstowe Basin and part of the known Bovey Basin succession.

The flora of the Bovey Formation comprises dominantly sub-tropical and tropical species, including palms, ferns, heathers and many plants which grew in swamps. These plants probably grew either in the basins or on the lowland areas surrounding the basins. However, the bulk of the lignite of the Bovey Basin is formed of *Sequoia*, derived from forests which presumably grew at higher altitude on the Dartmoor upland. G. C. Wilkinson (personal communication) writes that:

The upper part of the Bovey Formation is dominated by the fern spore *Polpodiaceaesporites* which infers damp or humid conditions, the tricolporate pollen *Tricolporopollenites* associated with trees, shrubs and herbs of broad affinity and the conifer pollen *Pityosporites* which is generally considered to represent the 'upland' flora (that is at a higher level than the typical flood-plain flora) and may be dispersed over long distances. The overall concept is one of a ubiquitous flood-plain flora dominated by shrubs/herbs and trees, and ferns, in association with other wet-loving plants and a stable 'upland' coniferous vegetation. There is no significant botanical variation throughout the time of deposition of the upper part of the Bovey Formation. The compositions of the early Oligocene floras of the Bovey and Hampshire Basins (as represented by the Hamstead Beds) are substantially different and reflect two distinct environments. Palynological studies are at present unable to infer an accurate time correlation between these two deposits.

The beginning of the modern geography of South West England, indeed the whole of Britain, can be detected in the early Tertiary. There is evidence of an upland ridge, based on the Cornubian batholith, from which rivers drained northwards, the most important flowing via Cadham, Petrockstowe and the Stanley Bank Basin to empty into a Celtic Sea. Possibly the Flimston deposit in Pembrokeshire was deposited by a river flowing southwards to the same sea. The Bovey Basin sediments were deposited by a southward-flowing river discharging into the English Channel area. Because the drainage of Devon was controlled by faults trending northwest–southeast, it is unlikely that rivers flowed eastwards from Dartmoor to deposit the Dorset ball clay strata, the clays of which are chemically different in some respects from the Devon ball clays and were possibly derived from weathering profiles on Mesozoic clays.

LOCALITIES

Palaeocene and Eocene Flint Gravels

9.1 *Tower Wood Quarry* (SX 877857): Tower Wood Gravel. This quarry lies at the northern end of Great Haldon and is the type locality for the member. It exposes 9 m of that gravel, the lowest 3 m of which are obscured by talus. The visible 6 m comprise unabraded flints up to 0.3 m in diameter in a matrix of clay with very little sand. The flints are peripherally frost shattered. A basal bed of gravel with quartz and schorl may be present beneath the gravel.

9.2. *Buller's Hill Quarry* (SX 882847): Buller's Hill Gravel. This is the best exposure in the Haldon Gravels. The quarry is the type locality for the gravel, with 2 m of it resting on up to 7 m of Tower Wood Gravel, although there has been much periglacial mixing of the two. The Buller's Hill Gravel is pale greyish-brown, rather darker than the underlying Tower Wood Gravel, and contains frost shattered and chatter-marked flints in a matrix of sandy clay. On the north face, head gravel has cut down through the junction in two places. On the east face, 3 m of Buller's Hill Gravel lies on 1 m of Tower Wood Gravel on Upper Greensand, but the junction is much confused by periglacial disturbance.

9.3. *Buller's Hill* (SX 885848): Buller's Hill Gravel. A gully between two minor roads provides the best section showing a clay body in the member. It is a structureless pale grey clay, mottled brown at its margins, lying within pale brown gravel with flint, quartz, and schorl pebbles in a patchy clayey sand matrix.

9.4. *Royal Aller Vale Quarry* (SX 877694): Aller Gravel. This is the type area for the Aller Gravel and provides excellent sections of up to 30 m of abraded flint gravel and sand with subordinate silt and clay. The gravels show an average westerly dip of about 7°. The strata show marked lateral and vertical variation and erosional surfaces; cross-bedding and channel structures are common. The adjacent Zig-Zag Quarry (SX 880690) also shows excellent exposures of Aller Gravel.

Bovey Basin

(Localities 5, 6, 7, 8, 9, 13 and 14 are owned by Watts, Blake, Bearne and Company PLC, and permission to visit is required from their office in Newton Abbot.)

9.5. *Preston Manor 'White Pit'* (SX 859759 to SX 859753): Abbrook Member. Grey sandy and silty clay, silt, sand and grey plastic clay, with minor lignitic clay, are exposed in this pit. Along the eastern side of the pit the underlying red mottled Lappathorn clays and sands can be seen. A sequence similar to that in the 'White Pit' is exposed in the John Acres Lane Pit (SX 862751).

9.6. *Rixey Park Pit* (SX 852762): Abbrook Member. Grey silty and sandy clay, grey plastic lignitic clay show breccias of pale grey and dark brown clay particles in a matrix of dark brown clay.

9.7. *Chudleigh Knighton Pit* (SX 842771): Chudleigh Knighton Clay and Sand Member, Purplish-brown smooth clay, grey very silty and slightly silty clay, grey silt, and coarse grey sand dip southwest. Horizons of intraformational breccia can be found, consisting of angular to sub-round clasts of grey clay up to 40 mm, in a matrix of grey clay. Siderite-cemented sand (sandrock) is present at some levels. North-northwest–south-southeast faults affect part of the sequence.

9.8. *Clay Lane Pit* (SX 844769) displays the interdigitation between grey brecciated clays of the Chudleigh Knighton Clay and Sand Member and brown clays, and lignites of the Southacre Clay and Lignite Member. The upper 12–13 m of the pit are affected by marked periglacial contortions, clay 'anticlines' being separated by pockets of alluvial gravel.

9.9. *Southacre Pit* (SX 855754): Southacre Member. This pit is the type area for the member, some 60 m thick in this area. The beds are lignite and brown clay, with some grey-fawn silty clay and rare sand dipping 10° southwest. The basal part of the Stover Member occurs in the west face of the pit. The Parks Seam, some 6 m below the top of the member, consists of 4.3 m of greyish-fawn clay with yellow fine laminations of powdery siderite and lignitised plant remains, some well preserved. Rootlets occur in this and other of the grey silty clay beds and the seam is brecciated in places. The Southacre Member with the Parks Seam containing very well developed vertical rootlets is also exposed in the Newbridge Pit (SX 846766).

(Newbridge Pit, and localities 10, 11, 12, 15 and 16 below, are owned by ECC Ball Clays Limited, and permission to visit must be obtained from their office in Kingsteignton).

9.10 *Broadway Pit (Teignbridge Pit)* (SX 860737): Southacre, Stover and Blatchford Sand Members. The pit exposes about 15 m of brown clay and lignite of the Southacre Member, overlain by about 5 m of Stover Member grey clay and sand, overlain by Blatchford Sand Member, all dipping 11° west. The Southacre Member, overlain by thin Stover and by Blatchford strata, is also seen at the East Golds Pit (Pinsent's Quarry), owned by Watts, Blake, Bearne and Company PLC (SX 861730), the beds dipping 10° west. Well developed periglacial disturbances in the top 9 m of the beds have been recorded from this pit.

9.11. *Ringslade Clay Pit* (SX 846726): Ringslade Clay. Exposed in the pit are purplish-grey and pale grey smooth clays with patches of lignitic material, grey silty clay, and mottled clay with fragments of Devonian slate. Devonian slate to the south of the pit overlies the clays and may have been thrust over them. Exposures of about 1 m of Aller Gravel on Devonian slate, overlain in turn by about 1.5 m of Staplehill Gravel, have been observed in the benches south of the open pit.

9.12. *Stover Clay Pit* (SX 844741): Stover Member. A small intermittently worked pit exposes Stover clays. Lignite is overlain by 3 m of dark brown, fawn and greyish-fawn clay which show brecciation and complex disturbed-bedding structures in places. The ball clay sequence is overlain by coarse granitic sand and grey sandy clay.

Petrockstowe Basin

9.13. *Westbeare Pit* (SS 513121): clay and lignite. Sections mainly along the southeastern part of the pit show a sequence of beds dipping at angles of up to 45° southwest. The topmost part of the sequence consists mainly of sandy clay with beds of highly lignitic, medium to dark brown clay with lignite at their bases. Below the lignite, rootlets can be seen descending into the very much more siliceous clays. The middle part of the sequence contains rather sandy clay and clayey sand with little or no brown clay or lignite. Some dark staining caused by finely divided marcasite is visible. The lowest sequence in the pit is again highly lignitic and contains seams of sand-free clay and further rootlet beds.

9.14. *Courtmoor Pit* (SS 519115): structural features. A shallow syncline trending north-northwest–south-southeast, involving clays with a variable sand content and with pale brown layers, can be seen on opposing southeastern and northwestern faces of the pit. In the northwestern corner, dark lignite-bearing clays dipping east-northeast are truncated to the west by a fault trending parallel to the syncline. This fault dips at 60–70° in an easterly direction, and appears to have acted as a pathway for solutions as it is mineralised with siderite.

9.15. *Stockleigh Moor Pit* (SS 535093): fault and sedimentary cycles. This pit straddles a monocline with an arcuate shape but is roughly parallel to the nearby marginal fault which trends northwest–southeast. The steep limbs of the monocline shows dips of up to 45° southwest. The succession displayed consists of a number of fining-upward cycles of grey clays with a few slightly carbonaceous bands and a fairly prominent band of red staining. The bases of the cycles are commonly characterised by coarse sand to fine gravel and are strongly erosional into the preceding cycle.

9.16. *Wooladon Pit* (SS 529078): sedimentary cycles and Tertiary weathering of Carboniferous rocks. Wooladon Pit lies in a small cuvette opening into the main Petrockstowe Basin over a buried ridge of Carboniferous rocks. The beds exposed are clay, some of which are highly lignitic. The sand content is very variable but several seams of almost sand-free clay are present. Several beds of fine sand occur at the bases of fining-upward cycles. These are erosive, and at certain times during development of the pit, sand-filled channels have been uncovered. At the base of the sequence, visible on the southern side of the pit, red-stained clay overlies a fine-grained yellow quartz gravel which rests, in turn, unconformably on Carboniferous shale and sandstone which have been tropically weathered during the Tertiary. This weathering has reduced the once-hard sandstone and shale to soft green sand and blue grey clay respectively.

Chapter Ten

The Tertiary Igneous Complex of Lundy

Lundy is an island in the Bristol Channel 20 km north-northwest of Hartland Point, and some 50 km south of the nearest points on the south Wales
coast. It is 5 km long from north to south with an average width of 1 km. The
island consists of a plateau sloping from a height of about 140 m in the south
to 90 m at North End, surrounded by a steep bevel or 'sideland' surmounting
20–30 m high cliffs, the bases of which are easily accessible only in a few
places (*Figure 10.1*).

The island is composed principally of Tertiary granite, only the Castle
Hill–Lametry peninsula area in the southeast being formed of Devonian
metasedimentary rocks. Dykes of mafic and intermediate igneous composition intrude both the granite and the metasedimentary rocks. All rock types
are exposed in the cliffs. The granite is readily accessible in quarries on the
east side of the island and in the upper parts of the cliffs almost all round the
island. The dykes are most easily reached around North End and from
Quarry Beach to the Lametry peninsula on the east side of the island. There
are few exposures inland except in the northern part of Lundy, as the plateau
is formed mainly on head deposits.

TERTIARY IGNEOUS COMPLEXES IN THE BRITISH ISLES

Apart from the Lundy complex and a few dykes in north Wales and northern England all the Tertiary igneous rocks occur in northern Ireland and
western Scotland, and it is from there that our knowledge of these rocks has
been mainly derived.

The results of Tertiary igneous activity can be regarded as composed of two
major elements, the lava plateaux and the central ring complexes. A third
element consists of a vast swarm of mafic dykes which extends over a considerable area and has a remarkably constant trend, varying only slightly
from northwest–southeast. This Tertiary igneous area forms part of the Brito-
Arctic or Thulean Province which extends to the Faeroes, Iceland, Greenland,
Jan Mayen and Spitzbergen. In Britain the volcanic activity lasted for only a
relatively short period, but in Iceland, for example, it has persisted to the
present day.

The intrusive complexes result from the second of the three phases which
constitute the complete Tertiary igneous cycle. This phase was preceded by
the outpouring of lavas and followed by the emplacement of regional dyke
swarms (*Figure 10.2*).

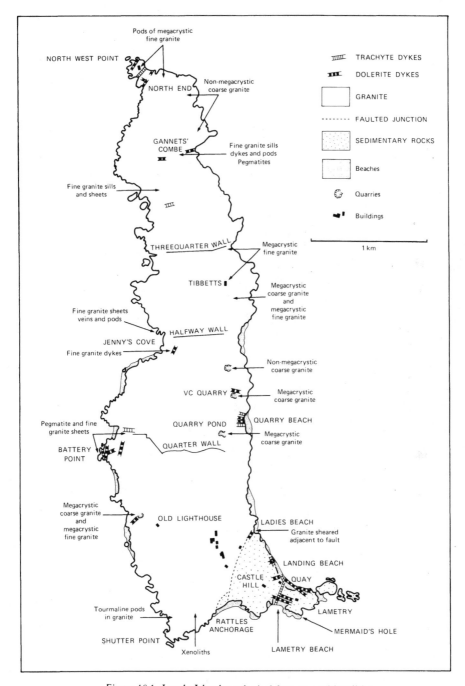

Figure 10.1. Lundy Island; geological features and localities

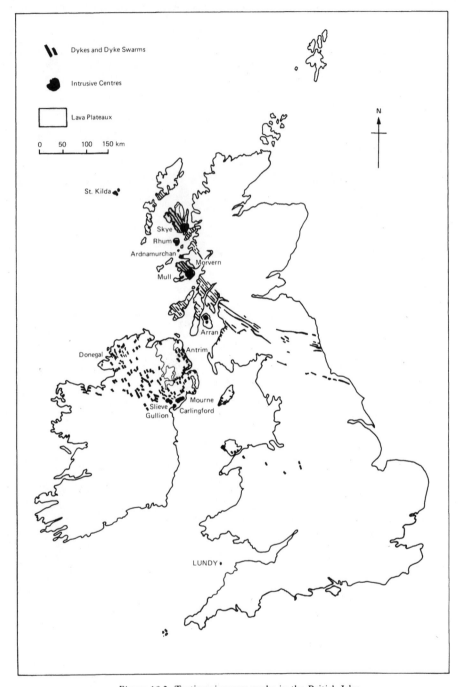

Figure 10.2. Tertiary igneous rocks in the British Isles

Although it was originally maintained that the lava plateaux had been built up by successive eruptions from dyke fissures, it is now generally accepted that the intrusive complexes represent the 'basal wrecks' of great central volcanoes from which the lavas erupted. On Mull a volcanic crater 10 km in diameter has been identified which was in existence during the latter half of the basalt phase. Lavas probably erupted from such craters which were almost obliterated by the subsequent intrusion of the plutonic complexes. The latter were present as centres of igneous activity and sites of magma reservoirs before emplacement of the dyke swarms, which, although related to the plutonic centres, arrived too late to be feeders for the lava plateaux.

Over 2,000 sq km of lavas remain on the Hebridean islands and adjacent parts of the Scottish mainland together with almost 3800 sq km in northeast Ireland. A thickness of about 1,800 m of lavas has been recorded on Mull, and it seems probable that the original thickness of Antrim lavas was similar. In several areas eruption commenced explosively, producing tuffs. Most of the lavas are of olivine basalt, intercalated on Mull with mugearite, olivine poor basalt and tholeiitic basalt. On Skye the basalts are followed by a succession consisting of trachyte and andesite, rhyolitic tuff and rhyolite. In Antrim the lower olivine basalt flows with minor mugearite and trachyte are separated from the upper olivine basalt flows by tholeiitic basalt or rhyolite.

The plutonic phase of the cycle, responsible for the central complexes, resulted in a range or rock types from ultramafic to silicic, but gabbroic and granitic types are dominant. In Skye an early peridotite was enveloped by the olivine gabbro which forms the Cuillin Hills. Subsequent emplacement of the Red Hills granites completed the sequence of major intrusions. A full range of rock types occurs as dykes, sills, composite sills and cone sheets. The simple nature of the intrusion sequence on Skye contrasts with the complexity on Mull and the Ardnamurchan peninsula, where basic and acid magmas were available simultaneously and igneous masses are arranged concentrically around intrusion centres. On Mull, two centres operated successively, and on Ardnamurchan three. In Arran the centre of igneous activity also migrated: emplacement of crinanite sills and quartz dolerites in the south of the island preceded intrusion of the north Arran granites, and the central ring complex, composed mainly of explosion breccia and fine-grained granite, lies between the early developed sites. Renewed activity in south Arran led to the intrusion of late quartz porphyry and pitchstone sills.

The Slieve Gullion and Carlingford complexes both exhibit gabbros, dolerites, granophyres and vent agglomerates, arranged in ring structures and elongated plugs. In the Mourne Mountains complex, five varieties of granite can be distinguished, three being emplaced successively at an eastern centre and two at a western centre, probably following the subsidence of blocks of country rock bounded by ring fractures. The intrusions are relatively thin and are underlain at no great depth by a basic mass.

The third and final phase of the cycle consisted of the injection of swarms of northwesterly trending mafic dykes which extend from the Outer Hebrides to northern England and from west Donegal to County Down and north Wales. The Scottish dykes are especially abundant in the vicinity of the plutonic complexes. Two contrasted suites of rocks are present: olivine dolerite–crinanite and tholeiite–quartz dolerite. The Skye swarm is composed mainly of the former suite and extends from South Harris to Loch Linnhe. Quartz dolerite dykes from the Mull centre reach South Uist and the few intrusions reaching northern England are composed of tholeiite or quartz dolerite also, but the major part of the swarm occurs northwest of the Clyde and consists of olivine dolerite and crinanite types. The Arran swarm, extending across Kintyre to Jura and Islay, includes members of both suites.

The northern Irish dykes trend north-northwest. In Antrim and north Down they are of olivine dolerite, in the Hillsborough area trachybasalt, while around the Mourne Mountains, olivine-free basalts and andesites predominate. An intense swarm concentrates in Donegal and another is associated with the Carlingford plutonic centre.

THE LUNDY GRANITE

General Features

The Lundy intrusion is composed mainly of a coarse-grained megacrystic granite, the groundmass having an average grain size of 1–2 mm. The proportion of feldspar megacrysts can attain 28 per cent although locally, for example in an old quarry near Halfway Wall Bay (SS 138456), the granite is non-megacrystic. Average megacryst lengths range from 18 mm to 30 mm; the maxima at particular localities ranging from 35 mm to 70 mm. Dollar (1942) recorded a syenitic variant of this granite which crops out at Shutter Point and near North West Point (*Figure 10.1*).

The other principal rock type is a fine-grained megacrystic granite. Typically the groundmass is even-grained with an average size of 0.1 mm, but some specimens show a range of about 0.03–0.3 mm. The proportion of feldspar megacrysts fluctuates from 9 per cent to 21 per cent and the average megacryst length from 19–24 mm, with a maximum recorded length of 50 mm. This fine-grained megacrystic granite crops out in many parts of the island although occurrences are concentrated in certain areas. The distribution accords generally with the area which Dollar (1942) showed, but it appears that he overestimated its extent by linking together occurrences which are in fact isolated. Irregular pods and patches of this rock type are found within the coarse granite and may be portions of an early-formed crust enclosed by subsequent advances of magma during the same period of intrusion. Alternatively this material may be granitised argillaceous and silty-argillaceous country rock (Edmonds *et al.*, 1968, 1975; Hawkes and Dangerfield, 1978).

Throughout the island fine-grained poorly megacrystic granites occur within the main granite outcrop. Biotite poor and biotite rich types occur as vertical, inclined or horizontal sheets, rounded pods or irregular masses. Both types locally develop pegmatitic borders which are generally associated only with the upper margins of horizontal or gently inclined sheets, although some bodies grade laterally into pegmatites.

The junction between the granite and the metasedimentary rocks trends north-northeast–south-southwest and may be mainly a fault, but with the remnants of a normal contact on the south coast (Edmonds *et al.*, 1979). The junction has been invaded by mafic dykes at Rattles Anchorage (SS 138437) and at Ladies Beach (SS 140442), where the granite is shattered and crushed for a distance of about 10 m from the junction. The boundary is not exposed inland but seems to be straight or slightly curved, showing that the plane of junction is nearly vertical. The remainder of the granite boundary lies beneath the sea, but the shapes of both the gravity and submarine contours suggest that the granite mass is roughly circular with a diameter only slightly greater than the length of the island (Bott *et al.*, 1958). Additionally, a circular positive aeromagnetic anomaly over Lundy confirms that the subsurface extension of the granite does not extend far beyond the island (Cornwell, 1971). The occurrence of xenoliths in exposures on the south and southwest coasts suggests closeness to the edge or the roof of the intrusion.

Dollar (1942) considered the granite to have the form of an elongate dome, possibly laccolithic, which trends north-northeast–south-southwest and is 8 km long and about 5 km wide. Bott *et al.* (1958) agreed with the laccolithic form of the intrusion, their gravity data being consistent with a body 1.6 km in thickness and approximately cylindrical in shape with a diameter of about 4.8 km.

Isotopic age determinations by Miller and Fitch (1962) and Dodson and Long (1962) have shown the Lundy granite to be about 52 Ma old and therefore of Eocene age.

Petrography

Both the coarse-grained and fine-grained megacrystic granites are composed of feldspar and quartz megacrysts set in a groundmass of potassium feldspar, quartz, plagioclase and varieties of mica. Almost all the feldspar megacrysts are potassic and staining of thin sections shows that these are Carlsbad twinned perthites. Their margins are commonly very irregular because the crystals have outgrown the areas of their original plagioclase hosts and incorporated groundmass feldspars. Inclusions of groundmass minerals with the megacrysts are common, and some megacrysts show the zones of quartz blebs described by Shelley (1966). Quartz megacrysts range in size from 3–10 mm, and typically consist of aggregates of several anhedral crystals cemented together along sutured margins by the addition of silica. Quartz megacrysts are of late formation: inclusions of groundmass minerals occur but are uncommon.

Groundmass potassium feldspars are patch and string or vein perthites, commonly Carlsbad twinned, which result from the replacement of pre-existing plagioclase crystals. Quartz is present as anhedral crystals with very irregular margins and commonly shows undulose extinction. Much of the mineral is interstitial. Plagioclase normally ranges in composition from oligoclase to sodic andesine. It is commonly Carlsbad and lamellar twinned and rarely pericline or chess-board twinned. Compositional zoning is rare. Plagioclase commonly shows some alteration to sericite, and patchy replacement by potassium feldspar can be observed. Biotite is a normal groundmass constituent, crystals occurring singly or in clusters. Locally it alters to muscovite or chlorite. A pale-brown faintly pleochroic mica, probably phlogopite, occurs in the coarse granite and in the fine granite a little muscovite is commonly present. The principal accessory minerals are zircon, garnet, tourmaline, topaz, apatite and an opaque iron oxide. Garnet and tourmaline can be distinguished in some hand specimens.

The fine-grained poorly megacrystic granites are composed mainly of potassium feldspar (perthite), quartz and plagioclase with a few perthite megacrysts up to 20 mm long and rare quartz megacrysts up to 10 mm in diameter. Plagioclase crystals range from albite to oligoclase and show Carlsbad, lamellar and pericline twinning. Some specimens carry pleochroic biotite which may occur with muscovite. Where muscovite is the only mica it is commonly accompanied by topaz. These rocks have xenomorphic granular texture and an average grain size of 0.4 mm.

A typical pegmatite from Battery Point (SS 128449) is composed of quartz-feldspar intergrowths with some biotite and muscovite. Other examples occur on crags above Jenny's Cove (SS 134457) and in Gannets' Combe (SS 135473). Dollar (1942) recorded greenish apatite, lithium mica and molybdenite in these rocks.

THE LUNDY DYKE SWARM

General Features

Edmonds *et al.* (1979) recognised the presence of over 230 dykes on Lundy. Mafic and intermediate types can easily be identified in the field. Within the former group aphyric, porphyritic and vesicular analcime bearing rocks can be distinguished; but, except for glassy varieties, the intermediate igneous dykes present a uniform appearance. Both types intrude the granite and the Devonian metasedimentary rocks.

The mafic dykes average about 1 m in thickness and are dark-grey to greyish-black or olive-grey to olive-black dolerites. They commonly weather to friable brown material, are readily eroded and can clearly be seen in the cliffs.

Only about 30 of the dykes are intermediate in composition. They range from medium-grained to glassy in texture, and are pale grey or greenish-grey trachytes and trachyandesites. On weathering they become light brown or buff in colour. Many bear small rectangular or acicular feldspar phenocrysts.

Most of the dykes are vertical or nearly so and appear to be fairly straight, although a few examples deviate abruptly through right angles. This feature is well displayed on the Landing Beach below the quay (SS 143438) and 230 m north-northwest of the quay (SS 142440). Some dykes bifurcate and several show braiding. The margins of the dykes are sharp, commonly glassy, and some dykes have vesicular interiors. Trachytic dykes may show flow-banding at their margins. Mafic dykes typically have close-spaced joints parallel to their walls with subordinate joints at right angles, while the trachytic dykes show wide-spaced transverse jointing. It seems that the dykes were emplaced easily as only a few show any marked disruption of the granite into which they are intruded. Where brecciation and granulation occur, the country rock is normally affected for no more than a few centimetres from the contacts. An exception to this is the 2.5 m wide trachyte dyke cutting the slates adjacent to the quay.

Field relations indicate that the dolerite and trachyte dykes were emplaced contemporaneously. On the western side of Lametry Beach (SS 142437) the 2.5 m trachyte dyke mentioned above strikes north-northeast and cuts west-northwest trending mafic dykes. Conversely, 230 m north-northwest of the quay (SS 142440) a mafic dyke, striking northwest and parallel to a trachyte dyke, turns sharply through 90° to cut the latter. Dollar (1942) noted a pronounced west-northwest trend to the dyke swarm.

Isotopic age determinations by the potassium-argon method on whole-rock samples of trachyte and dolerite dykes have given ages of 50 Ma and 54 Ma respectively. Within the limits of experimental error, the results are therefore indistinguishable from isotopic ages obtained from the granite of 50 to 55 Ma by Miller and Fitch (1962), and of 52 Ma by Dodson and Long (1962). Mussett *et al.* (1976) gave ages for Lundy dykes spanning the period 54–55 Ma, but stated that an age of 54 Ma from a mafic dyke is the most reliable.

Petrography of the Dolerite Dykes

The mafic dykes include olivine dolerites, analcime bearing olivine dolerites (crinanites,) olivine free dolerites and quartz dolerites. There is considerable variation in grain size and texture: most specimens contain a few phenocrysts and porphyritic varieties of each type occur.

The olivine dolerites are compact dark-grey to greyish-black rocks, but some porphyritic specimens have an olive-grey or olive-black groundmass. Most porphyritic specimens contain phenocrysts of olivine and plagioclase or plagioclase alone. Titanaugite phenocrysts are rare. Plagioclase phenocrysts are greenish-grey labradorite, commonly zoned and patchily chloritised. Olivine phenocrysts are generally replaced by serpentine or iddingsite although in some specimens the cores of the crystals are fresh. The groundmass is composed generally of labradorite, titanaugite, serpentinised olivine, opaque iron oxide and chlorite. In the aphyric varieties chlorite forms primary irregular and angular interstitial patches. Potassium feldspar occurs in a few specimens and apatite is an accessory mineral. Many examples, particularly the aphyric types, have a subophitic texture: others are intergranular.

A few per cent of analcime and other zeolites can be present in the olivine dolerites but where this is exceeded the dolerites pass into crinanites. These are dark-grey to greenish-grey fine-grained rocks which can be distinguished in hand specimen from normal olivine dolerites by the presence of amygdales filled with analcime. Porphyritic and aphyric types are found. Plagioclase and olivine occur as phenocrysts. Except for the presence of analcime in irregular interstitial patches and in amygdales, and the general absence of primary chlorite, the groundmass of the crinanites is similar to that of the olivine-dolerites. Subophitic texture predominates in the porphyritic types.

Several quartz dolerites occur on Lundy. They are fine, even-grained non-porphyritic olive-grey rocks with an irregular texture. The pyroxene appears to be augite, and with plagioclase comprises most of the rock. Quartz occurs as scattered interstitial anhedral grains with plagioclase and pyroxene as rare xenocrysts.

A few dykes contain neither quartz nor olivine. Rocks of this type are dark greenish-grey or black and generally non-porphyritic. The plagioclase ranges from oligoclase to labradorite in composition and the pyroxene is titanaugite. Chlorite is common, occurring interstitially, filling vesicles, and replacing plagioclase and pyroxene. Accessories include analcime, apatite, biotite and opaque iron oxide.

Petrography of the Trachyte Dykes

The intermediate igneous dykes are mainly quartz trachytes or quartz trachyandesites (latites) containing varying accessory mineral assemblages and showing a variety of textures. Additionally there are a few glassy spherulitic rocks, a few cryptocrystalline felsites, and several rocks exhibiting basic affinities.

The quartz trachytes and quartz trachyandesites are composed essentially of oligoclase, potassium feldspar and minor quartz. It appears that, when first formed, these rocks consisted mainly of oligoclase, and that subsequent potassium metasomatism resulted in the partial conversion of plagioclase to potassium feldspar and the formation of interstitial quartz. The total feldspar content is more than 70 per cent and ratios of potassium feldspar to oligoclase show a range in rock type composition from trachyte to trachyandesite. Quartz content ranges from 2.7 per cent to 8.6 per cent, averaging about 4 per cent.

Many of the dykes contain a few phenocrysts, but none can be termed porphyritic. The phenocrysts are generally potassium feldspars which because of their glassy appearance in hand specimen may be sanidine. In the ground mass oligoclase occurs mainly as elongated laths but with some equant crystals. It is commonly altered to chlorite. Staining of thin sections with sodium-cobaltinitrite reveals extensive replacement of plagioclase by a network of minute potassium feldspar crystals. Much of the quartz is present as interstitial anhedra, and most specimens contain on average 1 per cent of opaque iron oxide. A blue-green sodic amphibole is present in about half the specimens examined. Analytical work shows this to be arfvedsonite (Edmonds *et al.*, 1979). In a few specimens this mineral exceeds 6 per cent. Other accessory minerals include altered hornblende, a brown mica, pyroxene and apatite, and many specimens contain interstitial glass which can attain 20 per cent. These rocks display intergranular, orthophyric, variolitic and trachytic textures.

Two dykes near Mermaid's Hole illustrate a glassy facies into which the main groups of dykes may pass. They consist of streaky greenish-grey or bluish-grey rocks, composed of yellowish-brown glass, with phenocrysts of corroded sanidine, aegirine or aegirine-augite and opaque iron oxide. An analysed specimen from one of these dykes contains over 7 per cent K_2O.

Dykes in various parts of the island consist of fine-grained felsitic trachytes, which may result from a more acid fraction of the intermediate magma. They are buff or grey-coloured rocks with a few phenocrysts of altered and corroded potassium feldspar, probably sanidine. Their groundmass is composed oligoclase, partly replaced by potassium feldspar, quartz and a little opaque iron oxide, together with accessory amphibole and biotite.

Specimens from a few dykes are essentially of intermediate composition, exhibiting some mafic igneous features not shown by most of the trachytes. These rocks may provide a link with the mafic group of dyke rocks. Phenocrysts can be of plagioclase, potassium feldspar or augite. Groundmass feldspar varies within the oligoclase to andesine range. Titanaugite, serpentinised olivine and opaque iron oxide occur in some specimens and accessories include apatite, rare mica and arfvedsonitic amphibole.

PETROGENESIS AND TECTONIC SETTING

It now remains to attempt to explain why the rocks of Lundy are there as we see them today, but first it is necessary to place them in their tectonic setting and to examine briefly the geophysical evidence bearing on the question.

The island is situated on the basement platform of the Cornubian Massif, close to the southern edge of the Bristol Channel Trough where the latter

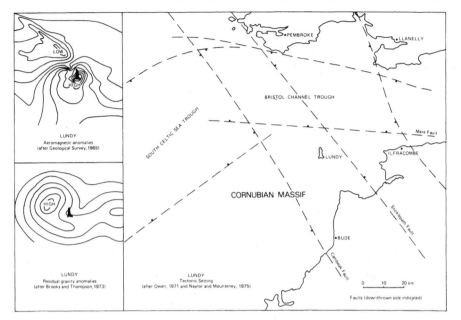

Figure 10.3. Lundy: tectonic and geophysical setting

joins the South Celtic Sea Trough. Both these sedimentary basins are down-faulted (Owen, 1971; Naylor and Mounteney, 1975), and at this, its extreme western end, the Bristol Channel Trough is dissected by a set of northwest–southeast trending faults which are extensions of the strike-slip faults seen on-shore (Dearman, 1963). Thus *Figure 10.3* shows that Lundy is on a horst, bounded on the northeast by the Sticklepath Fault, on the southwest by the Cambeak Fault and on the north by a westerly extension of the Mere Fault.

Both the troughs have been subsiding zones of sedimentation since Permian times. Throughout the Jurassic and early Cretaceous, sedimentation was mainly restricted to the troughs themselves, but due to the influence of spreading at the Mid-Atlantic Ridge, by the beginning of the Upper Cretaceous, only the Cornish area of the Cornubian Massif remained above sea level. Therefore the area of basement platform where the Lundy complex is now situated received a cover of Permian and some Cretaceous and Palaeocene sediments.

The main Lundy gravity high (Brooks and Thompson, 1973) has an arcuate form with the island at its centre of curvature. However, its main development is west of the island with a peak some 10 km west-northwest of Lundy. An approximately circular positive magnetic anomaly shown on the aeromagnetic map of Great Britain, Sheet 2 (Geological Survey of Great Britain, 1965) partly overlaps the gravity high without coinciding exactly with it, and reaches its peak over the southern part of the island. On this evidence, Brooks and Thompson attributed the Lundy gravity high to a major basic intrusion in the area west of the island, lying at shallow depth and between 2.5 km and 4.0 km in thickness.

A negative magnetic anomaly extends for some 30 km northwestwards from a point 10 km northwest of the island and was interpreted by Cornwell (1971) as being due to a Tertiary dyke or dyke swarm, dipping steeply southwestwards and at a shallow depth below the sea floor.

During the opening of the Atlantic Ocean the junction of the Bristol Channel and South Celtic Sea Troughs may have become a weak point in the crust which acted as a focus for the intrusion of a mass of basic magma some 10 km across; the Lundy granite representing either the mobilisation of a portion of crust by this basic magma, or conceivably an earlier granitic body. The Tertiary age of the Lundy granite is now clear, although it is just possible that a Variscan granite could have been re-constituted in Tertiary times if a large enough source of heat were available. Dollar (1942) entertained and then abandoned the idea of palingenesis on the grounds that no suitable source of heat could be identified. Now, however, the location, by geophysical methods, of a large body of mafic igneous rock in the vicinity means that the idea bears some re-consideration.

Whichever origin is correct, the silicic material, due to its lower density, was intruded above the main mass of mafic material. Because of the presence of slaty xenoliths, the presence of an intrusive contact on the south coast of the island, and the coarse texture of the granite, it seems that the granite was emplaced in Devonian metasedimentary rocks beneath a cover of Permian and Cretaceous rocks. After the solidification of the granite a renewal of activity led to faulting, including the formation of the granite boundary fault on Lundy and possibly early movement along the northwest–southeast lines described above. This event was closely followed by a final phase of intrusion which gave rise to the Lundy dyke suite and perhaps the elongated dyke swarm northwest of the island. The dykes on Lundy may be a southeasterly extension of this postulated swarm, considerably reduced in volume. Alternatively, the Lundy dykes could represent the highest portions of a late-stage set of dykes emanating from the main mafic pluton.

The principal movement along the northwest–southeast faults occurred after the emplacement of the Lundy dyke suite and led to the formation of the Lundy horst. Shearman (1967) suggested a late Oligocene date for this movement.

Chapter Eleven

The Quaternary

Although in parts of Europe sedimentation continued uninterrupted through the Tertiary into the Quaternary, in Devon there is a major gap in the geological record between strata belonging to these systems. The strata also present a contrast of environments of deposition: the youngest Tertiary rocks in Devon are the fluvial and lacustrine of a subtropical climate, whereas the Quaternary deposits represent a variety of environments, many of which reflect the cold climates that produced the ice sheets that covered much of northwest Europe during the Pleistocene. From the Eocene onwards, there was a downward trend in global mean temperatures that developed into the Pleistocene cold phases (known commonly as the Ice Ages). What caused the reduction in temperature is still uncertain; but whatever the cause, the temperature fluctuations during the Quaternary were much more extreme, and around much lower mean temperatures, than those that occurred previously.

Just where the Tertiary–Quaternary boundary should be recognised is still a subject for debate, although for a long time it has been conventionally defined at the base of the Calabrian Stage marine deposits in Italy. However, no strata of this age are known in Devon. Estimates of the age of the boundary have varied greatly, but recent study of deep-sea deposits has produced a figure of around 1.6–1.7 Ma (Berggren *et al.*, 1980). Within the Quaternary, the boundary between the Pleistocene and Holocene (Recent) Series is placed at the base of the Flandrian Stage, and dated at 10 ka (^{14}C) BP (10,000 years before the present, the present being taken conventionally as 1950 AD: all such dates in this account are defined on the basis of carbon-14 or other radiometric dating techniques).

The study of deep-sea sediments has also allowed the climatic history of the Quaternary to be determined with greater accuracy than was possible using land-based sequences. It is now known that there were at least eight glacial phases during the last 700 ka, and that during the whole of the Quaternary there were at least 17 cold phases (Bowen, 1978). This contrasts with older views that there were only four, five or at the most six major glacial phases during the Quaternary. Although to some extent the difference reflects variation in the definition of what actually constitutes a glacial or interglacial as compared with a stadial or interstadial (which are less extreme and/or shorter phases of temperature fluctuation), it is mainly due to the detail that the more complete oceanic record contains when compared with the gaps and ambiguities present in the land-based sequences. According to the oceanic record, the classical 'Ice Age' of Europe only occupied the last 850 ka of the Pleistocene, so that a greater number of climatic events have been compressed into a shorter time span than formerly believed. The new data hold out the twin prospects of reliable global correlations and a complete chronological record of the Quaternary.

Unfortunately, attempts to link in detail the land-based evidence with that of the deep-sea sediments face great problems, so that for the time being it is still necessary to rely on the classification and nomenclature for the British Quaternary shown in *Table 11.1*. Nevertheless, some authors have used north European terminology to describe Quaternary sequences in Britain, so that a correlation of the British and European Stages is also shown in *Table 11.1*, but it is highly speculative for that part of the sequence before the Ipswichian-Eemian Stage. No attempt is made here to indicate equivalence with the classical Alpine sequence, the security of which is increasingly doubted even in the Alpine region itself.

In Britain the last three major cold stages—the Anglian, Wolstonian and Devensian—produced important ice sheets, although only during one of the older stages did an ice sheet reach Devon. However, the proximity of the southern margin of the Devensian ice sheet (which reached its maximum development about 18 ka BP) in south Wales had a profound effect on Devon, producing a periglacial environment that resulted in the almost complete obliteration of features formed during the earlier glacial stages. Even the most recent glacial advance of all, which occurred in the Loch Lomond Stadial between about 11 and 10 ka BP at the end of the Devensian, and involved the growth of ice caps, valley and cirque glaciers in Scotland, and numerous small cirque glaciers (not shown in *Figure 11.1*) in the Lake District and in Wales as far south as the Brecon Beacons, had its periglacial impact on the uplands of South West England. This last cold phase was terminated about 10 ka BP by a period of rapid warming, which led to the climate of the present day.

Table 11.1 STAGES OF THE BRITISH QUATERNARY

Series		Stage		Climate		N.W. Europe
Holocene		Flandrian		t		Flandrian
Pleistocene	Upper	Devensian	c	g	p	Weichselian
		Ipswichian	t			Eemian
		Wolstonian	c	g	p	? Saalian
	Middle	Hoxnian	t			Holsteinian
		Anglian	c	g	p	? Elsterian
		Cromerian	t			
		Beestonian	c		p	
		Pastonian	t			
	Lower	Baventian	c		p	
		Antian	t			
		Thurnian	c			
		Ludhamian	t			
		Waltonian				

Abbreviations: c, cold g, glacial deposits known p, permafrost known t, temperate.
Based on Mitchell *et al.* (1973) and West (1977).

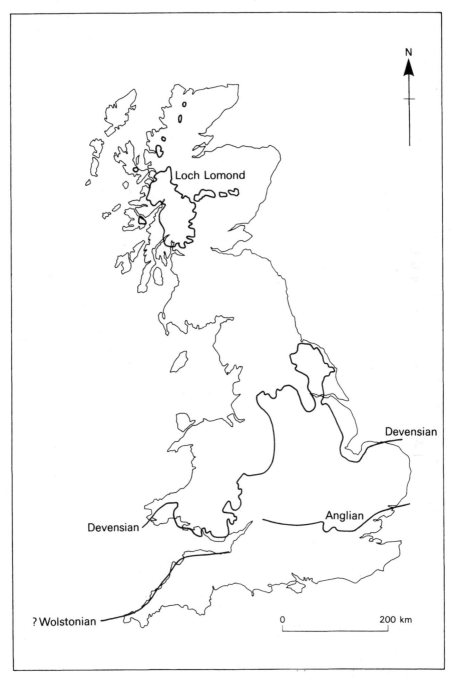

N

Loch Lomond

Devensian

Devensian

Anglian

? Wolstonian

0   200 km

Figure 11.1. Principal glacial limits in Britain, including the southernmost limit of glaciation

The climatic complexity of the Quaternary is shown not only by the Loch Lomond Stadial, but by a warmer phase, known as the Windermere or Late-glacial Interstadial, which preceded it between about 13 and 11 ka BP. Such interstadials also occurred earlier in the Devensian, which was therefore a complex glacial stage consisting of alternating stadials and interstadials, and it must be assumed that earlier cold stages were similarly interrupted by

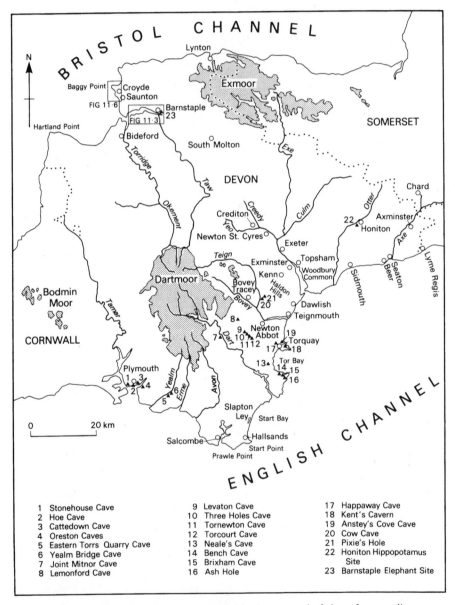

1	Stonehouse Cave	9	Levaton Cave
2	Hoe Cave	10	Three Holes Cave
3	Cattedown Cave	11	Tornewton Cave
4	Oreston Caves	12	Torcourt Cave
5	Eastern Torrs Quarry Cave	13	Neale's Cave
6	Yealm Bridge Cave	14	Bench Cave
7	Joint Mitnor Cave	15	Brixham Cave
8	Lemonford Cave	16	Ash Hole

17	Happaway Cave
18	Kent's Cavern
19	Anstey's Cove Cave
20	Cow Cave
21	Pixie's Hole
22	Honiton Hippopotamus Site
23	Barnstaple Elephant Site

Figure 11.2. Locations of places mentioned in the text, and of sites of mammalian remains

milder intervals. The Last (Ipswichian) Interglacial began about 128 ka BP, but the date of its upper boundary is still hotly debated, largely because of problems of definition. Some workers argue that it gave way to the Last (Devensian) Glacial about 118 ka ago, with interstadials interrupting the ensuing Early Devensian at about 100 and 80 ka BP, whilst others believe it more apt to incorporate these interstadials within the interglacial, thus placing the boundary at about 75 ka BP.

The pre-Ipswichian record is extremely hazy, and it is not even certain whether the maximal ice advance of the Pleistocene in western Britain occurred in the Wolstonian or in the earlier Anglian, as in eastern England. Whatever its age, this maximal ice sheet impinged on the northern coast of South West England but, apart from a small area in north Devon, there is no reliable evidence that any part of the county was ever covered by glacier ice, although this was disputed by Kellaway et al. (1975).

The following account attempts to summarise the evidence of Quaternary climatic vicissitudes as recorded in the landforms and superficial deposits of Devon, dealing in turn with glacial phenomena, periglacial phenomena, river terraces and drainage evolution, sea-level changes, and the non-marine fossil record. The chronological significance of those types of evidence not yielding direct age information is then assessed in a section on the Quaternary sequence, and finally a selection is made of localities where aspects of the Quaternary of Devon can be seen with relatively easy access. The locations of the most important places mentioned in the text are shown in *Figure 11.2*. All altitudes are in metres above or below Ordnance Datum (OD), unless otherwise stated, and all radiocarbon dates are uncorrected.

GLACIAL PHENOMENA

Glacial phenomena, that is the deposits and landforms that can be attributed directly to the former presence of glacier ice, include both material laid down directly by the ice (glacial till) and features resulting from the erosional and depositional activity of glacial meltwaters (fluvioglacial phenomena). These are very restricted in their distribution in Devon; so far as is known at present, such phenomena are confined to the north of the county where, in the Fremington–Hele area (*Figure 11.3*), their occurrence has been accepted for a long time and by a succession of researchers.

Glacial Deposits in the Fremington–Hele Area

The earliest account of the glacial deposits was by Maw (1864), and more recent investigators have included Mitchell (1960), Stephens (1966, 1970), and Wood (in Straw, 1974). Published descriptions of the deposits vary, both because of the different perception of different investigators, and because of the spatial variability characteristic of glacial deposits. The so-called Fremington till forms a continuous clay body that extends for 4 km between Fremington and Lake (*Figure 11.3*), with small, isolated till bodies to the north and west of the main mass. The maximum recorded depth of clay, in a well-boring at Roundswell, is 24 m (Maw, 1864). The principal exposures now are in Brannam's Claypit (SS 529317), the earlier workings at Clampitt (SS 527328) having been long abandoned. *Figure 11.4* shows the deposits revealed between 1967 and 1970, as recorded by Wood, whilst *Figure 11.5*

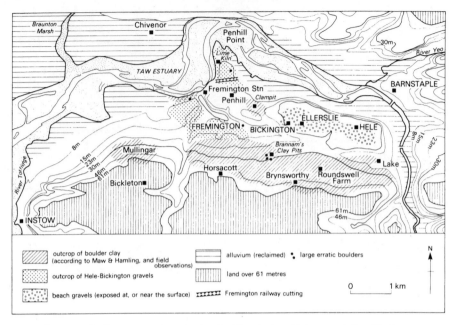

Figure 11.3. Quaternary deposits in the Fremington–Hele area (after Stephens, 1970)

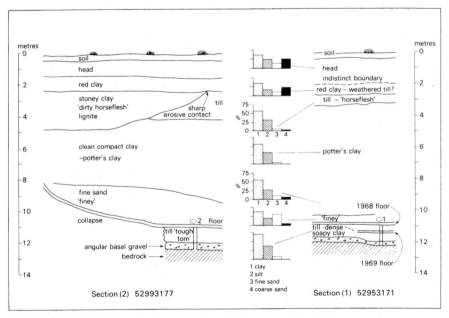

Figure 11.4. Sections in the south face of Brannam's Claypit, Fremington (after Wood, in Straw, 1974)

shows a composite section drawn by Stephens on the basis of several visits. A series of clays about 10 m thick is overlain by up to 2 m of yellow head (solifluction material) and is underlain by about 15 cm of gravel, beneath which is bedrock.

Consideration of the head properly belongs in a later section, because it is a periglacial rather than a glacial deposit, but it is described here for convenience. It has a relatively high clay content as compared with most other head exposures in north Devon, no doubt because of the underlying clays over which it has moved downslope, and from which it is partly derived. Ice-wedge casts and frost cracks are present in the upper parts of the head, and there is evidence that frost heaving has re-orientated some of the angular stones into vertical or near-vertical attitudes. Erratics and striated stones have been found in the head, presumably derived from the glacial clays in the immediate vicinity of the pit for, according to Stephens (in Straw, 1974), they are not found in the head away from the pit neighbourhood.

The glacial clays themselves include four or five distinct units. The uppermost clay unit, about 1 m thick and red in colour, is stony and contains both erratics and striated stones. Stephens described it as a weathered till, whereas Wood regarded it as a separate deposit, and not merely the weathering product of the underlying 'fresh till'. His reasons included the sharpness of the horizontal boundary between the red clay and the 'fresh till', and the fact that the former lies over a different deposit (a stony silt, which replaces the 'fresh till') in one part of the exposure. The base of the red clay coincides with the base of the overburden stripped off by the pit operators.

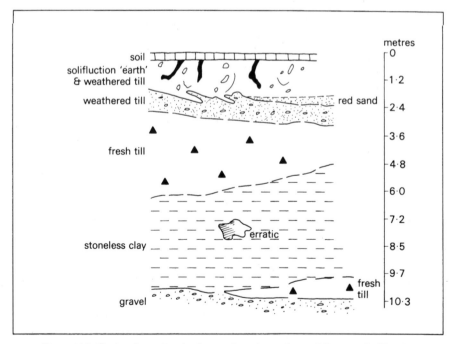

Figure 11.5. Composite section in the south and east faces of Brannam's Claypit (after Stephens, 1970)

Beneath the red clay two stony clays are separated by up to 6.1 m of 'stoneless' potter's clay. The upper, dark reddish-brown, stony clay is of variable thickness (up to about 2.5 m), is compact and tenacious, and contains subangular to subrounded pebbles, cobbles and boulders. In addition to local Devonian and Carboniferous grit, shale, sandstone and limestone, the rock types present include dolerite, granite, quartzite, andesite, tuff, gneiss and flint (Taylor, 1956; Wood in Straw, 1974). Shell fragments have also been found, together with a derived microfauna that includes 11 species of foraminifera. A high proportion of the foraminifera are damaged, and they are all typical Irish Sea types. The dark brown 'stoneless' clay is very fine-grained in its upper part, with a clay fraction (less than two micrometres) of 60 per cent or more, but it coarsens downwards, eventually to become a fine sand at the base. It has a pseudo-laminated structure, and although stoneless in the sense that it contains no layers or abundant mixture of stones, several small pebbles are found scattered through it, and a few large boulders have been found within it, including one of dolerite and one of granodiorite. The lower stony clay is broadly similar to the upper one, but is less stony and even tougher.

The basal gravel is rarely exposed during clay working because it contains water under pressure and would create difficulties if it were breached, but analyses by Wood showed it to have a median grain size of 3.4 mm, and to be poorly sorted. The gravel has been alternatively interpreted as a raised beach deposit and as a fluvioglacial sediment. The former view is based on its alleged similarity to, and/or stratigraphic continuity with, undoubted raised beach gravels exposed at the coast nearby, and the latter view is based on a denial of such relationships. As will be seen later, this difference of interpretation has played a vital part in explaining rival versions of the Quaternary succession in Devon, and in South West England as a whole. Kidson and Wood (1974) used both sedimentological and geophysical evidence to differentiate between the sub-till gravel and the raised-beach gravel at Penhill Point (*Figure 11.3*), and to demonstrate the fluvioglacial origin of the former. Stephens (1966), on the other hand, not only believed the basal gravel to be stratigraphically continuous with frost-disturbed raised-beach gravel at Fremington Quay (SS 512332), but also interpreted a stony clay resting on the latter gravel as glacial till, on account of its content of erratic and striated stones. The Hele gravels, capping the Hele–Bickington ridge at altitudes up to 55 m OD, are also considered to be of fluvioglacial (outwash) origin by Stephens (1970) and Kidson and Wood (1974), rather than marine as suggested by Mitchell (1960). They have been shown by borings to interdigitate with the Fremington till, and are likely to have been deposited by streams emanating from the same ice sheet that deposited the till.

Although the glacial deposits in the Fremington–Hele area have a limited distribution, their importance in the Quaternary of South West England is very great, for they provide the only widely accepted evidence that glacier ice ever reached the peninsula. The included erratics and marine microfauna indicate glacial transport of some till constituents from the Irish Sea Basin, and for reasons to be discussed later it is likely that the till and associated fluvioglacial gravels were deposited near the limit of the Irish Sea portion of a Wolstonian ice sheet. Problems remain in the detailed interpretation of these deposits, however. The depositional environments of the different till units are still not clear, particularly in the case of the 'stoneless' clay, which was considered to be lacustrine by Maw (1864), Stephens (1966) and Edmonds (1972a). Wood (in Straw, 1974) and Edmonds (1972a) considered the entire series of till units to represent ice-front oscillations, which the latter author correlated with the sequence of four river terraces he identified in the Taw and tributary valleys near Barnstaple, all but the lowest terrace having been formed whilst the edge of the Wolstonian ice sheet lay nearby. The highest (fourth) terrace was formed during a temporary withdrawal of the ice following the deposition of the basal till unit, and the third was graded to a lake dammed by a readvance of the ice, the water surface standing at about 30 m OD. The potter's clay was deposited in this lake. The second terrace was formed as the ice, resuming its advance, overrode the lake deposits and deposited the upper till(s). Such a detailed sequence of events must be regarded as tentative, given the amount and nature of the available evidence. An alternative interpretation of the glacial deposits is that the different layers merely reflect sequential changes in the conditions of glacial sedimentation, not necessarily involving ice marginal oscillations.

Glacial Phenomena Elsewhere

Till. Away from the Fremington–Hele area, glacial tills have been identified by Stephens (1966, 1970; in Straw, 1974) at a number of sites near Croyde (*Figure 11.6*) and at Westward Ho! (SS 423292). In all these cases the stony clay referred to as till rests on raised beach deposits, and is described as highly weathered, with erratic stones (including chert and mica schist at Croyde). Stephens (in Straw, 1974) himself doubted whether the tills were *in situ*, whilst other workers (such as Kidson and Wood, 1974) regarded them merely as components of the head.

Glacial Landforms. No clear examples of landforms fashioned directly by glacier ice have been recorded in Devon, even in the small area shown by the presence of till to have been occupied by ice. Just inside Somerset, however, on the southern fringe of Exmoor, is the impressive northeast-facing hollow known as the Punchbowl (SS 883345), which has been identified as a possible glacial cirque (D. Dalzell, personal communication). An alternative explanation is that the Punchbowl is a nivation hollow, formed in the manner of other enlarged valley heads discussed below. If its enlargement was the result of erosion by an isolated cirque glacier, then it is the only cirque so far identified south of the Bristol Channel, the closest ones to the north being in Glamorgan.

Giant Erratics. In addition to the erratic pebbles in the till and head deposits, large far-travelled boulders occur both in the till and exposed in coastal locations, mainly on old shore platforms (Taylor, 1956). The largest and most accessible coastal erratic boulders are a block of pink granite at Saunton (SS 44013787) and a block of granulite gneiss 1.4 m in size near Baggy House (SS 428400), whose weight has been estimated to be 5 tonnes.

The erratics in the till were manifestly carried by the ice sheet, but the mode of transport of the coastal ones has been disputed. One view is that they too were brought by the ice sheet, an interpretation that is supported by the similarity of rock types between the coastal erratics and those in the till: porphyry and dolerite are the commonest in both cases, and spilite occurs in

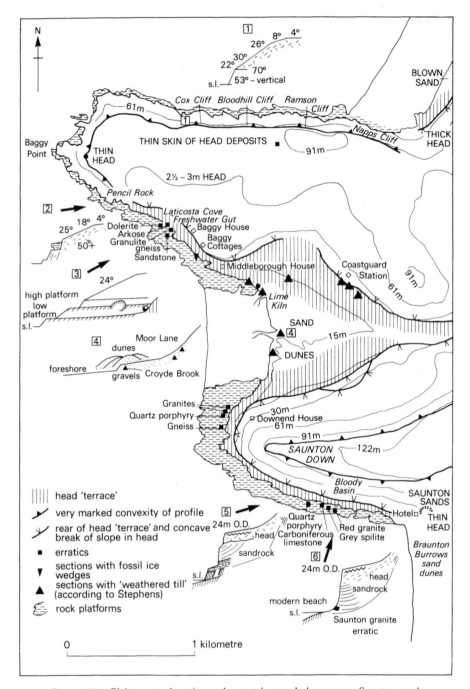

Figure 11.6. Pleistocene deposits and coastal morphology near Saunton and Croyde (after Stephens, 1970)

both. Furthermore, derivation from western Scotland is likely for many of the erratics from both till and platforms. However, another view is that the coastal platform erratics were delivered by icebergs calved from glacier ice farther north, an interpretation that is consistent with the widespread distribution of erratics on shore platforms on the Bristol Channel and English Channel coasts. In south Devon erratics have been recorded as far east as Prawle Point, but they occur as far distant as Sussex and also on the French coast.

These hypotheses are probably not mutally exclusive, the south coast erratics being best explained by ice-rafting. In north Devon, however, in addition to the blocks in the till an isolated block of epidiorite was found at about 80 m OD on Baggy Point promontory (SS 43564070) by Madgett and Madgett (1974) which can only have been placed in position by an ice sheet. So it is possible that both modes of transport have been involved, and even that individual blocks have been delivered initially by iceberg and subsequently moved by glacial transport, or perhaps the reverse.

Fluvioglacial phenomena and drainage diversion

The presence of the outer edge of an ice sheet in the Taw–Torridge estuary implies that it impinged on the coast elsewhere in north Devon, and that the drainage of the area beyond the ice sheet was affected, both by diversion of the non-glacial drainage and by meltwater issuing from the ice. Glacial diversion of the drainage of the lower Taw and Torridge basins was postulated by Edmonds (1972a), who suggested that the Wolstonian ice advance blocked the estuary and caused reversal of the northward flow of the lower Taw between Barnstaple and Chapelton, from where the Taw catchment water, augmented by meltwater from the ice, flowed westward toward Bideford (*Figure 11.7*).

Stephens (1966, 1970) postulated ice-marginal and/or submarginal meltwater channels, all now dry, at several points along the coastal cliffs of north Devon. Detached dry valley segments at Hartland Quay (SS 223248), Damehole Point (SS 226264) and St Catherine's Tor (SS 225242) hang above the sea at both ends, have flat floors underlain by 1–2.5 m of head, and are bounded by steep valley sides. They were regarded by Stephens as forming a series falling southwards from Hartland Point, but in fact their floors, both individually and collectively, decline northwards in altitude, and can more readily be visualised as remnants of a former river valley to which the present coastal streams were tributary, that was subsequently dismembered by marine erosion. This explanation, favoured by Arber (1911), is also applicable to other marginal channels postulated by Stephens at Clovelly Court (SS 310255), Sloo Farm–Worthgate (SS 373236) and the Valley of Rocks, west of Lynton. The latter, an impressive dry valley with a thick fill of head and with tors and other frost-riven features on its sides (Mottershead, 1977), was interpreted by Simpson (1953) as a remnant of a marine-dissected extension of the present East Lyn river. This interpretation was upheld, with significant modifications, by Dalzell and Durrance (1980), who postulated an Ipswichian date for the initial capture, followed by further marine dissection of the virtually dry valley system. Dismemberment of river valleys by marine cliff recession has also occurred in south Devon between Torquay and Teignmouth, and between Sidmouth and Beer.

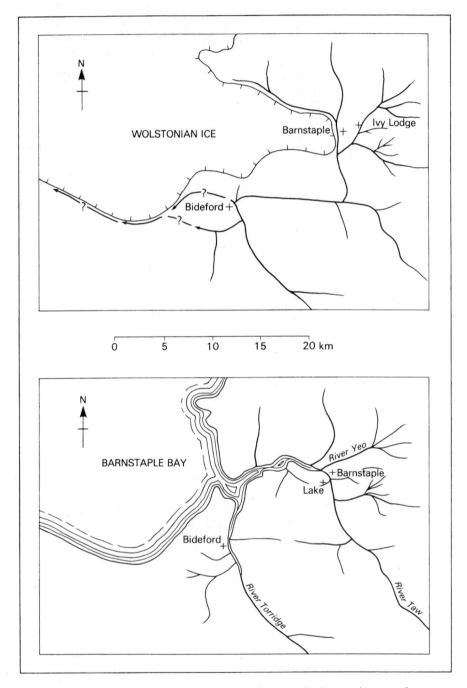

Figure 11.7. Possible glacial diversion of drainage in the Barnstaple area (after Edmonds, 1972a)

PERIGLACIAL PHENOMENA

In contrast to the glacial evidence, periglacial phenomena in Devon are both widespread and abundant, and there are very few areas that can reasonably be regarded as devoid of them. Until fairly recently, research into periglacial features was concentrated in the coastal areas and on Dartmoor; study of other areas was unsystematic and largely confined to investigation of chance discoveries. Since 1972, however, our knowledge has been increased greatly by the systematic study of periglacial phenomena in the Bovey Basin, east Devon, on Exmoor and in the Tamar area, by T. Gouldstone, N. R. Piggott, D. Dalzell and D. J. Miller respectively, and references to those areas in the following account owe much to their unpublished work.

Erosional Periglacial Features

Tors. One of the most dramatic aspects of the scenery of Dartmoor is the presence of tors—upward-projecting masses of naked bedrock either crowning the hilltops (summit tors) or arising from steep valley sides (valleyside or hillslope tors). The hillslopes descending from both types are often strewn with blocks of rock, known locally as clitter, that have been derived from the recession of rock faces by differential weathering and mass movement. The tors are thus residual features, and occur on both the granite and the metamorphic aureole. Two main models have been advanced for the origin of the granite tors.

The two-stage model advocated by Linton (1955) involves deep weathering of the granite under warm-temperate or tropical climatic conditions during the Tertiary, followed by stripping of the weathered material by solifluction and other forms of mass movement in the Quaternary. According to this interpretation the tors consist of the surviving corestones of relatively coherent granite exposed by removal of intensively weathered material, and the distribution of tors is related to variations in joint spacing in the bedrock: the tors occur where the joints are most widely spaced, and where the intensity of chemical weathering has therefore been least. The angularity of some tors is attributed to limited periglacial frost action.

The single-stage model advanced by Palmer and Neilson (1962) is that the tors are purely palaeo-arctic forms resulting from frost shattering and riving of exposed bedrock (*gelifraction*) in conjunction with downslope transfer of the shattered debris by cold-climate solifluction (*gelifluction*). Rapid rock-face recession by these means during cold phases of the Pleistocene has caused widespread lowering of the ground surface, the present tors being the last remnants of higher ground surfaces. Palmer and Neilson denied that chemical weathering has contributed significantly to the decomposed granite or 'growan' that undoubtedly exists (for example at Two Bridges, SX 596752), which they ascribed to pneumatolysis as in the case of the major china clay bodies (Chapter Five). They claimed that the tors and rotted granite are spatially dissociated, rather than closely associated as stated by Linton. Any tendency for tors to be associated with wider joint-spacing is consistent with the greater effectiveness of gelifraction, the more closely spaced the planes of weakness are. This last point is well demonstrated by the tors and associated features on the metamorphic aureole, and on sedimentary rocks in the Valley of Rocks near Lynton.

Although Waters (1964, 1965, 1971) attributed to gelifraction and geli-fluction an importance in granite tor formation far greater than that implied by Linton, he nevertheless accepted the occurrence of selective, subsurface weathering in the Tertiary and in the interglacial periods of the Quaternary. The suspicion that the true explanation incorporates elements of both main rival hypotheses has grown with recent work, which has concentrated mostly on the central problems of the origin of the growan and the relation of tors to joint-spacing variations. As indicated by Brunsden (1964) growan has been formed by three main processes whose products are not always easy to dis-tinguish: pneumatolysis, chemical weathering, and mechanical disintegration by frost action. Eden and Green (1971) used textural and mineralogical evi-dence to distinguish between the products of the first two, and to establish the occurrence of the second, though it is largely restricted to the main river valleys and the margins of the granite outcrop. Since this distribution accords with that of most of the granite tors, Eden and Green supported a two-stage origin of the latter, with weathering under rather cooler (subtropical to tem-perate) conditions than envisaged by Linton. In contrast, Doornkamp (1974) concluded from an electron microscope study of quartz-grain surface textures that the weathered growan was produced mainly by mechanical disintegra-tion, thus favouring a wholly periglacial two-stage origin of the tors. Gerrard (1974, 1978) has confirmed that tors are located where joint spacings have permitted differential weathering and erosion.

It seems advisable to conclude that the final phase of tor formation was related to extensive stripping of regoliths in conjunction with frost-shattering processes in a periglacial environment, and that both exhumation and modifi-cation of two-cycle tors and the formation of one-cycle tors could have occurred at the same time in different parts of Dartmoor. Tors on non-granitic rocks·showing no evidence of deep weathering, as in the Dartmoor aureole (for example at Cox Tor) in north Devon (such as the Devil's Cheese-wring, Valley of Rocks) and in south Devon (such as Signalhouse Point, SX 771355) are less ambiguously the products of gelifraction and gelifluction, and in some cases are associated with cryoplanation features.

Cryoplanation (*Altiplanation*) *Terraces.* These are erosional bedrock benches at the foot of frost-riven scarps, or around tors. They often occur in staircases, and result from the parallel retreat of the scarps under the combined action of

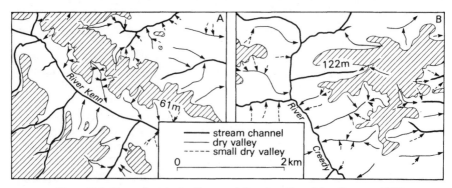

Figure 11.8. Dry valleys in the Kenn and Creedy valleys (after Gregory, 1971)

frost shattering and nivation, the detrital products of which are transported across the benches by gelifluction and sheetwash. These processes are most effective on resistant but well-jointed and well-bedded rocks, and benched hillslopes attributable to them have been recognised at several locations in the Dartmoor aureole (D. J. Miller, personal communication), including Black Hill (SX 765786), Brent Hill (SX700617), Peek Hill (SX 555700), Whitchurch Common (SX 528745), Cox Tor (SX 531762), Smeardon Down (SX 522782), White Tor (SX 543786), Cudliptown Down (SX 532795), Brent Tor (SX 471804), Gibbet Hill (SX 503812), White Hill (SX 534837), Great Nodden (SX 538874), Southerly Down (SX 535878), Lake Down (SX 540890), Sourton Tors (SX 543897) and East Hill (SX 597940). The best developed features are probably at Cox Tor, where the highest cryoplanation terrace is backed by the frost-shattered scarp in metadolerite 5 m high that forms the tor itself. Good examples also occur on Exmoor (D. Dalzell, personal communication), on Trentishoe Down (SS 628479) and Holdstone Down (SS 620479), where polygonal patterned ground is present on some of the steps.

Nivational Forms. The term nivation is applied to a suite of localised and powerful erosional processes that operate on susceptible lithologies beneath and near snowbanks, and include frost shattering, gelifluction and slopewash. The principal nivational landforms in Devon, cryoplanation terraces apart, are large nivation hollows, rounded or bowl-shaped valley heads, overdeepened niveofluvial valleys, and dells, all of which result from enlargement of depressions which grew more actively the larger they became.

Very few large nivation hollows have been identified so far in Devon, though this may reflect the lack of detailed fieldwork over large parts of the county as much as a dearth of such features. A large hollow facing northeast on the western flank of the West Okement Valley (SX 556897) is a striking example of a probable nivation hollow. Much more common are the deeply incised, enlarged rounded or bowl-shaped valley heads whose dimensions are out of keeping with a purely fluvial origin by headward erosion and/or spring sapping. Such features are widespread in west, east and north Devon, as also are overdeepened niveofluvial valleys, of which the Tor Wood valley (SX 542894), near the nivation hollow just mentioned, is an excellent example. Also widespread, though much smaller, are *dells*—shallow, commonly bowl-shaped depressions or furrows, usually dissecting the sides of larger valleys, though in places quite separate from the present stream pattern. They commonly have two heads, which may be steep-walled; they indent river and niveofluvial terraces, as in the Culm and Creedy Valleys; and in places they form groups of parallel furrows or grooves. They commonly contain head deposits on their floors and slopes. The origin of dells is usually ascribed to niveofluvial processes on either permanently or seasonally frozen ground. The only published maps of such features in Devon are those of Gregory (1971), who mapped them, together with other types of dry valley, in the Exminster–Kenn and Crediton–Newton St Cyres areas (*Figure 11.8*), and in the catchment of the River Otter (*Figure 11.9*), but their distribution is much more widespread than yet revealed in published work. They occur on lithologies ranging in age from Carboniferous to Pleistocene, and including both permeable and impermeable rocks.

Dry Valleys. Whereas the smaller dells are usually distinctive in form, and readily evoke a periglacial origin, some of the larger ones are difficult to

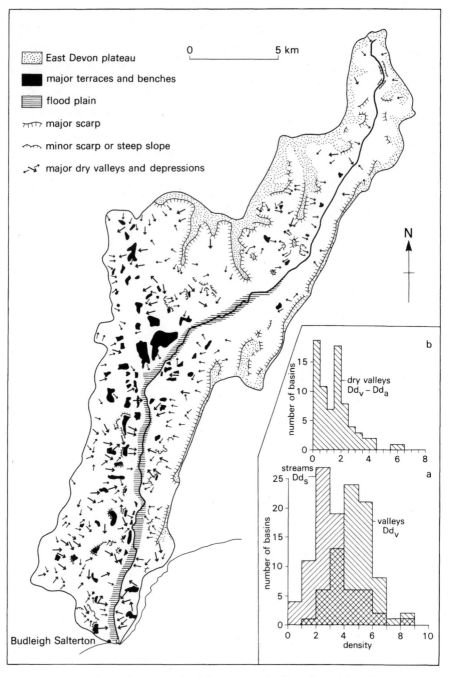

Figure 11.9. Dry valleys and related features in the Otter Basin (after Gregory, 1971)

distinguish from dry valleys that may have formed in non-periglacial environments. Gregory (1971), using both morphometric and mophological criteria, identified two distinct types of dry valley in the Otter basin: the dells as already described, and larger dry valleys up to 30 m deep, up to 1 km long, and with steep sides (up to 20°). The larger ones commonly have a flat floor underlain by 1.5–3.5 m of head and slopewash material. Whereas the dells are discontinuous and in places separate from the present stream network, the larger dry valleys extend the line of the latter, though in some locations they hang above stream-occupied valleys to which they are tributary. A chronological distinction is also implied by the fact that the larger dry valleys dissect the high-level and middle terraces of the Otter Valley, and descend to (but never below) the lowest terrace, whereas the dells often dissect the latter. According to Gregory, the dells functioned during the last cold phase, whereas the larger dry valleys became dry before the last periglacial phase due to lowering of the water table caused by the deepening and opening out of the Otter basin during the successive cold and warm stages of the Quaternary. It is clear from Gregory's work, and other work in progress, that there is scope for detailed study of dry valleys and dells in other parts of Devon.

Erosional Slope Forms and Valley Asymmetry. During each periglacial phase, gelifluction stripped slopes of the weathered material (regolith) they had acquired in the preceding warmer phase, and selective gelifraction and gelifluction tended to emphasise lithological differences, thereby producing and exaggerating irregularities in slope profiles. Even where significant lithological variations are absent, steep rectilinear slopes now devoid of regolith, save for a thin veneer of scree, probably testify to the effectiveness of periglacial mass-wasting processes which largely removed the frost-shattered debris. Such slopes are well exemplified in the lower Heddon Valley (SS 655493), and are a conspicuous element in the northern coastal cliffs. Beyond such cases, interpretation of erosional forms in a periglacial context is hazardous, for there is no one slope form or asemblage of slope forms that is unique to the periglacial environment or distinctively periglacial in nature. Asymmetric valleys, with one slope markedly steeper than the other, and with no lithological contrast to explain the difference, have often been regarded as diagnostic of periglacial conditions. Among several examples in Devon is the Farley Water Valley (SS 752445), whose steeper slope faces northeast. Unfortunately, one cannot argue unequivocally for a periglacial explanation for valley asymmetry, because the many factors that produce it operate in a wide variety of morphoclimatic conditions, though they may often be more pronounced in periglacial zones than elsewhere.

Depositional Periglacial Features

Screes and Blockfields. Frost shattering of exposed bedrock surfaces, and comminution of the riven blocks by frost action and other forms of rock weathering, produce screes and blockfields. However, they are not unambiguous indicators of Pleistocene periglacial activity unless it can be shown either that they do not form at the present day, or that the present rate of formation is incompatible with the volume of material in them. Although some frost-riving occurs today in upland Devon during winter, it seems clear that the impressive amounts of rock waste on the partly and sometimes largely vegetated screes that occur down to low altitudes in several north Exmoor valleys are largely or even wholly a relic from former cold conditions. Similarly relict are the blockfields or clitter spreads of Dartmoor which, although presently being modified by weathering processes, testify to former conditions of severe frost shattering of bedrock and relatively rapid downslope transfer of the blocks by rafting on the finer-grained soliflual material or head beneath.

Head. The term *head*, first used in a geological context by De la Beche (1839), has since been applied widely to periglacial slope deposits in southern England. An attribute commonly mentioned in published descriptions of head in Devon is its very common occurrence, which has been confirmed by recent work. It occurs wherever the slope was sufficient to allow gelifluction and frost creep to take place, which can be on gradients as low as 1°, as has been shown by work in present-day periglacial environments. It is therefore prob-able that only the flattest summit platforms and interfluves escaped significant modification by periglacial mass wasting.

In present-day periglacial areas the upper half metre or so of water-satu-rated material (the active layer) moves downslope in spring and summer at typical measured rates of 1–5 cm per year, the rate varying according to slope angle, aspect, vegetation and other factors. The legacy of such processes in Devon, which is profound, varies according to the nature of the bedrock, the depth and character of the previous regolith, and the nature of the pre-existing topography. The steeper upslope areas in several places were largely stripped of regolith and exposed to rapid denudation by frost action, whilst at lower levels the pre-existing topography was largely smothered by soliflual material, as attested by the long, smooth slope profiles so characteristic of large parts of the county.

On Dartmoor the depth of head varies, but reaches 5–6 m in a number of places. According to Waters (1964, 1965, 1971), the layers of soliflual debris exposed in several pits record two distinct cold phases (*Figure 11.10A*). 'Dur-ing the earlier cold phase successive horizons of the pre-existing weathering profile were removed from the upper parts of slopes and deposited in reverse order lower down as the main head ... The later cold phase was characterised by the downslope transfer of boulders and blocks of sound bedrock, detached from the newly-exposed tors, and their deposition as the upper head' (Waters, 1965). There is no evidence of the intervening warmer phase. Green and Eden (1973), by contrast, found no evidence of widespread inversion of the weathering profile. The bedded growan seen between the weathered granite and overlying head in several exposures (for example at SX 659902), and attributed by Waters to downslope transfer of growan by surface wash before deposition of the head, was interpreted by Green and Eden as the result of local displacement of the underlying *in situ* growan by the head passing over it (*Figure 11.10B*). They thus envisaged contemporaneous movement of the bedded growan and head, and challenged Water's identification of two separ-ate heads, suggesting that most of the coarse debris on the lower slopes is in the lower part of the head, and was derived from local basal sources rather than from the clitter spreads and upslope rock exposures.

In south Devon, head deposits of stones and boulders in a matrix of sand and clay reach thicknesses of over 8 m in several coastal exposures between Plymouth Sound and the Dart estuary. The fine cliff exposures in Wembury Bay (SX 511485) show an old shore platform at 3–4.5 m OD overlain by up to 9 m of massive and largely uniform stony head whose constituents are derived

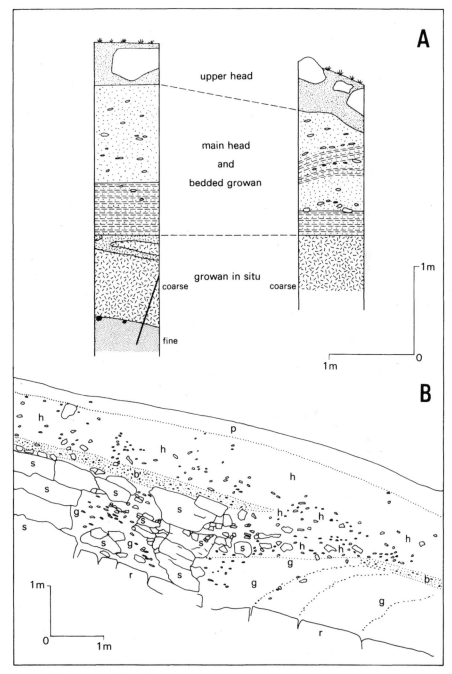

Figure 11.10. Slope deposits on Dartmoor: A, according to Waters (1964); B, according to Green and Eden (1973)

mainly from the underlying Dartmouth Slate, though other local rocks are represented. The uniformity of the head is interrupted in places by discontinuous sand horizons and local concentrations of larger blocks of bedrock. The schist-derived coastal head deposits between Start Point and Hope Cove have been studied by Mottershead (1971), who noted that they are best preserved where the coast is relatively protected from marine erosion, such as Hope Cove (SX 673398), Salcombe Harbour and Lannacombe Bay (SX 816369), where maximum thicknesses of 27–33 m are exposed. The head apron extends from the steep, tor-crowned bedrock slopes of an old cliff-line to form a blanket over old shore platforms, from which it is being stripped in places by marine erosion (*Figure 11.11*). The maximum width of the head apron is just under 300 m, at Langerstone Point (SX 782354). In general, the depth of head, the surface gradient of the apron or terrace and the calibre of the material all diminish away from the old cliff-line. There is no evidence of any inversion of the weathering profile; on the contrary, there are often indications that the entire regolith has moved *en masse*. There is crude bedding in some places, however, suggesting repeated shallow flows of material. There is also a finer material than the main body of head, either lying above the latter, or forming a lens within it. Mottershead suggested that this finer 'Upper Head' may result from slopewash, though it also contains a large proportion of silt that may represent a windblown component.

Coastal head aprons are also prominent in north Devon, producing long, sweeping, concave lower hill slopes that are particularly striking between Barnstaple and Saunton, and around Croyde Bay. Along the northern fringes

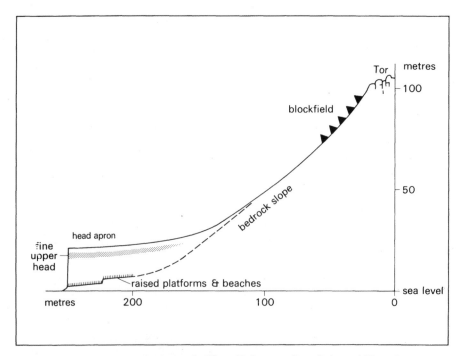

Figure 11.11. Generalised coastal cliff profile between Start Point and Hope Cove

and valleys of Exmoor, however, many slopes have been stripped almost entirely of regolith. Cliff exposures between Saunton and Baggy Point (*Figure 11.6*) show that the head deposits, varying in thickness up to 20 m, rest in different places upon one or more of the following: old shore platforms, erratic blocks, raised beach deposits, cemented blown sands. The head consists mostly of angular slate fragments, in some places with no finer material, and in others set in a sandy matrix. As on Dartmoor, there has been debate as to how many solifluial phases are recorded in the head stratigraphy; Stephens (1966, 1970) claimed that two heads (representing different glacial periods) are present, and Kidson (1971) maintained that there is only one. The nature of the evidence that has evoked such conflicting interpretations is exemplified by an exposure at Middleborough, in Croyde Bay, which shows a layer of coarse, angular slate fragments underlain and overlain by finer material consisting of smaller slate fragments in a sandy matrix. Such variations can readily be explained in terms of one cold phase, during the onset of which the down-slope transfer was mainly of previously weathered material; this being followed by a period when the stripped rock faces, exposed to severe frost action, yielded abundant gelifracts (frost shattered fragments), after which there was renewed transfer of finer material when the climate had ameliorated sufficiently for frost-riving of bedrock to give way in importance to chemical weathering and the comminution of existing rock fragments. Although the fossil ice-wedges in the coarse, stony layer may imply renewed severe cold following a cessation of downslope movement, as suggested by Stephens, there is no evidence to support his view that the ice-wedge formation and overlying head were the only elements in the sequence to have formed in the Devensian, which included enough climatic fluctuations to accommodate all elements of the head stratigraphy.

Loess. Wind-blown silt, known as loess, is an important component of the material deposited in periglacial environments: their relatively dry climates and skeletal vegetation cover allow ready deflation of fine material from fluvial and fluvioglacial deposits. Silty drift deposits, usually under 0.5 m thick but locally up to 1.5 m thick, occur extensively on the east Devon plateau, and in the dry valleys dissecting the Devonian limestone plateau inland from Torbay, and sporadically in other areas, including Woodbury Common, the Haldon Hills and Dartmoor. The distribution, particle size characteristics and mineralogical composition of the material suggest that it consists mainly of loess blown from distant sources to the east, probably during the coldest phase of the Devensian (Harrod *et al.*, 1973).

Periglacial Structures

Fossil periglacial structures recorded in Devon include vertical stones, involutions and cryoturbation structures, ice-wedge casts, patterned ground, and certain bedrock structures.

Vertical Stones. These result from differential frost heave rotating and tilting stones until they stand on end within a matrix of finer material. Repeated frost heave cycles also cause the progressive upward migration of stones. Vertical stones are sufficiently distinctive to be regarded as periglacial indicators, and have been recorded from several exposures in east and north Devon, in the Bovey Basin, and in the top 25–45 cm of head on the western flanks of Dartmoor.

Involutions and Cryoturbation Structures. Deformations that result from cryostatic pressures in unfrozen material trapped between the downward migrating freezing layer and perennially frozen ground beneath are known as involutions; these can also result from differential pore-water pressures when ice-rich sediments thaw, a mechanism that requires neither cryostatic pressure nor permafrost, and can operate in non-periglacial environments. The term involution implies some uniformity and regularity of spacing of the structures. Where relatively amorphous periglacial disturbances are present, the term cryoturbation structure is better employed. Involutions are still visible in a largely infilled gravel pit (SX 596762) west of Two Bridges (Waters, 1961, 1964), at Peartree Point (SX 819366) in south Devon (Mottershead, 1971), and together with cryoturbation structures in several exposures in east Devon.

Ice-wedge Casts. The former presence of frost cracks or fissures caused by thermal contraction of the ground may be preserved as casts. When recognised they are valuable palaeoclimatic indicators, because they form only in areas where the mean annual air temperature is −6°C or lower. Their palaeoclimatic value is limited, however, by problems of recognition, for they have often been confused with desiccation cracks, tree-root casts and load structures. Such problems make it impossible to give a definitive list of genuine ice-wedge casts in Devon, but they have been reported from localities on Dartmoor, Exmoor, in the Bovey Basin and in north and east Devon. The presence of fossil ice-wedge polygons near Newton St Cyres is suggested by differential crop markings on aerial photographs; these polygons are at an altitude of about 76 m OD, and the ice-wedge casts near Croyde are within a few metres of the present shore, suggesting that permafrost in Devon was not confined to the upland areas.

Patterned Ground. The arrangement of the constituent materials of the regolith into more or less distinct patterns, including circles, polygons, nets, steps and stripes, constitute the phenomena of patterned ground. Their precise origins are uncertain, but the processes involved include cracking (whether by desiccation, dilation or frost), frost sorting, differential frost heaving and mass wasting. Among the clearest patterned ground phenomena still visible in Devon are the stripes in the clitter downslope of the Staple Tors (SX 543760), Great Mis Tor (SX 563770), Sheeps Tor (SX 565683) and Hen Tor (SX 594654) on Dartmoor, and the impressive stone nets and polygons on the cryoplanation terraces of Holdstone Down (Exmoor) referred to earlier. It is not certain whether the spreads of earth hummocks (turf mounds) at Cox Tor (SX 532762) and Hay Tor (SX 759771) are a relic of patterned ground features, or due to more recent intricate dissection of the thick soil cover or to some other cause. The palaeoclimatic significance of such features is less clear than that of fossil ice-wedges, not only because of genetic uncertainties, but also because active patterned ground is present today as far south as the uplands of northern Britain. Nevertheless, it is almost certain that the Devon features are a relic from colder conditions in the past, and that more widespread patterned ground is elsewhere concealed beneath the vegetation mat.

Bedrock Structures. There are several occurrences of bent-over strata (outcrop curvature) in Devon where steeply dipping weathered bedrock is overlain by head, and the uppermost ends of the inclined rock strata have been dragged downslope by the overlying mass of moving soliflual waste. Examples are visible at Outer Hope Cove (SX 675401) where the bedrock is schist

(Mottershead, 1971), and in Laughter Tor Quarry south of Bellever (SX 657764) on the Dartmoor granite (Green and Eden, 1973), and occur quite commonly at the top of coastal cliffs in both south and north Devon. More problematic are the large scale post-depositional contortions in the Oligocene Bovey Formation exposed in the Southacre (SX 854754), Longmarsh (SX 852757) and East Golds (SX 860730) pits north of Newton Abbot. These were attributed by Dineley (1963) to cryoturbation, but Gouldstone (1975) thought this unlikely because it implies an improbably thick (up to 15 m) active layer to account for the depth of disturbance, and instead suggested an origin by the growth of ice lenses within the permafrost. Similar structures have recently been described from Clay Lane Pit (SX 844769) by Jenkins and Vincent (1981).

RIVER TERRACES AND DRAINAGE EVOLUTION

The major published works dealing with Devon river terraces are those of Green (1949) and Brunsden (1963) in the Dart valley, and Kidson (1962) in the Exe. Unfortunately, these studies treated river terraces in the wider context of denudation chronology and drainage evolution, at a time when such investigations were based on a number of assumptions that are now known to be either false or highly dubious. One such assumption was that river terraces were necessarily graded to sea levels higher than the present, and that rises and falls of sea level inevitably cause, respectively, aggradation and incision (or 'rejuvenation') whose effects are ultimately transmitted throughout the entire river basin. A further assumption was that former long-profiles of rivers could be reconstructed as smooth, graded curves, down-valley extrapolation of which would enable the altitude of contemporary sea level to be determined, and thus allow correlation of terraces between different (even distant) river basins.

In reality, large scale aggradation, such as is apparent in several Devon valleys, is quite likely to have been a cold climate phenomenon, bearing in mind the large amounts of soliflual debris that move downslope towards rivers in periglacial areas, and therefore to have occurred during periods of low world-wide sea level. Moreover, changes from aggradation to incision may well have been responses to the diminution of sediment supply caused by thick vegetation growth on the valley sides during warmer phases, so that downcutting along part of a river's course may well have corresponded with relatively high sea level. Difficulties in visualising such relationships readily disappear when account is taken of the landward and seaward coastline migration that accompanied changes of sea level, and that the latter affected directly only the lower parts of river courses. Extrapolation of river profiles and terrace gradients has rightly fallen into disrepute with the realisation that river long-profiles are characteristically irregular, with 'knick points' that often owe more to rock type variations than to 'rejuvenation'. Floodplain gradients also vary considerably between different rivers, so that terrace gradient and altitude are most unreliable parameters for inter-basin correlation of river terraces.

In short, river terraces are much more a reflection of climatic changes and the associated changes in geomorphic processes and river regimes than of sea-level changes, and it follows from the above discussion that many of the conclusions about terrace chronology and correlation reached in the papers on the Dart and Exe Valleys referred to above must be regarded with suspicion.

Apart from work by Gregory (1971) in the Otter Valley, referred to earlier, the only more-recent published terrace work is that of Edmonds (1972a and b), Stephens (1970; in Straw, 1974) and Green (1974). Edmonds has mapped terraces in the Taw and tributary valleys (as mentioned earlier) and in the South Molton area, where two high-level terraces trend east–west across present drainage lines, but in conformity with the trend indicated by the terrace system farther west towards Barnstaple. In the entire terrace system, six levels have been identified, of which the lowest is attributed to the Ipswichian, the next three to the Wolstonian, and the two highest to pre-Wolstonian times.

In the lower Axe Valley, the surface altitude of a particularly impressive terrace declines downvalley from 61–65 m OD at Chard Junction to about 46 m at Broom and about 16 m near Seaton (Stephens, 1970). Large gravel pits at Chard Junction (ST 342044) and Kilmington (SY 275974) show about 10–12 m of crudely bedded and poorly sorted coarse gravels and sands, severely cryoturbated in the top 2 m or so, and with a thin capping of silty material of presumed colluvial or aeolian origin. At least another 10 m of gravel have been proved beneath the floor of the Chard Junction pit. The gravel is clearly fluvial. In addition to local chert, flint, greensand and chalk, erratic pebbles of Palaeozoic rocks comprise up to 10 per cent of the stones. Stephens used these erratics as evidence of a glacial source, suggesting that the main Axe terrace was the product of glacial meltwaters overflowing through the Chard gap (a col with its floor at 84–88 m OD) from proglacial 'Lake Maw', supposed by Maw (1864) to have been ponded in the Somerset Levels area by the ice that deposited the Fremington till. A fundamental difficulty with this hypothesis is to envisage relatively sediment-free lake overflow waters depositing such a large volume of coarse material: the latter could have been derived only from the col itself, so that the erratics remain unexplained. According to Green (1974) the far-travelled material, which is present also in the gravels of the upper Axe and neighbouring valleys, is derived from early Tertiary gravels that occur within the present basins of these rivers, and not from a glacial source.

The Axe terrace gravels also contain Palaeolithic chert and flint implements, large numbers having been recovered from both the now-disused Broom pit (ST 326020) and from Chard Junction. The ovate chert hand axes include both sharp-edged and water-worn specimens, and are of Early–Middle Acheulian type, which indicates a possible age range of Middle Hoxnian to Wolstonian. The terrace cannot, therefore, be older than Hoxnian, and if it is accepted that some of the implements have not undergone significant fluvial transport, then it cannot be younger than Wolstonian.

In addition to the Dart, Exe, Taw, Otter and Axe Valleys mentioned above, river terraces are well developed in the Tamar, Bovey–Teign, Creedy–Yeo, Culm and Lyn basins, and offer a promising field of future research in the quest for a more complete understanding of the Quaternary sequence. As in past studies, though, their true value in identifying former sea levels will be strictly limited, and confined to cases where terraces can be traced unequivocally into marine features, to which attention is now turned.

SEA-LEVEL CHANGES

The air of controversy that pervades most aspects of the Quaternary of Devon, discussed so far, is also strong in the field of relative sea-level changes, evidence of which includes both marine phenomena above, and intertidal and terrestrial phenomena below, present sea level. There is also convincing evidence that many of the shore features in the present intertidal zone are of considerable antiquity, and have merely been modified by recent shore processes. The term *relative sea-level changes* refers to changes of sea level relative to the land, without regard to whether world sea level or local (or regional) crustal movements have been responsible.

Ancient marine phenomena at and above present sea level

Evidence of past relative sea levels higher than and near the present level includes marine platforms resulting from erosion of bedrock, and former beaches of depositional origin. Loose terminology plagues the literature on these features, the term 'raised beach' often being applied misleadingly to rock platforms. Moreover, the altitudes given are of variable accuracy, and related to various, and not always accurately specified, datum lines. This may be of little consequence in the case of older features that cover a considerable altitude range, but even the younger and less variable features have only rarely been accurately heighted.

The highest ancient marine platforms claimed to be of Quaternary age were identified during the work on the Tertiary and Quaternary denudation chronology referred to earlier. Balchin (1952) in the Exmoor area, Green (1949) and Brunsden (1963) in the Dart valley, Kidson (1962) in the Exe valley, Waters (1960a and b) in east Devon and on Dartmoor, and Orme (1964) in the South Hams and on southern Dartmoor, between them identified a large number of planation surfaces at elevations up to 580 m OD, and dating back to the early Tertiary. There is no need here to become involved in the often convoluted arguments about whether the higher surfaces are of marine or subaerial origin, because it has been widely agreed that the highest feature of possible Quaternary age is a marked and widespread surface with a shore feature at about 210 m OD. This has been correlated with a feature in South Eastern England that bears a marine fauna of Red Crag type, of Waltonian (Lower Pleistocene) age. Marine platforms have been recognised on morphological criteria at between 9 and 16 stages between the 210 m feature and present sea level. The criteria used include bevelled hilltops, flats, and bluffs at the back of spur flattenings, and the most persistent levels in the areas listed above are at about 180, 130, 100, 85, 45, 15, 7, 4, and 0 m OD, of which the four lowest are usually much more distinct than the others and are mostly found in present coastal locations where their relationships to other landforms and to Quaternary deposits are commonly exposed.

Kidson (1971, 1977) grouped the lower-level shore platforms of South West England into four altitude ranges, in terms of cliff-notch heights related to OD: 18–20 m, 6–9.5 m, 3.7–5.5 m, and 0 to −6 m. He referred (1971) to the great confusion over use of the term '25-foot' platform, 'applied with little discrimination to surfaces from sea level to well above 10.7 m (35 feet)', and arising from a failure to recognise the existence of more than one feature.

Since this term has been applied to features at different altitudes or of different ages at different places, any connotations of contemporaneity or equivalent altitude are illusory, so the term is meaningless and should be abandoned.

Lower-level platforms occur both singly and in staircases of up to three members at several locations between Plymouth Sound and Hope's Nose (Torquay), at elevations of about 7.4, 4.3 and 0 m OD (Orme, 1960). Among the best locations for examining these features, together with the fossil cliffs backing them and the head deposits draped over them, is the coast between Prawle Point and Start Point and at Rickham Sand (SX 753368). They are also well developed in Barnstaple Bay and Croyde Bay, their upper limits having been measured (Kidson, 1971) at 8–8.5 m, 5–5.5 m and −1.5 m OD at Westward Ho! (SS 423292) and Saunton (SS 438379 and 432385), and lie in the ranges 10.7–15 m, 5.5–7.6 m and 0–6 m OD between Saunton and Baggy Point (Stephens, 1970). A platform in Lee Bay (SS 693492) has an altitude of 7 m OD (Dalzell and Durrance, 1980). Although the lowest platforms are within the present intertidal zone, and some of the higher ones are affected by storm waves and spray, the antiquity of them all is demonstrated by the presence on them at numerous locations of one or more of the following: erratic blocks, raised beach deposits, cemented blown sands, head deposits. The significance of these relationships for dating and the Quaternary sequence is discussed later.

Raised beach deposits are mostly scattered in protected locations on the landward parts of shore platforms. The main south coast locations are at Hope's Nose (SX 950637), Thatcher Rock (SX 944628), Churston Cove (SX 919570) and Shoalstone (SX 936568) around Torbay, between Hallsands and Start Point, and between Start Point and Prawle Point. In all these cases the beach deposits lie on platforms between about 11 m and 4 m OD, whilst at Plymouth Hoe (SX 477536) they are on platforms at about 17–18 and 6.5 m OD (Masson-Phillips in Zeuner, 1959; Orme, 1960). In the north, raised beach deposits are preserved at Lee Bay (on the platform noted above) and at Woolacombe (SS 454442), between Pencil Rock and Middleborough in Croyde Bay (*Figure 11.6*), along the foot of Saunton Down promontory, and at Westward Ho! (SS 423292), resting on platforms at altitudes between about 4 m and 14 m OD. The deposits extend up to about 20 m OD in places, and have been described by several authors (such as Stephens, 1966, 1970; Stephens and Synge, 1966; Edmonds, 1972a). The deposits on both coasts are variable in character, both between sites and even in the same exposure, from coarse rounded pebbles and cobbles to fine sand, with or without comminuted shell fragments. Chalk flint occurs in both raised and modern beaches. The finer materials are commonly cemented, especially in Croyde Bay and at Saunton where the beach materials are overlain by considerable thicknesses of aeolian sands ('dunerock') which are bound by a calcareous cement, and together with the cemented beach sands comprise the so-called 'sandrock'. Detailed sedimentological studies of these deposits by Greenwood (1972) has enabled the distinction of the beach and aeolian components of the sandrock, which probably represent a regressional sequence, the blown sands having been derived from intertidal sands stranded by marine regression.

The fossiliferous nature of the Tor Bay raised beach deposits is well known, the faunal assemblage at Hope's Nose including 17 species of marine mollusca, whilst 43 species have been identified at Thatcher Rock. Commonest species at Hope's Nose are *Ostrea edulis, Mytilus edulis, Patella vulgata* and *Cardium edule*, the latter being common also at Thatcher Rock, though out-numbered by *Cardium echinatum*. The climatic connotation of the fauna at both sites is temperate, though slightly cooler than the present day at Thatcher Rock. At Saunton and Croyde, 25 species of marine mollusca have been identified, the most abundant being *Purpura lapillus, Littorina littorea, Patella vulgata, Cardium edule, Mytilus edulis, Mya truncata* and *Ostrea edulis*. Again, a temperate climate is indicated, at least as warm as the present day. The cemented blown sands contain terrestrial gastropods.

In most cases the raised beach deposits and any dunerock are overlain by head, and at Sharpers Cove (SX 786357) and Gorah Run (SX 791364) the basal layers of head are interbedded with beach material (Mottershead, 1971), suggesting that either solifluction or renewed mass movement of previously deposited head was under way during beach deposition in those localities.

In view of the scattered nature of their occurrence and the fairly wide altitude range they cover (a precise statement of which is prevented by the shortcomings of the available height information), it is not possible to state how many different former sea-level stands are represented by the raised beaches in Devon, and the stratigraphic relations do not allow an undisputed answer.

Intertidal and terrestrial phenomena below present sea-level

Evidence of past relative sea levels lower than the present level includes buried and submerged shore platforms and their backing cliffs, buried and submerged river channels, and buried and submerged intertidal and terrestrial sediments and organic materials.

Buried and/or Submerged Shore Platforms and Cliffs. These have been ident-ified by geophysical, boring and diving techniques. An extensive buried rock platform occurs at about −10 m OD under Saunton Sands (MacFarlane, 1955), and a broad off-shore rock shelf at between −9 m and −15 m OD has been reported in Start Bay. Off Plymouth, a submerged 10 m high cliff occurs with its base at about −42 m to −43 m OD, and fronted by a gently sloping smooth platform. A similar feature occurs at the same altitude in Start Bay (Hails, 1975; Kelland, 1975), and an ancient shoreline has been postulated to be present in Tor Bay by Clarke (1970), who also identified a buried cliffline with a base at about −48 m OD off Berry Head. Donovan and Stride (1975) claimed on the basis of detailed Admiralty surveys that three degraded cliffs occur around Devon and Cornwall with bases at −38 m to −49 m, −48 m to −59 m and −58 m to −69 m OD respectively, to which Wood (1976) added platforms at −18 m and −26 m OD, though these authors ascribed to them a late Tertiary age rather than Quaternary.

Buried and/or Submerged River Channels. These have been discovered beneath the estuaries of the Taw–Torridge, Tamar, Erme, Dart, Teign and Exe rivers. The buried channel of the Taw–Torridge attains a measured alti-tude of −24 m OD near its mouth, although its deepest part is estimated to reach a level of −31 m, and it has a gradient of about 1 in 600 over a distance of 8 km (Durrance, 1974). The rock floor of the Tamar submerged channel descends to lower than −40 m OD in places where tidal and/or fluvial scour have kept it clear of sediment. The buried channel of the Erme, which reaches −27 m OD at its mouth, has an average gradient of about 1 in 140 over a distance of 3.2 km, whilst rockhead levels of -36 m and -43 m OD have been recorded in the buried channel of the Dart, which has an average gradient of about 1 in 350 over a distance of 10 km. The most detailed knowledge of buried channels and related buried terraces in Devon has un-doubtedly been obtained in the Teign and Exe Estuaries, using a combination of seismic and borehole evidence (Durrance, 1969, 1971, 1974). The floor of the buried channel of the Teign Estuary drops in level from -7.8 m OD at Hackney to -20.5 m OD at Teignmouth, at an average gradient over the 6 km of about 1 in 470. The Exe has two sets of buried channels: an older rock-cut, gravel-filled feature descending to below -50 m OD and a younger set cut in both bedrock and gravel, partly re-excavating the gravel fill of the older channel, to a level of about -30 m OD. Both sets of channels contain buried terraces. The gradient of the younger buried channel over the 10 km between Topsham and Dawlish Warren is about 1 in 540. According to Clarke (1970) the (younger) Exe buried channel continues southwards off Torbay to a level of at least -46 m OD after receiving the Teign buried channel as a tributary (*Figure 11.12*).

The ages of the buried and/or submerged features discussed so far cannot be established unequivocally, and hypotheses regarding their ages are best dealt with later in the wider context of the Quaternary sequence.

Buried and/or Submerged Intertidal and Terrestrial Deposits. Organic material contained within these may allow direct assessment of age to be made. The best-known organic remains in this category are the 'submerged

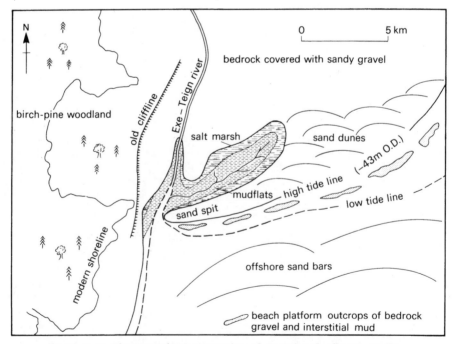

Figure 11.12. Diagrammatic reconstruction of coastal and offshore conditions, south Devon, about 9,500 years B.P. (after Clarke, 1970)

forests' that crop out both in and beyond the present intertidal zone at many locations around South West England. In some cases these beds are being removed by marine erosion, and in others they are normally covered by intertidal deposits, being occasionally exposed at low tide after periods of stormy weather. They include peat, leaf-litter, roots, seeds and tree stumps, the latter commonly in the position of growth and rooted in thin soils on either bedrock or earlier Quaternary deposits. The main Devon occurrences are at Saunton and Westward Ho! in Barnstaple Bay, in Tor Bay and in Start Bay.

The most frequently studied intertidal site is that at Westward Ho! (Churchill and Wymer, 1965). Here a kitchen midden containing Mesolithic flints and many intertidal mollusca rests on pollen-bearing estuarine clay at an altitude of -2.4 m OD, and is overlain by an oak fen-wood peat, 0.3 m thick, from which seeds of saltmarsh plants have been recorded. A thin seam of clay rests on the midden in places. The peat has yielded a radiocarbon date of about 6,500 BP (Q-672, Table 11.2). Taken together, this evidence suggests a Mesolithic occupation site in existence at the high water mark of spring tides about 6,500 radiocarbon years ago, and the fact that the high tide level was then at -2.4 m OD compared to its present level of about 4 m OD implies a net rise of relative sea level of about 6 m over the last 6,500 years, assuming that the tidal range has not varied significantly over this period. In 1970 severe storms exposed thicker (0.6 m) peat at a higher level (about 1 m OD). The peat contained tree roots in situ, wood from which was dated as about 5,000 BP (St 3402, Table 11.2).

The pollen-bearing estuarine clay beneath the peat at Westward Ho! is underlain by a blue clay that contains no foraminifera, diatoms, mollusca, pollen or seeds, and this in turn rests on beach cobbles that are underlain by head (Stephens, 1970); the latter two deposits are cryoturbated, with upturned stones in the beach deposit and polygonal arrangement of the stones in the head. There are thus two ancient beach deposits at Westward Ho!: the one

Table 11.2 RADIOCARBON DATES RELEVANT TO FLANDRIAN SEA-LEVEL CHANGES IN DEVON

Location	National Grid Reference	Material	Altitude (m OD)	^{14}C date	Laboratory Number	Reference (Radiocarbon)		
						Volume	Year	Page
Beesands	SX 821410	Peat	−4.3	4300 ± 50	SRR-164	21,	1979,	214
Beesands	SX 821410	Peat	−4.6	4770 ± 50	SRR-165	21,	1979,	214
Blackpool	SX 855478	Wood	c.0.0	2540 ± 70	SRR-318	21,	1979,	215
Hallsands 70VC (offshore)	SX 825390	Peat and wood	c.−18.2	8110 ± 60	SRR-237	21,	1979,	214
Newton Abbot	SX 849730	Wood	−1.0	3332 ± 70	SRR-163	16,	1974,	249
North Hallsands	SX 818389	Peat	c.−1.0	1680 ± 50	SRR-317	21,	1979,	214
Slapton Ley	SX 823439	Peat	−0.2	1810 ± 40	SRR-492	21,	1979,	226
Slapton Ley	SX 823439	Peat	−1.8	2890 ± 50	SRR-493	21,	1979,	226
Teignmouth (offshore)	SX 996705	Peat	−23.8	8580 $^{+830}_{-755}$	NPL-86	8,	1966,	341
Topsham	SX 962883	Wood	−3.0	3300 ± 120	Birm-533	17,	1975,	261-2
Topsham	SX 962883	Wood	−3.6	3910 ± 130	Birm-534	17,	1975,	261-2
Topsham	SX 962883	Wood	−4.1	3990 ± 120	Birm-600	18,	1976,	250
Westward Ho!	SS 433295	Peat	−2.1	6585 ± 130	Q-672	6,	1964,	126-7
Westward Ho!	SS 432296	Wood	c.+1.1	4995 ± 105	St-3402	14,	1972,	331

just described, which is exposed periodically on the present foreshore, and has undergone frost disturbance in a cold climate and been buried beneath later sediments, including the Flandrian 'submerged forest'; and another, the raised beach referred to earlier, which rests on a rock platform at about 8–9 m OD and is overlain by head. These two old beaches have figured in attempts to unravel the Quaternary sequence, as is shown later.

In Tor Bay, Clarke's (1970) sedimentological, foraminiferal and pollen studies enabled him to recognise non-marine, intertidal and inshore-marine deposits in cores from the present sea bed. Although the scope of the pollen work was sufficiently limited for Clarke himself to urge caution in its use, he compared his results with radiocarbon-dated Flandrian pollen zones for the purpose of dating approximately the progress of the landward migration of the shoreline during that part of the Flandrian transgression during which relative sea level rose from −43 m to −17 m OD. Clarke's tentative diagrammatic reconstruction of conditions prevailing about 9,500 radiocarbon years ago, when he estimated mean sea level to have been about −43 m OD, is shown in *Figure 11.12*, illustrating various types of off-shore and coastal depositional environments, the approximate positions of the high and low tide lines of that time, and the present shoreline. According to Clarke's estimates, the rise of sea level between about 9,000 and 7,000 BP was about 1.5 m per century, and involved the landward migration of the shoreline by an average distance of about 7.6 m per year. To any inhabitants of those shores it must have been a fact of life that the sea advanced upon low-lying, probably fertile land, and a folk memory of the 'Lost Kingdom of Lyonesse' may have come down to us today from just such a cause.

The progress of the Flandrian transgression was marked in Start Bay by the steady landward migration of a barrier-estuarine-lagoon complex during the past 8,000 years (Hails, 1975; Morey, 1976), at a rate which was rapid between 8,000 and 5,000 BP, but which slowed down greatly after that. There was greater coastal stability after about 3,000 BP, and the present freshwater Slapton Ley was established behind the shingle ridge by about 1,000 years ago, and has been free from significant marine incursion since then. Six radiocarbon dates on freshwater peats and wood that were buried by marine deposits or submerged were used by Morey, together with Clarke's data, to construct a sea-level curve. Unfortunately, the precise relationship of each peat or wood sample to its contemporaneous sea level is unknown, so Morey's curve merely records the positions of the dated samples in relation to Ordnance Datum. They are listed in *Table 11.2*, and are plotted on *Figure 11.13* in relation to local mean high-water mark of spring tides (expressed as 'sea level'), on the assumption that the vegetation must have been growing no lower than the high water mark of the time. Also plotted is the curve produced for the Somerset Levels by Kidson and Heyworth (1978), allowance having been made for the tidal differences between the two areas. No allowance can be made, however, for compaction of the sediments in the case of the Devon data, which may partly explain the apparent discrepancies between the Devon and Somerset evidence. Kidson and Heyworth used evidence from pollen analysis, macroscopic plant remains, diatoms and foraminifera, together with 65 radiocarbon dates, in a study that is far more detailed than any work carried out in Devon. They allowed for compaction of the sediments, and claim to have established the relationship of their dated samples to contemporaneous sea level. Their Somerset Levels curve agrees very closely with an even more detailed one for Cardigan Bay (Kidson and Heyworth, 1978), suggesting a lack of local or regional tectonic movements during the period in question. In other words, their curve is thought to be eustatic, representing world-wide sea-level fluctuations.

A significant feature of the Somerset Levels curve is that it has been constructed as a smooth curve, on the assumption that the oscillations suggested by the data are the result of minor variations in the relative rates of sedimentation and sea-level rise. This contrasts with results from northwest England and Southern Sweden (Mörner, 1969), which suggest a markedly oscillatory eustatic sea-level rise during the Flandrian (*Figure 11.13*) reflecting the known oscillatory behaviour of world climate and of glacier volumes during the same period. It may also be at variance with evidence from Topsham, where a temporary exposure (SX 962883) in the alluvial flat bordering the Exe Estuary revealed a waterlain deposit suggested by its sedimentary properties, pollen analysis and the included beetle fauna to be of freshwater fluvial origin. Three pieces of driftwood within the deposit, covering an altitude range of −3.0 m to −4.1 m OD, yielded radiocarbon dates ranging from about 3,300 to about 4,000 BP (*Table 11.2*). This evidence suggests that local high water mark about 4,000 radiocarbon years ago was no higher than −4 m OD, and that

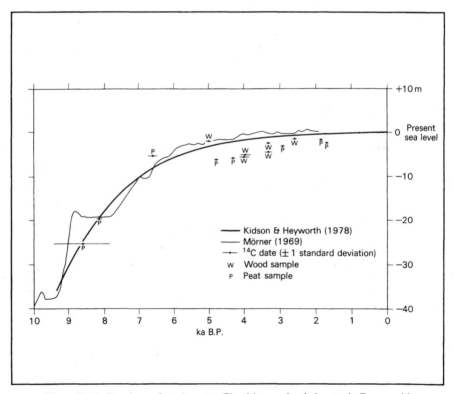

Figure 11.13. Dated samples relevant to Flandrian sea-level changes in Devon, with eustatic curves from Somerset and south Sweden

about 3,300 radiocarbon years ago it was no higher than -3 m OD (the present high water of spring tides at Topsham being 2.25 m OD), implying a significantly lower mean sea level at those dates than is suggested by the Somerset Levels curve (*Figure 11.13*).

Finally, driftwood incorporated in fluvial gravels at East Golds Pit, Newton Abbot, has been dated at about 3,300 BP (SRR-163, *Table 11.2*). The wood was at an altitude of -1.0 m OD, and was about one metre below the contact with overlying estuarine muds (Gouldstone, 1975). The inference is that local high-water mark was below -1 m OD about 3,300 years ago.

THE NON-MARINE FOSSIL RECORD

Flora

Studies of the Quaternary non-marine flora in Devon have been limited in number and, apart from one study of material from the 'Honiton Hippopotamus Site' (discussed later), have been confined to deposits of Flandrian age. Some of these studies have been from the buried and/or submerged deposits already discussed, and the only published work that remains to be considered is that of Simmons (1964) on Dartmoor. Fortunately, the Late Devensian and Flandrian vegetation history of Bodmin Moor has been relatively well studied and dated (Brown, 1977), allowing inferences to be made concerning conditions across the Tamar.

A generalised pollen diagram for the Flandrian vegetation history of Dartmoor (Simmons, 1964) is reproduced as *Figure 11.14A*, in which the timescale is approximate and based on analogy with dates of pollen zone boundaries in other areas, and not on local radiocarbon dates. The prehistoric vegetation history was summarised by Simmons in terms of four phases (*Figure 11.14A*), the first of which (A) was a period dominated by open heath, with sedge swamps and willow and birch thickets in suitable hollows. The response by the vegetation to the markedly improving climate gave rise in phase B to the immigration of deciduous trees led by hazel, followed by oak and elm, until the upland was mostly or possibly entirely covered by forest (the position of the then treeline having not been unequivocally determined). In phase C came the beginnings of disforestation at high altitudes, and by the end of this phase blanket bog was well established. The relative responsibility of climatic change and man for this early disforestation is not clear, but Simmons interpreted small scale fluctuations in the pollen record associated with the presence of charcoal at one site as resulting from Mesolithic alterations of woodland fringe areas. Recent pollen work in the Okement Valley in northern Dartmoor (Caseldine and Maguire, 1981) has shown similar fluctuations in the pollen record that have been radiocarbon dated to 7,360 $\pm$ 65 BP. There seems little doubt, therefore, that Mesolithic man was having an effect on the forest cover possibly as early as the eighth millennium BP, though there is no doubt that Neolithic man had a much greater impact, towards the end of phase C. After the sporadic clearances followed by regeneration that characterised the Neolithic, phase D saw more continuous clearance, with little or no regeneration. This clearance gathered pace through succeeding cultural periods to produce the largely man-induced vegetation cover of today.

The Bodmin Moor work (*Figure 11.14B*) is useful in the present context in that it extends our knowledge of vegetation history back into the last few

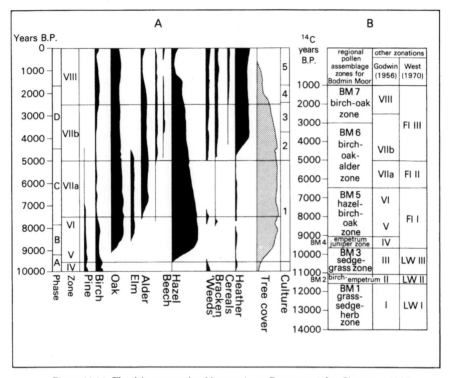

Figure 11.14. Flandrian vegetation history: A, on Dartmoor (after Simmons, 1964);
B, on Bodmin Moor (after Brown, 1977)

thousand years of Late Devensian time (the Lateglacial), and has provided radiocarbon dates over the period from 13,088 ± 300 to 6,451 ± 65 BP that allow much greater chronological precision than is available from the Dartmoor evidence. According to Brown's detailed pollen work (1977), limnic sediments began to accumulate shortly before about 13,000 BP, when cold climatic conditions are indicated by the dominant vegetation of open grass heaths, snow beds and flushes. Amelioration of climate then allowed juniper scrub to invade, and the tree birches expanded their distribution about 12,000 BP, the birch pollen maximum occurring towards the end of the Lateglacial Interstadial at between 11,553 ± 280 and 11,069 ± 220 BP. About the latter date a marked deterioration of climate set in, giving rise to the solifluction of upland soils and the development of grass sedge mires. This was the last cold phase of the Devensian, known as the Loch Lomond (or Younger Dryas) Stadial. The Hawks Tor Lateglacial site (SX 147744) has considerable importance in the late Quaternary history of South West England because solifluction deposits there rest upon organic Lateglacial Interstadial deposits, proving that solifluction was active during the Loch Lomond Stadial. Unfortunately, an unconformity at the base of the Flandrian organic deposits has prevented a significant carbon-14 date being obtained for the Devensian/Flandrian boundary on Bodmin Moor, the oldest Flandrian date obtained being 9,654 ± 190 BP. In the early Flandrian, the spread of tree birches and

willow in the valleys, with *Empetrum* and juniper on the hillsides, was followed by the replacement of the latter two genera before 9,000 BP by hazel,
with oak spreading in very soon afterwards. According to Brown, oak, birch
and hazel, although the dominant woodland genera, probably colonised only
the more sheltered upland sites.

Fauna

Three circumstances have conspired to make Devon one of the most important counties in Pleistocene faunal studies. Firstly, Devon is located cen-
trally in the zone across which Pleistocene mammalian species migrated in
response to the climatic vicissitudes, with northern species such as the woolly
mammoth, woolly rhinoceros and reindeer móving southwards into Devon
and the continent during the glacial periods, and with warm species such as
the straight-tusked elephant, narrow-nosed rhinoceros and hippopotamus
spreading northwards into and beyond Devon during the interglacials.
Secondly, the presence of caves in the Devonian limestones to the south and
southeast of Dartmoor provided suitable locations for the accumulation of
mammalian remains, whether by animals falling down open shafts, or by
animals (including Man) using caves as refuges in which they have left their
own remains and those of their prey, or by deposition of remains in cave
stream deposits. Furthermore, the alkaline conditions in the limestone caves
have favoured the preservation of even the smallest bones. Thirdly, Devon
was the scene of much exploratory work in the nineteenth century, partly
because of the circumstances just noted, but especially because of the presence
of enthusiastic pioneers in the study of cave deposits. The first scientific study
of a bone cave in Britain was by J. Whidbey at Oreston, near Plymouth, in
1816, and a whole succession of investigators has worked on the cave faunas
right through to the present day (Sutcliffe, 1969; in Straw, 1974).

The locations of the caves and other sites where Pleistocene mammalian
remains have been found in Devon are shown in *Figure 11.2*. The probable
stratigraphic range of the deposits in the principal bone caves is summarised,
together with the significant mammalian species, in *Figure 11.15*, from which
it is clear that the majority of the bone deposits are probably of Devensian
age. Only one cave (Kent's Cavern) contains evidence of fossiliferous deposits
of pre-Wolstonian age. A possible reason for this lack of Middle Pleistocene
or older fossiliferous cave deposits, apart from the general one of destruction
by denudation, is that the caves were probably below the water table for
much of the Lower Pleistocene at least, since they all occur at elevations
below 100 m OD.

The three most important caves in terms of the significance of their deposits
for elucidating the Quaternary sequence are Kent's Cavern, Tornewton Cave
and Joint Mitnor Cave.

Kent's Cavern (SX 93456415) has deposits covering the longest timespan,
although they pose serious chronological problems because the earliest deposits (the 'Breccia') appear to be much older than the next younger fossiliferous
layer (the 'Cave Earth'). The 'Breccia' contains Lower Palaeolithic hand axes,
suggesting a pre-Wolstonian age, whilst the fauna of the 'Cave Earth' and
'Black Band', which includes woolly mammoth, woolly rhinoceros and reindeer, is associated with an Upper Palaeolithic industry consistent with a

STAGE		KENT'S CAVERN	TORNEWTON CAVE	JOINT MITNOR CAVE	OTHER DEVON CAVES	SIGNIFICANT MAMMALIAN SPECIES
	HOLOCENE	'Black Mould' — — — — — — — — 'Granular Stalagmite' — — — — — — — —	'Diluvium'		? NEALE'S CAVE ? HAPPAWAY CAVE	Domestic animals Wolf Brown bear ? Lynx
UPPER PLEISTOCENE	DEVENSIAN GLACIATION	'Black Band' — — — — — — — — 'Cave Earth'	'Reindeer Stratum'		ANSTEY'S COVE CAVE ASH HOLE BENCH CAVE BRIXHAM CAVE CATTEDOWN CAVE COW CAVE LEMONFORD CAVE LEVATON CAVE ORESTON CAVES TORCOURT CAVE YEALM BRIDGE CAVE	Woolly mammoth Woolly rhinoceros Reindeer Horse Brown bear Spotted hyaena Narrow-skulled vole, *Microtus gregalis*
	IPSWICHIAN INTERGLACIAL	?'Crystalline Stalagmite'	'Hyaena Stratum'	Bone Deposit	EASTERN TORRS QUARRY CAVE	Straight-tusked elephant Narrow-nosed rhinoceros Hippopotamus Red deer Fallow deer Brown bear Spotted hyaena
	WOLSTONIAN GLACIATION		'Glutton Stratum'		? Lower deposits of COW CAVE	Woolly rhinoceros Reindeer Brown bear Glutton (wolverine) Hamsters, *Cricetus cricetus* & cf. *Allocricetus bursae* Steppe lemming, *Lagurus lagurus* Snow vole, *Microtus nivalis*
MIDDLE-LOWER PLEISTOCENE	HOXNIAN ANGLIAN CROMERIAN	? Breccia				Sabre-toothed cat Cave bear Vole, *Pitymys gregaloides*

Figure 11.15. Probable stratigraphic range of deposits in principal Devon bone caves (after Sutcliffe, in Straw, 1974)

Devensian age. There is thus a major hiatus in the fossil record, covering much of the Wolstonian and Ipswichian, during which the 'Crystalline Stalagmite' accumulated.

Tornewton Cave (SX 818673) was used successively as a bear lair, a hyaena lair, and a human occupation site. It not only has the most complete Upper Pleistocene sequence in any British cave, but is also the only known locality to have presumed Wolstonian ('Glutton Stratum') and Devensian ('Reindeer Stratum') deposits separated by an interglacial deposit ('Hyaena Stratum') containing hippopotamus remains. Furthermore, study of the abundant rodent faunas in the cold phase deposits has shown them to be markedly different, which has wider stratigraphic implications because the cold phase faunas are otherwise indistinguishable, and would have been confused had each occurred at a different site. Comparison of the Ipswichian fauna of the 'Hyaena Stratum' with Ipswichian faunas at sites on the Thames terraces has led Sutcliffe (1976) to the view that there are too many distinct mammalian assemblages to include in one interglacial, and that there are at least two distinct warm phases included under the name Ipswichian. This view would have very important implications for our understanding of the Quaternary sequence, but alternatively the faunal differences between the sites may be explained by accepting a change in fauna through a single Ipswichian inter-glacial which paralleled the postulated vegetational change (Mayhew, 1976).

Although the remains of animals that fell down an open shaft into *Joint Mitnor Cave* (SX 743644) are confined to two relatively thin layers, they constitute the richest Ipswichian mammalian assemblage from any British cave, with at least 18 species represented, all of interglacial character. In addition to the species listed in *Figure 11.15*, the fauna includes wolf, fox, wild cat, cave lion, badger, wild boar, giant deer, bison, hare, water vole and field vole.

A major frustration for Quaternary stratigraphers and chronologists working in Devon is that there is no direct link between the fossiliferous cave deposits and landforms and deposits outside the caves. The closest approach to such a link is the relationship of the cave system of which Joint Mitnor Cave is a part to the terraces of the River Dart: the caves occur in the limestone underneath a terrace flat whose altitude is about 83 m OD, and must have been below the water table when the Dart flowed at that level. The terrace is therefore presumably pre-Ipswichian in age, but one cannot go further in view of the doubts expressed earlier about the likely validity of existing interpretations of the Dart terrace sequence.

Two of the mammalian sites in *Figure 11.2* are not cave sites. The *Honiton Hippopotamus Site* (ST 162006) was discovered in 1965 when road construction sectioned a depression in the Triassic bedrock, revealing deposits rich in mammal bones and peaty organic material, overlain by head. There were indications that the bones were initially embedded in a peat horizon and then transported, possibly as part of a mudflow. The mammal remains include at least 17 different individuals of hippotamus, together with remains of straight-tusked elephant, giant ox and red deer (Turner, 1975). The faunal assemblage suggests an Ipswichian Interglacial age, but pollen analysis from the bone deposit showed sparse tree pollen, the percentages of herbaceous types being much higher than are usually found in interglacial deposits. This apparent anomaly, which has also been encountered at other Ipswichian sites that have yielded rich faunas of large herbivorous mammals, may be partly explained in terms of concentrations of these mammals causing local extension of herbaceous vegetation by their grazing, browsing and trampling activities; and partly in terms of extreme over-representation of herbaceous types in the pollen record caused by pollen deposition in hippopotamus dung.

The *Barnstaple Elephant Site* (SS 562330) was discovered in 1844 in a brick-clay pit, the site of which is now a built-up area whose surface elevation is about 10–11 m OD (Arber 1977). The remains, found under 4.5 m of clay, include the teeth of two straight-tusked elephants together with portions of elephant's tusk and a vertebra. The presence of elephant suggests an interglacial age, and the site was mapped by Edmonds (1972a) as part of a terrace attributed to the Ipswichian Interglacial.

THE QUATERNARY SEQUENCE

Attempts to reconstruct the sequence of events in the Quaternary of Devon from the evidence summarised above have been fraught with difficulties and beset (and stimulated) by controversy. Among the major difficulties are the fragmentary nature of the evidence and the lack of reliably dated and widely and unambiguously identifiable marker horizons. Such shortcomings in the quality of the evidence are causes of complaint amongst Quaternary scientists working in many areas, but the problems are particularly acute in an area that lay beyond the limits of the last glaciation, and almost all of which lay beyond the limit of maximum glaciation, and whose relief and topography have not been favourable to the widespread development of stratigraphic continuity. Controversy has in some cases been fuelled by fundamental disagreements about the very nature of the evidence, let alone its significance. This is particularly the case in north Devon, which has a few more pieces of the Quaternary jigsaw puzzle than other parts of the county, and which will therefore be considered first.

Several coastal cliff exposures around Barnstaple Bay and in the Taw estuary show raised beach deposits overlain by head, and resting on one or more shore platforms. In some cases cemented blown sand (dunerock) intervenes between the raised beach deposits and the head, and in places both raised beach and blown sand contain scattered angular stones derived from the cliffs behind as the deposits accumulated. Near Saunton and Croyde giant erratics rest on shore platforms, one at Saunton being trapped between platform and raised beach material, and the large one at Croyde between platform and head. The head deposits are variable in composition, as noted earlier, and it is sometimes apparent that different phases of soliflual movement are represented in the head stratigraphy.

The succession as just described is both simple and widely accepted, and may be summarised thus:

Head deposits
Blown sand
Raised beach
Giant erratics
Shore platforms

There are major disagreements about how the Fremington till relates to this coastal cliff sequence, and about how many Pleistocene stages are represented. The two main rival views of the Quaternary sequence in north Devon are summarised in *Table 11.3*. Both schemes agree about the Wolstonian age of the till, even though there is no unequivocal evidence that it is of this age rather than of some earlier date, such as Anglian. Beyond that, the two schemes differ with regard to virtually everything else.

According to Stephens (1966, 1970; in Straw, 1974) the main raised beach and associated blown sand, whose fossil contents indicate a temperate (interglacial) fauna, is of Hoxnian age, a belief based largely on three lines of evidence: the alleged marine origin of the gravel beneath the till at Freming-

Table 11.3 RIVAL INTERPRETATIONS OF THE QUATERNARY SEQUENCE
IN NORTH DEVON

	Stephens (and Mitchell)	Kidson (and Zeuner)
Devensian	Upper cryoturbated head	Head
Ipswichian	Weathering horizon, and lower raised beach at Westward Ho!	Raised beach
Wolstonian	Tills and Lower Head (incl. Fremington Clay)	Fremington till and Giant erratics
Hoxnian	Main raised beach	
Anglian	Giant erratics	
Earlier Pleistocene	Shore platforms	Shore platforms

ton; the alleged stratigraphic continuity of that gravel with undoubted raised beach gravel exposed at the coast; and the alleged presence of till above raised beach materials in some coastal exposures. The only Ipswichian beach material according to Stephens is that exposed on the present foreshore at Westward Ho! Kidson and Wood (1974), however, demonstrated by using borehole, geophysical and sedimentological techniques that the sub-till gravel is unlikely to be marine, and that it has no stratigraphic continuity with the gravels at the coast. They also regarded the alleged tills in the coastal exposures as components of the head. Acceptance of these latter views tends to undermine the more complex scheme, because it removes the need to regard the raised beach(es) as older than Ipswichian, and to assign the giant erratics and the till to different glacial periods. There are then no compelling reasons for recognising two separate glacial periods within the head deposits that lie over the raised beach materials, especially given the climatic vicissitudes known to have characterised the Devensian. Similar reasoning applies to the head deposits along the south coast and inland, where there have also been disagreements as to how many soliflual phases were involved, and in how many cold periods. Evidence of significant solifluction even at low altitudes in the Devensian is provided by the head over the Ipswichian Honiton Hippopotamus deposits, and at higher levels the occurrence of Lateglacial Interglacial deposits between heads on Bodmin Moor proves that more than one soliflual phase occurred during the Devensian, the last one in the very last (Loch Lomond) stadial period.

There are no reasonable grounds for believing that each major stratigraphic element has to be allocated to a different stage of the Quaternary, and conversely there is no reason why every stage must be represented everywhere, although the chances of survival of evidence must become greater towards the present day. Taking this into account, there seems little doubt that the simpler scheme of Zeuner and Kidson (*Table 11.3*) offers the more economical explanation of the available information, and does not postulate events for which there is no clear evidence, and on those grounds it is probably more sound scientifically than the more complex scheme of Mitchell and Stephens. Nevertheless, no stratigraphic element older than the Flandrian has yet been unequivocally dated. The age of the fossiliferous raised beach is beyond the range of radiocarbon dating, and Andrews *et al.* (1979) attempted to resolve the question of its age by applying amino acid racemisation studies on the limpet *Patella vulgata* as a means of correlating the Saunton raised beach with other interglacial raised beaches in South West England and Wales. Unfortunately the results do not resolve the problem, for they suggest the presence of material of two different ages in the Saunton sample, one age being greater than at any other site studied. This may indicate that the beach is the product of two distinct high (interglacial) sea levels, or that contamination has affected some valves (Andrews *et al.*, 1979).

The possibility that there were two pulses of high sea level during the Ipswichian has been suggested by the evidence of both beach deposits and mammalian remains in Minchin Hole Cave, Gower (Bowen, 1977), and Hollin (1977) has argued from the evidence of Ipswichian interglacial sites in the Thames estuary area that there was an early mean sea-level stand at about 7 m OD, followed by a middle period when sea level fell below its present level, in turn followed by a rise to above 14 m OD late in the interglacial.

Such suggestions, together with the likelihood that evidence of more than one level would survive from the fluctuating sea-level record of an interglacial, makes it easy to accept the idea that both raised beaches at Westward Ho! might date from the same interglacial, and that the altitude variations amongst the raised beach deposits in south Devon noted earlier do not necessarily signify that material from more than one interglacial is present.

The age of the shore platforms in and above the present intertidal zone remains problematical, and one can say little more than that they probably all pre-date the giant erratics. This applies to both north and south coasts, where it is clear that even the 'present' intertidal platform is merely being exhumed from beneath a cover of Pleistocene head and raised beach deposits, and trimmed. It is not even clear whether these platforms were all formed at times of interglacial or interstadial high sea level, as is often assumed (Kidson 1971; Sissons 1981), for as pointed out by Stephens and Synge (1966), both the formation of a platform near present sea level and the rafting of erratics onto it could have been accomplished during a glacial period if the edge of an ice sheet was sufficiently close to cause isostatic depression of the earth's crust. Furthermore, it has been argued that the glacial/interglacial fluctuations of world sea level through the Quaternary have been superimposed on a longer term secular downward trend, so that even the cold phase sea-level 'lows' of the early Pleistocene were at or near present sea level.

The only evidence whose chronological significance remains to be discussed occurs below present sea level—the buried and/or submerged platforms and the buried channels. It is difficult to date these erosional features except in relation to the more recent deposits that rest on or in them, and since the latter are either themselves difficult to date, or are much younger than the features they occupy, then dating becomes a matter for conjecture. A further complication is that both platforms and channels may have been re-occupied by the agencies that formed them, so that they are really composite features of multiple ages, like the 'present' intertidal platform. The well-developed platform at -42 m to -43 m OD that occurs widely around southwest England represents a significant erosional phase, which Sissons (1979) has suggested might be the same as that responsible for the Main Lateglacial Shoreline in Scotland. The latter, which is at much higher altitudes than -42 m owing to glacio-isostatic recovery, is thought to be the product of severe frost action combined with wave action during the Loch Lomond Stadial. A probable Late Devensian age for the younger buried channels of the Exe Estuary, and the buried channel of the Teign Estuary, has been inferred by Durrance (1974), who argued on the basis of their similar gradients, depths and seismic transmission velocities that the buried channels of the Erme and upper Taw-Torridge Estuaries are of similar age. A Late Devensian age was also suggested for the younger Dart buried channel, but the older and deeper Dart and Exe channels were regarded by Durrance as of Early Devensian age.

It is clear from the foregoing discussion and the summary of evidence that preceded it that the Quaternary sequence in Devon can be only very sketchily reconstructed from the evidence so far available, and that there are fundamental ambiguities or shortcomings in almost every type of information available. As Stephens (1970) has observed, 'Argument and counter-argument... will continue, but for the moment much remains unproven'.

LOCALITIES

11.1. *Westward Ho!* There are two distinct sites: the cliff section and the foreshore.

   The *cliff section* extending from near Rock Nose (SS 419290) to east of the holiday camp (SS 424291): this shows a raised shore platform at an altitude of 8–9 m OD overlain by a raised beach deposit, which in turn is overlain by head. The platform, cut in Culm sandstone, is separated by a cliff from the 'present' intertidal platform at 2–3 m OD. The raised beach deposit, 0.5–4.5 m thick, consists of Culm sandstone cobbles in a sandy matrix. It lacks erratic material and is unfossiliferous. Erosion of the raised beach deposit provides some of the material for the modern storm beach and pebble ridge. The head, 2–3 m thick, is of angular slaty material overlain in places by a stony clay. The latter has been interpreted by Stephens (1970) as a till, whereas Kidson and Wood (1974) regarded it merely as a facies variation in the head. According to Stephens the raised beach is of Hoxnian age, the coarse head and the 'till' are Wolstonian, and colluvial material over the stony clay is Devensian. According to Kidson and Wood the raised beach is Ipswichian and the overlying head Devensian.

   On the *foreshore* (SS 433295), the submerged forest and underlying deposits are occasionally exposed after the beach has been scoured by storms. The Flandrian peat rests on a Mesolithic kitchen midden which in turn rests on estuarine clay. The latter deposit is underlain by sterile blue clay, below which are cryoturbated beach cobbles underlain by cryoturbated head. The whole suite of deposits rests on the intertidal shore platform, which is therefore largely relict. The beach deposit is thought to be Ipswichian, the underlying head Wolstonian, and the cryoturbation of both beach and head is thought to have occurred in the Devensian.

11.2. *Brannam's claypit, Fremington* (SS 529317): Wolstonian till. Permission to enter this claypit should be obtained from the proprietor of Brannam's Pottery, Barnstaple. Care is needed to avoid treacherous soft ground in the clay workings. The working face shows about 10 m of glacial clays, including tills and glaciolacustrine clays, overlain by up to 2 m of head. The clay series is underlain, below the floor level of the pit, by about 15 cm of gravel resting on bedrock. The clays contain erratics of a wide range of rock types, as well as shell fragments and damaged foraminifera. The entire till series is presumed to be Wolstonian in age, though the overlying head may be Devensian.

11.3. *Saunton Down End* (SS 438379): at this site, which is approached most conveniently from the Saunton Sands car park, can be seen all of the major elements of the Pleistocene coastal stratigraphy of north Devon. Upon a shore platform cut across steeply dipping Devonian slates at about 4.5–5 m OD rests a large erratic boulder of pink gneissose granite which is overlain by raised beach material consisting of pebbles in a cemented matrix of sand and comminuted shell fragments. The raised beach deposit, up to 2 m thick, includes much erratic material, and is fossiliferous, some 25 species of marine mollusca indicative of a temperate (interglacial) climate having been identified here and at Croyde. The raised beach material passes upwards into 'sandrock', which is mainly cemented aeolian sands and comminuted shell fragments, exhibiting dune bedding. The blown sands contain shells of terrestrial gastropods in addition to marine shell fragments, and are about 9 m thick. They are overlain by about 21 m of head. According to Stephens (1966, 1970) the erratic was emplaced during the Anglian, the raised beach and blown sands are of Hoxnian age, and the head includes both Wolstonian and Devensian elements. According to Kidson (1971, 1977; in Straw, 1974) and Kidson and Wood (1974), the erratic was emplaced during the Wolstonian, the raised beach and blown sand deposits are Ipswichian, and the head is all Devensian.

11.4. *Croyde Bay to Baggy Point.* The southern margin of the headland that terminates seaward in Baggy Point bears Pleistocene deposits resting on at least three distinct shore platforms. Three particular sites are selected from the many accessible exposures: Middleborough, Freshwater Gut and Pencil Rock. All can be covered in one walk from the car park near the hotel at Middleborough.

Middleborough (SS 429398). An exposure in cryoturbated head shows a 0.9 m thick layer of coarse angular slate fragments underlain and overlain by finer material consisting of smaller slate fragments in a sandy matrix. Fossil ice wedges occur in, and are confined to, the coarse layer.

Freshwater Gut (SS 428400). A large erratic boulder of granulite gneiss rests near the inner limit of a shore platform at 5.5–7.6 m OD. It is overlain by head, which has been largely removed from around the boulder by marine erosion. This is a good site for studying the morphology of the old shore platforms and their recent modification by the sea.

Pencil Rock (SS 423402). A 1–1.5 m thick deposit of well-rounded raised beach shingle in a matrix of coarse sand, marine shells and shell fragments rests on a rock platform at about 13.5 m OD, and is overlain by sandrock, which reaches to well over 30 m OD. The lowest few metres of the sandrock are marine and the rest aeolian. Angular rock fragments, derived from cliff falls, are scattered through the sandrock, which otherwise consists largely of cemented sand and shell fragments. The multiple shore platform levels are well seen at and from this locality, which is also a good place from which to view the slope forms associated with the coastal head apron around Croyde Bay.

11.5. *The Valley of Rocks*, west of Lynton (SS 705495). This impressive dry valley, the result of dismemberment by coastal cliff recession of a once more-extensive East Lyn river, has a thick fill of head, and spectacular tors and other frost-riven rock features on its sides.

11.6. *Cox Tor* (SX 531762). A staircase of at least five cryoplanation terraces is well developed on the metamorphic rocks of the Dartmoor aureole, the highest bench being backed by the tor itself, a 4 to 5 m high frost shattered scarp in metadolerite. Spreads of earth hummocks occur on all of the benches, but are best developed on the highest two. From Cox Tor, a good view is gained of the granite Staple Tors (SX 543760) to the east, and of the stripes in the clitter spreads downslope of them.

11.7. *Joint Mitnor Cave* and the *William Pengelly Cave Studies Centre,* Higher Kiln Quarry, Buckfastleigh (SX 743644). The quarry and its caves are administered as a field centre by the William Pengelly Cave Studies Trust Limited. Access to Joint Mitnor Cave and the cave studies centre can be obtained through the secretary of the Trust. The cave studies centre contains exhibits and information relating to the caves, including Joint Mitnor Cave, and the latter still contains a demonstration section through the cave sediments showing the entire stratigraphic sequence, with some of the mammalian bones left *in situ.* This cave has yielded the richest known Ipswichian Interglacial mammalian assemblage from any British cave.

11.8. *Prawle point* (SX 773350). The stretch of coast between Prawle Point and Start Point (SX 830370) exhibits well the ancient shore platforms and the apron of head deposits that extends over them, and is being stripped from their seaward edges. Raised beach deposits intervene between platform and head in places. The old cliff line over which the head is draped is also well observed from the neighbourhood of Prawle Point.

Chapter Twelve

Industrial Minerals

The absence of coal in the Carboniferous Culm Measures in Devon meant that the Industrial Revolution largely by-passed the county, despite its quite substantial mineral resources. Consequently it remains, on the whole, unspoilt by large scale industrial development. Leaving aside the important metalliferous mineral resources dealt with in Chapter Six, there are still extensive deposits of other minerals that contribute to the economy of the county. This chapter considers, therefore, the kaolinitic clays—both china clay and ball clay deposits are worked—and building materials (sand and gravel, crushed stone and building stone, and brick clay).

KAOLINITIC CLAYS

Devon is the only county in Britain that contains high quality deposits of both china clay and ball clay within its boundaries. These are used extensively both at home and abroad in a variety of applications. For example, the survival of the ceramic industry depends on a continuing supply of kaolinitic clays from Devon. It may seem surprising therefore that the county does not itself rank as one of the main areas for pottery manufacture, for although there are a number of small specialist workshops in Devon (both traditional and recently-established) the British pottery industry is firmly centred on Staffordshire. This is because of the need for coal during the dramatic expansion of the ceramic industry in the 18th century—it was cheaper to transport the clay to the coalfield rather than coal to the clay-bearing areas, as, surprisingly, larger amounts of coal were needed, and there are no commercial coal deposits in Devon. Although modern kilns now rarely use coal for firing, this historical predominance of the Staffordshire coalfield is still firmly established.

The clays occur geologically in two district groups: the china clays which result from kaolinisation of the granite in southwest Dartmoor, and the ball clays which are of sedimentary origin and occur in basins of Tertiary age at Bovey Tracey and Petrockstowe (*Figure 12.1*).

The predominant mineral in both china and ball clays is kaolinite, but other mineralogical factors produce fundamental differences between the two types and also between individual clays of each type. Before describing the occurrence of the clays it is worthwhile considering the two most important properties that are major influences on the behaviour of the clay products. These are firstly that ball clay contains larger amounts of micaceous material

291

than refined china clay, and secondly that there are differences in the crystal-linity of the kaolinite between china clay and some types of ball clay. The mica content of a clay is very significant in terms of the way in which the clay fires—in particular its vitrification characteristics—and in terms of the unfired (green) strength. To understand the effects of crystallinity, however, requires some knowledge of the structure of kaolinite.

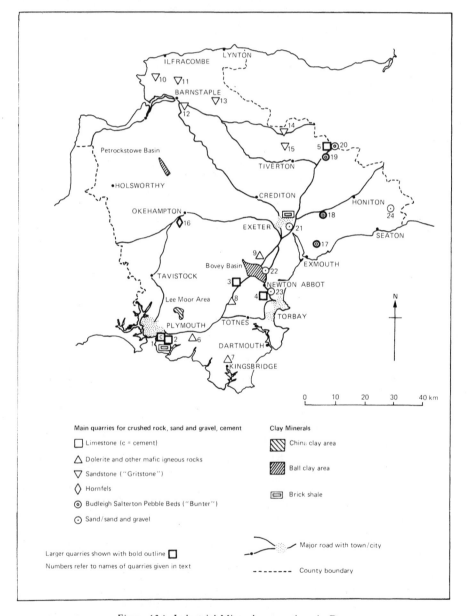

Figure 12.1. Industrial Minerals: operations in Devon

In china clays, the crystallinity of the kaolinite is generally described as well-ordered, whereas most ball clays contain a disordered kaolinite. The structure of both types is similar except that the latter is disordered principally by random layer displacement parallel to the crystallographic b-axis. On a scale from 0 to 2, crystallinity indexes derived from X-ray diffraction traces (Hinckley, 1962) for most china clays lie between 1.0 and 1.3; ball clays from the Bovey Basin range from 0.1 to 0.9 and from the Petrockstowe Basin around 0.1–0.3.

Primary kaolin deposits in Britain are all associated with granite masses. Formation of kaolinite from hydrothermal alteration of the granite after intrusion is the most common mode, though there is some evidence that weathering has also played a part in the process. During hydrothermal alteration, fluids migrated through zones of weakness in the granite and altered the feldspar to kaolinite but left the other constituents (mainly quartz and mica) unchanged. Kaolinisation processes seem to be concentrated in specific areas where fluid migration was preferentially taking place. Thus much of the Dartmoor granite is not deeply kaolinised. The amount of kaolinite present in the china clay deposits is very variable, and deposits with as little as 8 per cent of the mineral may be worked. In southwest Dartmoor, however, kaolinisation has been profound, and the percentage of kaolinite in some places is as much as 30 per cent.

Weathering processes also cause the alteration of feldspar to kaolinite, and much of the surface granite of Dartmoor does show kaolinised feldspar. But the extent to which this process has contributed to the main zones of kaolinisation is not clear. Isotopic evidence (Sheppard, 1977) suggested that cool meteoric water was an agent in the process, particularly in the formation of the St Austell china clay deposits—although this may still be related to a hydrothermal circulation system.

The kaolinite in ball clays has a diverse origin. Being sedimentary, these clays are secondary deposits and in the Bovey Basin much of the kaolinite was derived from china clay in the Dartmoor granite and from weathering of the granite exposed at the surface during Tertiary times. The Petrockstowe Basin, however, lies further from the granite outcrop and derived its poorly ordered kaolinite from the leaching of chlorites in the weathering mantle of the Carboniferous shales (Bristow, 1968). The poorly ordered clay in the Bovey Basin was probably also partly derived from a comparable source, though as the crystallinity index of the kaolinite from Dartmoor would probably not be affected by transportation, it seems that some part came also from a granite weathering mantle.

Differences in the particle size-distribution within the clays also contributes to the contrasting characteristics of china clay and ball clay. Kaolinite is comparatively less fine-grained in many refined china clays, with only about 40 per cent less than 2 micrometres (0.002 mm) in size so that the material can more truly be described as a kaolinitic clayey silt. Ball clay, on the other hand, is very fine-grained, particularly when derived from a weathering mantle. Some ball clays may have 85–90 per cent of kaolinite particles less than 2 micrometres in size, and some is so fine that 50 per cent is less than 0.5 micrometres. Settlement of such fine clays during sedimentation by normal processes is difficult to envisage, as Brownian motion would keep them in suspension. It is therefore believed that deposition took place in an acidic environment resulting from the release of humic acid during the decay of organic matter. Flocculation of the clay particles would then have been extensive, giving rise to *en masse* settlement of the flocculated clay, often with entrained wood fragments.

Because china clay and ball clay differ in these ways, characteristic properties are imparted to the various clay types. Thus china clay is very white-firing and has a natural whiteness in the raw state, but the material is essentially non-plastic and, ceramically speaking, very weak. In contrast, the fine-grained disordered kaolinite of the Petrockstowe Basin is very strong and plastic, but fires to a cream or light grey colour. In the Bovey Basin, however, the kaolinite particles have a mixture of mineralogy and particle size, including particles which have been derived from actual china clay deposits. As these particles are generally associated with very fine-grained or colloidal carbon the ball clay acquires a greater strength and plasticity than is found in china clays. The well ordered kaolinite also gives predictable properties to individual seams of the Bovey Basin ball clays and thus explains why this deposit is at present the world's most important source of clay for the ceramic industry and is, perhaps, unique.

The China Clays

Since the emplacement of the granite, Dartmoor has been subjected to some 250 Ma of erosion and denudation, and consequently some of the early kaolinisation zones which may have occurred near the roof of the granite have now been removed. On Dartmoor the main area of kaolinisation stretches from Shaugh Moor (SX 560633) through Lee Moor (SX 572629) towards the village of Cornwood (SX 605597), and the china clay is currently being worked by English Clays Lovering Pochin and Company Limited (ECLP) and Watts, Blake, Bearne and Company PLC (WBB). At Redlake (SX 640665), northwest of South Brent, workings have been defunct for many years, although traces of this small area of kaolinisation can still be found.

In Britain, the Shaugh Moor–Cornwood area of Dartmoor is second in importance only to the kaolinised parts of the St Austell granite in Cornwall in terms of the amount of china clay produced. The deposits are worked by directing high pressure jets of water onto the stope faces, washing out all loose material, including the kaolinite, quartz, mica and any fine tourmaline which is present. This flows in a stream to a sump at the lowest point in the pit, and is pumped to the refining plant. First the quartz is removed by spiral classifiers or by simple processes of sedimentation, and much of the resulting sand is sold to the building industry in the Plymouth area. Secondly, the china clay slurry passes through a series of hydrocyclones to remove the mica, which goes to waste. The china clay is then filter-pressed and dried.

Apart from the ceramic industry, an extensive market for the china clay is in the paper industry, where its raw whiteness makes it important both as a coating agent and as a filler, but for paper coating the average particle size has to be reduced, and the whiteness improved still further by bleaching. Because china clay is a fine, white, chemically inert material, a major use in industry as a whole is as a filler or extender for a variety of products. In fact approximately 90 per cent of the total china clay production of Devon is used in the paper and ceramic industries, but significant markets exist in manufacturing fields such as rubber and plastics, paint, insecticides, fertilisers and many other commodities. Production of china clay in Devon exceeds 500,000 tonnes per year, 80 per cent by ECLP and 20 per cent by WBB, and although most of ECLP production goes to the paper industry, WBB sell only 10 per cent to this market, with the majority of their production going to the ceramic industry. Of the total tonnage produced approximately half is exported, mostly through the port of Plymouth.

The Ball Clays

The ball clays of Devon occur in two areas, the Bovey Basin and the Petrockstowe Basin, both of which are associated with the Sticklepath Fault Zone, aligned across the county in a northwest–southeast direction. Although this is a wrench fault system, localised areas adjacent to faults were subjected to subsidence, thus providing sedimentary traps where clay, silt, sand and lignite collected. The geology of the Bovey and Petrockstowe Basins is described in detail in Chapter Nine, but here most attention will be given to the Bovey Basin because the post-depositional history of the Petrockstowe Basin has been very complex, creating characteristic features such as high dips, usually between about 30° and 45°, faulting (sometimes reversed), and minor folding. Unconformities, often with significant angular discordance, impose further complications on borehole interpretation. These features make exploration and exploitation of the Petrockstowe Basin rather problematical.

Apart from their genetic affinity to the Sticklepath Fault Zone, there are very few similarities between the Petrockstowe and Bovey Basins. The Bovey Basin has, on the evidence of a gravity survey, been shown to extend to a depth of approximately 1300 m (Fasham, 1971). However, because the upper strata of the Bovey Basin overlap all the earlier deposits, and to date boreholes have only penetrated to about 310 m, the strata of the lower section are not known. The upper strata consist basically of sands, clays and lignite. It is thought that the surface of the Bovey Basin some 50 Ma ago was not dissimilar to its present topography, with a level area at the foot of the steep slopes of Dartmoor. The surface conditions, however, were probably those of marsh, swamp and transient lake. Fast-flowing streams, laden with sediment and vegetable matter, flowed off the slopes of Dartmoor until their velocity was sharply reduced on reaching the comparatively level Bovey Basin. In the northwestern part of the basin an outwash fan developed, resulting in the largely sandy sediments found in this area today, but beyond this fan a turbid lake received plant remains and clay particles in suspension. Under acid conditions, set up by the decay of organic material, flocculation of the clay particles occurred with settling *en masse* producing seams of clay. Consequently these seams are often carbonaceous and may themselves contain plant remains (Vincent, 1974). The shore of the lake furthest from the inflow, that is in the southeastern part of the basin, was a backswamp environment where floating vegetation collected to form lignites. Thus, at any one time, sands, clays and lignites were being formed simultaneously in different parts of the Bovey Basin. Moreover the amount of subsidence also varied throughout the basin, with maximum subsidence occurring in different locations at different times. These sedimentary environments therefore continually migrated backwards and forwards over the area.

During the sedimentation of the Bovey Formation, the source of kaolinite changed. The oldest of the commercial ball clays was probably derived from the breakdown of slates and shales overlying the granite. These clays (the Abbrook Member) show varying degrees of disorder of kaolinite and have greater strength and plasticity than the later clays. In addition, they are comparatively free from carbonaceous material. Later, however, there was a climatic change resulting in a luxuriant vegetation cover that contributed a great increase of lignitic material to the deposits. These younger, very carbonaceous seams, comprising the Southacre Member, clearly exhibit a sand-clay-lignite relationship in which increasing quantities of sand occur towards the northwest and increasing quantities of lignite towards the southeast. The clays of the Southacre Member are carbonaceous and brown in colour, but contain resedimented china clay particles, and are the only truly white-firing ball clays so far discovered. China clay deposits, possibly from the Lustleigh valley to the northwest, probably contributed kaolinite to the Southacre Member which accounts for the high orders of crystallinity, sometimes as much as 0.9 on the Hinckley Index, exhibited by the clays.

Although the clays of the Southacre Member in the Chudleigh Knighton area (SX 845774) show good crystallinity, they are comparatively coarse-grained and exhibit the weak, non-plastic, but white-firing characteristics of china clay. Further to the south, however, the Southacre Member becomes much more carbonaceous, and the strength and plasticity of the clays are improved. Above the Southacre Member the beds of the Bovey Formation are generally sandy but they do contain some clays of medium crystallinity and medium grain size, although they have neither the white-firing attributes of the Southacre Member nor the strength and plasticity of the Abbrook Member. On the western side of the Bovey Basin, these beds overlie and largely conceal the outcrop of the earlier, commercially useful horizons. Consequently, the working of the ball clays is largely confined to the eastern outcrop. A typical example of a production area is thus found at Preston Manor (SX 857755), where the seams dip at about 12° to the west, and the white-firing carbonaceous clays of the Southacre Member are separated from the underlying plastic clays of the Abbrook Member by about 20 m of lignitic material.

In the past, when labour was cheap and overburden removal difficult, the most effective way of working the clays of the Bovey Basin was by underground methods. Today, the position is reversed: mechanisation has made the removal of large quantities of overburden economically viable, but labour has become progressively more expensive. In addition, much larger annual tonnages are now produced, which are more easily realised by opencast development, and therefore only 12.5 per cent of the ball clay from the Bovey Basin is now produced from underground mines. The total production from this basin is approximately 450,000 tonnes per year and of this 70 per cent is exported. As with china clay, two companies, WBB and ECC Ball Clays Limited are responsible for working the ball clay but, in contrast to china clay, WBB is responsible for more than 80 per cent of the production.

The name ball clay originates from the old method of working. The clay was cut into cubes about nine inches square, each one weighing about 13 kg to 15 kg which, with the normal handling processes, rapidly assumed a roughly spherical shape. Clay was sold in this form by the 'ball' and hence the name. Today, however, ball clays are extracted mechanically by dragline, ace-shovel or underground mining machine. Even the pneumatic spade which was used from 1930 until 1970 has virtually disappeared, except for limited underground use. At present, approximately 70 per cent of the total production is sold simply in shredded form, where the clay has been broken down into pieces ranging in size from 2.5–6.5 cm (1–2.5 inches). The shredded clays are then carefully blended to produce the properties required for its particular usage. The remainder is sold in dried, ground form (sometimes coated for fertiliser use), mainly bagged, but sometimes via bulk tankers for the home market. Although there are considerable markets in the fertiliser and filler industries, ball clays are dominantly used in the ceramic industry and are an important constituent in tableware, sanitaryware, wall tiles and electrical porcelain.

Of the total tonnage of ball clay produced in Devon each year, about 66 per cent is exported, mainly to European markets, but it is also sold to many other parts of the world, including the United States, India, Pakistan, East Asia and Japan. For European destinations, the clay is shipped either through the Port of Teignmouth (300,000 tonnes per year) or Bideford (15,000–20,000 tonnes per year). For more distant countries, deep water ports are necessary and, although Plymouth and Fowey are useful in this respect, most of the distant exports are containerised (in shredded or bagged form) and shipped via the ports of Liverpool, Southampton, Swansea, Newport, London, Hull and Felixstowe.

For the home market, road haulage is the main form of transportation used, the majority going direct to Stoke-on-Trent and the Staffordshire Potteries area.

CONSTRUCTION MATERIALS

Crushed Stone, Sand and Gravel

Construction and maintenance of houses, schools, hospitals, factories, roads, bridges, dams etc. in Devon requires about 5 million tonnes of aggregates each year, or the equivalent of around 5 tonnes per year for each person living in the county.

Naturally, the aggregate materials need to be as cheap as possible and, because transport makes up a substantial part of the delivered price, they are usually quarried at no great distance from the place where they have to be used. It is not usually economic, except in the case of specialised materials or where local supplies are very limited, for aggregates to be transported more than about 50 km from the quarry. Most of Devon's requirements are, therefore, obtained from within the county and only a small proportion of production, amounting to less than 20 per cent, is 'exported' beyond the county boundaries and this is mainly sent to neighbouring areas of Somerset and Cornwall. Very little aggregate is 'imported' from other counties into Devon.

The materials quarried for aggregates in Devon occur as widely varying types of deposits with widely different compositions and properties. The quality of aggregates is governed by a series of British Standard tests designed to assess the resistance of the materials to crushing, impact, abrasion, polishing etc., and also to assess size gradings and particle shape by means of British Standard sieves. Other properties with varying importance depending upon actual end-use are density, colour, insulating properties (heat, sound etc.), fire and heat resistance, resistance to chemical action, porosity and frost resistance, coefficient of expansion, flexural strength etc. Clearly there is a wide range of properties to consider and different rocks have different advantages and disadvantages. Within fairly broad limits, most customers can adapt their working methods to suit the aggregates most readily available, but having done this the main requirement is for overall consistency.

The location of quarries to produce aggregates is governed by three main considerations: the availability of deposits giving acceptable properties when processed; the ease of quarrying; and the proximity to the sites where the aggregate is needed. In certain areas, good quality aggregate may be difficult to produce or perhaps not available at all. A local builder or contractor may, therefore, have to adapt his methods to using poorer quality materials, which will probably increase his costs in other ways, or alternatively he may have to pay a higher price for difficult working or for longer distance transport of the good materials. Fortunately, few areas of Devon have this problem because good quality deposits are widespread in the county.

The major aggregate deposits quarried in Devon are shown in *Figure 12.1*, with a list of the principal operating quarries given in *Table 12.1*. Limestone, and quartzite from the Triassic pebble beds are the most important in terms of annual production, with dolerite, sandstone, hornfels and sand and gravel produced in lesser amounts, although very important locally.

Table 12.1 PRINCIPAL AGGREGATE QUARRIES IN DEVON

Type of Deposit	Symbol on figure	Age	Quarry/Quarrying Area (Numbered on *Figure 12.1*)	
LIMESTONE	L	Devonian	1	Plymstock*
	L	Devonian	2	Moorcroft
	L	Devonian	3	Linhay Hill
	L	Devonian	4	Stoneycombe
	L	Carboniferous	5	Westleigh Area
DOLERITE and other	D	Devonian	6	New England
mafic igneous rocks	D	Devonian	7	Torr
	D	Devonian	8	Whitecleaves
	D	Carboniferous	9	Trusham
SANDSTONE	G	Devonian	10	Vyse
('Gritstone')	G	Devonian	11	Little Silver
	G	Carboniferous	12	Venn
	G	Devonian	13	Bray Valley
	G	Carboniferous	14	Highleigh
	G	Carboniferous	15	Holmingham
HORNFELS	H	Carboniferous	16	Meldon
BUNTER PEBBLE BEDS	B	Triassic	17	Blackhill
	B	Triassic	18	Rockbeare Area
	B	Triassic	19	Hillhead
	B	Triassic	20	Whiteball
SAND/SAND AND GRAVEL	S	Permian	21	Bishop's Court
	S	Cretaceous	22	Babcombe Copse
	S	Eocene	23	Aller Area
	S	Quaternary	24	Kilmington

* Limestone quarried for cement manufacture.

Limestone

Workable limestone deposits are found in the Devonian of south Devon and the Lower Carboniferous of east Devon. These deposits contribute more than half of the county's aggregate requirements.

On the eastern outskirts of Plymouth, well-bedded medium to pale grey Middle Devonian limestone is worked for aggregate and cement manufacture in two large quarries (SX 510543 and SX 525540). Numerous old workings in the area are evidence of the long history of limestone extraction to supply the city. Because the urban area has grown out around and in places on top of the limestone deposits, large amounts of the stone have been 'sterilised' from being worked, and that which is still available is subject to strict environmental constraints to minimise blasting vibrations, dust, noise and visual intrusion. The alternative would be, however, to bring construction aggregates for Plymouth from quarries farther afield, adding substantially to cost and inconvenience.

Similar limestones are worked in the Ashburton-Newton Abbot-Torquay area. Again, the growth of Torbay has resulted in large deposits being sterilised and most of the quarries in the beautiful coral limestones traditionally used for building in the older parts of the towns are now completely disused.

Lower Carboniferous limestone is extensively worked in the Westleigh area of east Devon near the Somerset border, west of the M5 motorway and main line railway (ST 065175). Again, numerous small disused quarries in the area show the long history of limestone working. The main deposit at present being worked, between the villages of Westleigh and Holcombe Rogus, is an elongate inlier surrounded by Permian red mudstones and breccias and trending north-east–south-west, which is the direction of a prominent sharp anticline, or series of folds *en échelon*. The limestone is well and evenly bedded with some chert and intervening shaly bands. The coarse-grained limestone shows many of the features of deposition by powerful turbidity currents (Thomas, 1963a and b).

The lateral age equivalent of the Westleigh Limestone Group has long been worked in the area around Bampton (SS 960220), about 10 km to the north-northwest of Westleigh, where the sequence comprises interbedded chert, limestone and mudstone again sharply folded along axes trending east-south-east–west-northwest. The proportion of limestone in the deposits of the Bampton area is much less than in those of Westleigh.

In terms of production and reserves, limestone is without doubt the most important of the aggregate minerals in Devon and there are several reasons for this. First of all, the main deposits which are worked are substantial in size and are conveniently placed in relation to the main centres of population where most aggregates are required. Secondly, the deposits are relatively easily worked with no great thicknesses of overburden to be removed and tipped. Limestones in general are relatively easy to quarry, crush, and process when compared with some other aggregates, and the products are suitable for a wide range of uses. Limestone cannot, however, be used in bitumen-coated road surfacing because of the ease with which it polishes under the tyres of traffic. Limestone is also usually excluded from use in the construction of concrete dams such as at Wimbleball on the southern margin of Exmoor.

Dolerite and other Mafic Igneous Rocks

In contrast to limestone the mafic igneous group provides aggregates which are ideally suited for road surfacing because the wearing and polishing properties are generally good. The products are also suitable for most other aggregate applications. However, quarrying is generally more difficult and therefore more costly than for limestone, because the deposits tend to be smaller and much less regular in shape, sometimes with large inclusions of shaly material. The overburden is thick and irregular in many cases, occurring in the form of brown weathered dolerite with shales, particularly around the margins of the deposits. Blasting and crushing are generally more difficult due to the harder nature of the rock.

The brown weathered rock which is found where the mafic igneous rocks outcrop is often a major obstacle in the working of such deposits, and it is generally much more of a problem in South West England than with similar deposits elsewhere in the country. This is usually explained as due to the absence of any Pleistocene glacial scouring, which elsewhere has either removed most of the surface weathered material or has exposed deposits which were originally protected from any extensive weathering. Sections through the brown surface material usually show many of the features of a tropical weathering profile, and it appears that much of the brown stone results from the tropical climate which persisted for a long period during the early Tertiary about 50 Ma ago, at about the same time as the kaolinitic ball clays were being formed.

The thickness of the weathered brown stone often varies rapidly and irregularly. Brown discoloration sometimes penetrates deeply along prominent joints into the otherwise unweathered 'blue' stone and although this does not diminish the physical properties of the stone in any way, the appearance of brown fragments in the best quality aggregate can give rise to complaints from some customers and has to be avoided. The disappearance of the brown stone with depth is usually very gradual, and it is often necessary for a great deal of what appears to be best quality blue stone to be discarded, or sold as a second quality mixed material, because separation cannot be done economically. The economic limit of the thickness of brown stone which can be stripped varies with individual quarries but it would seem that about 15–20 m is probably the maximum.

The mafic igneous bodies which are worked occur in a broad arc swinging around the southern half of Dartmoor but extending well away from the margin of the granite to Torr Quarry near Kingsbridge (SX 745484). There the dolerite intrusion is sill-like in form but up-ended so that the quarry is long, narrow and deep. The deposit is about 700 m long but no more than 100 m wide at the surface.

In contrast, the deposit worked at New England Quarry (SX 597545), about 10 km east of Plymouth, is more irregular in outline although it is somewhat elongate parallel to the strike suggesting a sill-like form. This deposit is particularly well located in relation to Plymouth to complement the limestone deposits in applications such as main road surfacing where limestone cannot be used.

Similarly, the quarries at Whitecleaves, Buckfastleigh, (SX 737656) and Trusham in the Teign valley, (SX 848808), are well placed in relation to the

Newton Abbot–Torbay area. The Trusham deposit is one of an extensive series of sill-like bodies intruded into Lower Carboniferous shales and there are numerous old quarries in similar deposits around the area.

In terms of production tonnage, the quantity of mafic igneous rock quarried in Devon is small, and in the order of only 10 per cent or so of the county's total aggregate requirement.

The quantity of other igneous rocks quarried for aggregate in Devon is minimal. Contrary to the popular view, granite, especially coarse-grained varieties such as occur on Dartmoor, is not 'hard' in the aggregate sense, indeed its crushing strength is often so poor that it crushes beneath the roller during road construction. Furthermore, whilst its good polishing characteristics may be ideal for monumental purposes, this property is not good where skid resistance is required. Coarse-grained granite is used for aggregate in some counties where better quality materials are not available, but in Devon there is, as yet, no need to resort to using it.

Sandstone

Sandstone, or gritstone as it is usually known commercially to distinguish it from many softer materials used as building stone, is worked in numerous quarries in the northern and western areas of the county where the deposits have a widespread distribution and where there are no workable deposits of limestone or other aggregate sources. The nearest major limestone quarrying area at Westleigh is 50–70 km distant from the main centres of population in north Devon around Barnstaple and Bideford. Therefore, although the working and processing of gritstone in this area tend to be difficult and costly, there is a strong cost incentive because alternative materials have to be transported in over long distances.

The deposits which are worked are of Devonian and Carboniferous age, usually from the Baggy Sandstone and the Crackington Formation respectively, which outcrop in bands striking roughly east-west. The amount of true gritstone which occurs in the deposits is variable, as there are usually some interbedded shales (known locally as shillet), which can amount to a high proportion of the total in some places. This is only one of the exceptional working problems peculiar to this area giving rise to substantial quantities of second-grade material. Although some of this may be usable, equally often it needs to be tipped as waste, adding to costs and causing disposal and environmental difficulties. Further problems are caused by the highly abrasive nature of the gritstone which results in costly wear and maintenance problems on both fixed and mobile plant.

Nevertheless, despite working difficulties and high costs, good quality aggregates are produced from the north Devon deposits and they can be used throughout a full range of applications. Gritstones of this type have very good polish resistance, because they are mildly metamorphosed greywacke sandstones, generally regarded as the best naturally-occurring materials for road surfacing at locations where skid resistance is of prime importance.

Hornfels

The large quarry at Meldon (SX 570925) operated by British Rail extracts hornfels from the metamorphic aureole of the northern edge of the Dartmoor granite. The rocks are mainly Lower Carboniferous shales and cherts with some minor mafic igneous intrusions. Hornfels is not commonly quarried for aggregate, firstly because it is not a very common rock type, and secondly because it tends to be very tough and brittle, resulting in working problems and very angular and sharp-edged fragments. However, this extreme toughness is turned to advantage in the main product from this quarry which is rail ballast, and the good attrition properties are very valuable in the ballast used for main line tracks in these days of high-speed trains.

Triassic Pebble Beds

The narrow outcrop of the Triassic ('Bunter') pebble beds (Budleigh Salterton Pebble Beds and Uffculme Conglomerates) extends almost due northwards from Budleigh Salterton across east Devon to the Somerset border and beyond to the Bristol Channel coast. The full thickness of the deposit in Devon can be up to 35 m with distinct boundaries with the strata above and below. Almost invariably where the deposit is worked, the natural upper boundary with the overlying Otter Sandstones is not seen because of erosion and, therefore, the full thickness is not present. The pebble beds rest sharply on the Littleham Mudstones and the deposit is usually worked down to this boundary.

In Devon the pebble beds are made up predominantly of rounded pebbles up to about 25 cm diameter, but mainly in the range 5–10 cm, set in a buff to dark-red sand and silt matrix. Occasional thin sandy beds and lenses occur in which pebbles are absent, and some beds are entirely composed of silty clay, but such beds usually make up less than 10 per cent of the full thickness at any one place. The pebbles are predominantly of quartzite, including the pink and liver-coloured varieties derived from Ordovician sources which occur especially towards the southern end of the outcrop. Cementation of the deposit is minimal in Devon, but in the vicinity of the Somerset border, abundant beds of iron-cemented conglomerate appear, and the deposit cannot be worked any farther northwards.

The working of the pebble beds requires techniques intermediate between hard-rock quarrying and sand and gravel extraction, but explosives are not needed because of the lack of cementation. The material can be dug from the face using heavy excavating equipment. Wet processing is used to separate silt as a waste product and this presents some difficult disposal problems. Of the usable products, more than 40 per cent is usually sand; both fine sand for building mortar and plastering and coarse sand for concreting are produced. The remainder is coarse quartzite aggregate which is screened to various size gradings after crushing of the larger pebbles. The hardness of the quartzite gives rise to abrasion problems in the crushing and processing plant. The aggregates produced are of excellent quality and find a wide range of applications, including road surfacing because of the good polishing and wear resistance of the quartzite.

The southernmost area of quarrying along the pebble beds outcrop is at Blackhill on Woodbury Common (SY 030855) where a conservation balance has to be maintained between the amount of the deposit that can be extracted and the amount to be left in the ground in the interests of landscaping and restoration, and to safeguard the groundwater supply in the Otter Valley to the east.

Centred on Rockbeare Quarry (SY 060945) is a group of workings including those at Foxenhole near West Hill (SY 077947) and at Venn Ottery Common (SY 065913). Excavated materials from the outlying quarries are transported to the central plant for processing and distribution.

Close to the Somerset border, and approaching the point where iron-cemented bands start to present production problems, the deposit is worked at Hillhead (ST 065135) and at Whiteball (ST 090185) quarries. In this area the pebbles are of a greater variety than farther to the south and include some gritstone, limestone and quartz. Two small pits were worked in this area, near Burlescombe, to supplement the surge in requirements during the M5 motorway construction but these have now been completely restored to agriculture.

The pebble beds provide around 15 per cent of the aggregates used in Devon, mainly for requirements in the Exeter and east Devon area. Road surfacing materials are sold over a wider area, particularly along the south coast into Dorset and Hampshire because other aggregates with good wear and polish resisting properties are not available for this purpose from the younger rocks in those counties.

Sand and Gravel

Several miscellaneous workings of unconsolidated deposits may be included in this group.

Permian sands from the Clyst Sands, mainly wind-blown dune sand, are worked at Bishops Court (SY 965915) on the eastern outskirts of Exeter for building sand used in plastering, mortar, etc.

The Cretaceous Upper Greensand is worked for sand and gravel at Babcombe Copse (SX 870765) to the north of Kingsteignton, whilst in the area around Aller (SX 880690), south of Newton Abbot, sands and gravels up to 25 m in thickness and considered to be Eocene in age are worked in two pits. The Tertiary Haldon Gravels on the high ground between Exeter and Newton Abbot have not been worked to any great extent because they contain very little sand. Cenomanian sand is worked at Wilmington in east Devon. Near Axminster, the Axe Valley gravels of Quaternary age are worked at Kilmington (SY 275975). These gravels are particularly valuable because of their exceptional thickness for this type of deposit. Up to 17 m occur at this location and even thicker deposits are known upstream close to the border with Somerset and Dorset.

Dredged Aggregate

Sea and estuary dredged sands and gravels are brought in to several ports in Devon, both on north and south coasts, for local distribution. These supply up to 5 per cent of the county's requirements.

Secondary Aggregate

Waste materials such as mine and quarry spoil, power station ash, etc., are known as secondary aggregate. These can be suitable for certain construction applications, either as-produced, or with some processing, to substitute for primary quarried materials. Unfortunately, if processing is necessary, the cost saving over primary materials is rarely significant and can often be more than offset by the extra transport or other additional costs that may be involved.

Waste from china clay working on the southern edge of Dartmoor is the most abundant waste material which is used as aggregate in Devon. It occurs in the form of rock (*stent*), sand and micaceous residue. Large tonnages have already been tipped and are added to each year. Substantial amounts are used locally for aggregates, especially the sand which is used for building purposes and in concrete. The colour and texture can be an advantage where a particular architectural finish is required, but there are also disadvantages such as the irregular particle shape which leads to a higher cement demand in concreting. Transport costs restrict more widespread usage and the sand is not generally used in Devon outside the Plymouth area.

Building Stone

Many of the older buildings in Devon were built of stone, and the type of stone locally available governed the character of many villages from the grey gritstone of west and north Devon, to the white limestone of the Tor Bay area to red sandstone and purple volcanic rocks in mid Devon. Very little stone is now worked; the small amount which is worked is for repair work on walls and houses or for fireplaces. Much of it is produced as a by-product from major quarries at Westleigh, Venn in North Devon and Lummaton near Torquay, though ornamental limestone is worked specially at Elliotts Quarry (SX 590519) near Yealmpton. Merrivale Quarry (SX 546752) near Princetown still produces granite facing slabs, but much of its work is on stone from other sources.

A small quarry was opened up in 1979 at Dunscombe (SY 160887), near Branscombe, to serve the Exeter Cathedral restoration programme. Sandstone from the Upper Greensand is quarried from one of the sites which supplied the original fabric of the cathedral, and some churches nearby, in the twelfth to fifteenth centuries.

Brick Shale

Three brickworks are operating at present, using shale from two long-established quarries.

At Pinhoe (SX 955945) near Exeter, Carboniferous shales are worked to produce a distinctive red-coloured brick, on site, and also to supply a nearby works. At Steer Point (SX 545502), south of Brixton near Plymouth, Middle Devonian shales are used to make a wide variety of colours of bricks by selective quarrying, stockpiling, testing and blending.

The brickmaking operations in Devon are relatively small in comparison with some works elsewhere in the country. Devon is not traditionally a high brick use area. However, the combination of good raw materials with local techniques and strict quality control has led to the bricks having a high reputation at the top end of the market with sales throughout England and Wales and as far afield as Scotland.

Conclusions

The variety of materials worked to provide construction materials in Devon spans the whole geological spectrum. There age ranges from Palaeozoic to Holocene and they include igneous, sedimentary and metamorphic rock types, from old, hard and indurated to new, soft and unconsolidated. In the past there have been many small workings, but the modern trend is towards fewer but larger quarries, where mechanisation is used to full advantage. This is not only for economic reasons but because of increasing environmental constraints. In this way the disturbance can be confined to distinct areas which generally have a long history of quarrying.

Minerals cannot be worked without disturbing the ground, and without causing some inconvenience and change. The working of minerals requires a conscious effort by the operators to minimise the effects of these disturbances and perhaps this is easier in the aggregate industry than with many other minerals because substitution of source of supply and adaptation to using different rocks for the same purpose can be achieved more readily. However, like all other minerals, aggregates can only be worked at locations where suitable deposits are found. Devon is fortunate in having a broad choice due to its natural variety in geology.

Chapter Thirteen

Hydrogeology

Water is an important economic resource in any district, especially so in Devon where, during the summer months, its population is enlarged by thousands of tourists. Although Devon is not commonly thought of as having much industry, large quantities of water are also needed for industrial processes including paper making, mineral processing and dairying throughout the year. Though water falls free from the heavens, its collection, storage, purification and distribution cost money, as does the exploration for new sources to supply the county's growing demand.

Water may, therefore, be regarded as a mineral which can be extracted and sold; obviously, it differs from virtually all other minerals in being liquid and renewable. Indeed, much of the county is supplied with water from surface sources, rivers with storage reservoirs on them, but in many places smaller or larger sources have been developed in the groundwater domain, which means all water obtained from below the ground surface whether from unconsolidated sediments or fractured hard rocks. In a book on geology, this sector is necessarily of more interest, and will be treated in detail.

The *hydrological cycle* is the route and processes involved in the dynamic circulation of water on our planet, and concerns the atmosphere, rivers, lakes and seas, and the solid earth beneath. Hydrogeology is concerned with that part of the cycle which involves transport through the subsurface zone, particularly those aspects which relate to the movement of groundwater within saturated rock.

A proportion of the rain that falls on an area infiltrates into the soil and percolates down through unsaturated rock until it reaches saturated rock. Once there, it travels through it as groundwater until it flows out again into springs or wells. There is, therefore, a natural dynamic balance between the amount of rainfall in an area which enters water-bearing rocks, the amount of groundwater held in storage within them, and the amounts that re-emerge, issuing from springs and streams. It is important to remember that groundwater is perpetually flowing through the rock, albeit at very slow rates of movement in most cases, this circulation continuing whether or not rain is actually falling.

Water-bearing rocks, known as *aquifers*, act both as conduits for transmission of groundwater and as reservoirs for its storage. A knowledge of their characteristics is important in determining whether or not a formation can be used as a water resource, and if so how much water can be extracted from it on a continuing basis.

The most important hydraulic properties of an aquifer are its *permeability* and its *storage coefficient*. The former indicates the rate at which water can move in or out of the rock, the relative 'resistance' to flow through it. However, the hydrogeologist commonly uses a related characteristic, *transmissivity*, since this determines the flow which can be induced through the formation as a whole, taking the saturated thickness of the aquifer into consideration, and has obvious use in determining the yields from wells.

The storage coefficient, on the other hand, indicates the volume of water that can actually be removed from the aquifer. It is obviously related to the void space or *porosity* of the rock, but is a measure of how much of the pore space can actually be exploited for water supply. Some voids, for instance, may not be in connection with others and these would not therefore contribute to water supply. The drainable pore space is often termed the *specific yield*.

The shape and size of voids in any rock obviously determine its hydraulic properties. There are basically two kinds, intergranular porosity, the interconnected spaces between grains in sandstone and gravel, for instance, and fracture porosity, where fissures, joints and fractures provide small but significant pathways for movement and storage of water in hard rocks such as granite and Palaeozoic sedimentary rocks. Limestone and chalk both form major aquifers where fracture porosity is well developed. Percolating groundwater often dissolves the rock walls of fissures in limestone, producing very high permeabilities.

In Devon, the major aquifers of interest to any hydrogeologist contemplating large scale exploitation of groundwater almost wholly occur in the eastern half of the county. They consist of granular sedimentary deposits mainly in the Mesozoic, and include dune-bedded and fluvial Permian sandstone, Triassic conglomerate and sandstone, and the Cretaceous Upper Greensand. Elsewhere in the county Tertiary and Quaternary sands and gravels, though of limited extent, also form locally important aquifers, as at Taw Marsh on the Dartmoor granite, and at Westward Ho! on the coast.

Fissured aquifers are less significant in Devon. The Chalk, the major aquifer in southern England, occurs only as isolated outliers, in most cases draining freely into the underlying Upper Greensand. Beer and Chard are the only areas where it has any significance for water supply. Middle Devonian limestones are disappointing aquifers, partly because of their restricted outcrop but also because they have a low coefficient of storage despite having a high permeability. As a result they drain quickly after recharge and the amount of groundwater retained in storage is not sufficient to make large scale development feasible.

The aquifers in east Devon are separated by Permian, Triassic and Jurassic clays and mudstones. These are *aquicludes*, rocks which are almost incapable of transmitting water. Though boreholes in them often produce water, it is derived from more permeable horizons which occur within them, sandstone being the most common.

Few rocks in Devon are *aquifuges*, that is, rocks that are completely impermeable. Solid granite belongs in this category. Nevertheless, even on the Dartmoor granite it is commonly found that weathering and fissuring has created some water bearing capacity to the zone within the top few metres of the surface. The Palaeozoic sedimentary and metamorphic rocks and associated igneous intrusive rocks which occupy the rest of Devon are often termed the 'minor aquifers'; compared to the major British aquifers such as the Chalk and the Triassic 'Bunter' Sandstone (the Otter Sandstone in Devon) they can be considered to be insignificant, but the lowly amount of groundwater storage which occurs within the fissure systems which permeate these rocks is sufficient to be of local importance for small-scale private supplies.

EXPLOITATION METHODS

As might be expected from the above, the methods employed for extraction of groundwater and the way in which these resources are managed both differ fundamentally from the conventional methods of mineral exploitation (South West Water Authority, 1976). It is most important that it is not 'mined' in the conventional sense—that is, exploited at a rate which exceeds its rate of natural replenishment, unless this is carried out under controlled conditions. The most obvious way of doing this is simply to intercept the groundwater as it discharges from strata by constructing some form of catchpit or collecting chamber at springs. The groundwater resource then 'manages' itself. The rate of groundwater discharge at the spring fluctuates naturally as the amount held in storage varies, being highest in the winter and spring when recharge is occurring, falling throughout the summer and autumn when recharge is cut off. Any rainfall is absorbed in the soil zone, making up the moisture deficit that builds up from the high level of evaporation and transpiration occurring in these months (South West Water Authority, 1977, 1978). Seasonal fluctuations of the rest-water level in wells reflect the variation in groundwater storage that occurs as a result of this cyclical flow process. *Figure 13.1* compares typical groundwater level fluctuations at a well in the Triassic sandstone aquifer at Longmead in the Otter Valley with the monthy rainfall data for the same period, showing the proportion of the latter which contributed to groundwater recharge.

Wells and boreholes are the most commonly used 'artificial' method of abstracting groundwater. They have the advantage that the amount that is abstracted can be varied as required, up to the reliable maximum yield. Though a well and a borehole are essentially similar, the former is commonly a large diameter hand-dug shaft which only penetrates a few metres below the water table; boreholes, on the other hand, are usually of smaller diameter and are drilled, usually penetrating considerably further into the aquifer and thus tapping a greater saturated thickness of strata through which drainage can be made to occur.

Boreholes used for small private supplies are generally simply constructed, usually being drilled by percussive methods to a depth of 25–50 m and lined with slotted mild steel pipe of 15 cm diameter. However, where running sand is a problem, a more complex design is required using precision slotted stainless-steel screens, outside which a gravel pack is placed to support the formation. This allows high inlet water velocities without causing ingress of sand. The major public supply boreholes in the Triassic and Permian aquifers are of this type.

Though groundwater abstraction by this method appears straight-forward it must be borne in mind that continuous pumping at a high rate will even-

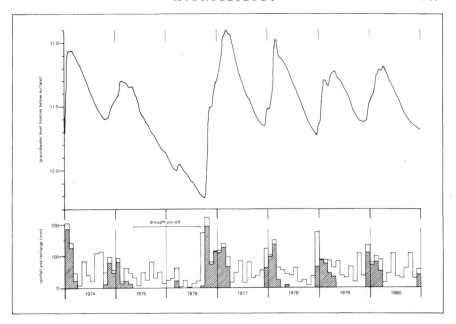

Figure 13.1. Groundwater-level changes in a well in the Triassic sandstone aquifer at Longmead in the Otter valley, compared with monthly rainfall and aquifer recharge

tually alter the dynamic flow balance in the aquifer by inducing flow to the well at the expense of its discharge elsewhere. Thus the development of groundwater resources has to be controlled and managed, so that deleterious effects such as the drying up of nearby springs or wells can be avoided. This is carried out by the South West Water Authority by the use of a licensing system which was initiated for this purpose under the Water Resources Act of 1963.

Before a new groundwater abstraction licence can be granted the Water Authority have to be satisfied that it will not affect the rights of any other water users in the area. Hydrogeological studies that are carried out include the assessment of local groundwater resources and the rate of natural replenishment and the conduct and analysis of pumping tests, from which the hydraulic properties of the water-bearing strata can be determined and the long term effects of the proposed abstraction can be forecast.

Legislation has also been drawn up to help prevent the pollution of groundwater. This is important for a number of reasons. Firstly, the slow rates of movement of groundwater mean that it may take many months or years for the effects of pollution to become evident at wells or springs; secondly, any effects may take years to dissipate; lastly, once pollution has occurred it can be extremely difficult and expensive to remedy the situation. For instance, if a borehole source becomes polluted with oil which has been disposed of in shallow strata nearby, it may have to be completely abandoned and replaced by a new well sunk outside the affected area.

GROUNDWATER RESOURCES AND WATER SUPPLY DEVELOPMENT

The development of groundwater sources for public water supply depends on the costs of their construction and operation, compared with the costs of alternative methods of providing the water. Only in the major aquifers of east Devon are yields sufficiently high to make them viable propositions: in that part of the county lying to the east of Exeter and Tiverton, the public water supply is wholly derived from groundwater sources developed in the Triassic and Cretaceous aquifers. These include spring catchpits as well as boreholes. New sources of supply are exclusively of the latter type, located using modern scientific techniques of hydrogeological investigation. The yields of such sources commonly exceed two million litres (450,000 gallons) per day.

Further west, groundwater resources are more limited and are only utilised for public supply in a few localities; sources are situated in the Permian and Cretaceous aquifers northwest of Dawlish, in the Permian strata of the Crediton trough northwest of Exeter, in river gravels beside the River Dart at Totnes, and beside the headwaters of the River Taw at Taw Marsh on the Dartmoor granite. Elsewhere in the county, supplies are provided by surface water sources, that is by river intakes and upland reservoirs, although the water authority retains a number of small local groundwater sources for emergency use. The balance between groundwater and surface water abstractions in the county is evident from the data on licensed sources shown in *Table 13.1*.

In rural areas of Devon, however, groundwater from both the major and the minor aquifers is used extensively to provide small private supplies for domestic or agricultural use. Springs are sometimes tapped because of their cheapness but, as they are susceptible to pollution, wells and boreholes are generally preferred. The reliable yields necessary to satisfy these needs rarely exceed 7,000 litres (1,500 gallons) per day. Luckily, there are few areas in the county where this amount of water cannot be obtained from a small-diameter

Table 13.1 RESOURCE ESTIMATES FOR THE MAJOR AQUIFERS IN DEVON

Geology/Aquifer		Location	Outcrop Area sq km	Theoretical Annual Replenishment (from rainfall)		% Exploitation by Licensed Sources (1979)
				mm	million cubic metres	
Permian	Sandstone	Dawlish–Exminster	21.7	340	7.4	15‡
		Clyst Valley	44.5	280	12.5	5
		Brampford Speke–Rewe	28.0	330	9.2	30*
		Knowle–Bow	9.3	400	3.7	15‡
		Cullompton	8.2	435	3.6	5
		Tiverton Halberton	15.7	480	7.5	5
Triassic	Pebble beds and sandstone	Culm Valley	32.0	420	13.4	5
		Otter/Tale Valley	88.8	370	32.9†	50‡
Cretaceous	Greensand and chalk	Total Outcrop	371	545	202.0	5

‡Includes major public supply source(s) of potable water.
*Includes major public supply sources available for river regulation purposes.
†Resources significantly augmented by river recharge in the vicinity of certain public supply boreholes.

borehole of suitable depth; the exceptions are locations on bedrock granite or poorly fissured slates. The widespread reliance on the services of water diviners in the county to locate such small private sources is intriguing: it probably arises because most people lack any appreciation of the mode of occurrence of groundwater in the water-bearing strata in their locality, combined with the apparent endorsement of the method by its 'success'. After all, when a borehole is drilled and water is encountered, the diviner can claim he has been proven right. The old tales which one hears of 'underground streams' often following weird and wonderful courses, sometimes being derived from rocks as far away as Cheddar or Paris, are similarly misplaced. Only in the Devonian limestones such as those at Ashburton and at Torquay are any true underground streams encountered. These occupy fissures in the limestone which have been enlarged through solution of the rock by the flowing water. However, given the paucity of information on groundwater in the minor aquifers and the difficulty of acquiring it, it is no wonder that people rely on those who claim special powers which enable them to locate 'underground watercourses' and even pay them for this service.

In the following account of available ground-water resources and the current state of their development in Devon, the water-bearing strata are considered in the order of their importance for the provision of large scale supplies. Resource estimates for the major aquifers are given in *Table 13.1*. Typical well yields and groundwater quality data for the major water-bearing strata in the county are given in *Tables 13.2 and 13.3*.

Table 13.2 YIELD DATA FROM LICENSED BOREHOLES IN DEVON

Geology	'Typical' Licensed Yield cubic metres/day	Range of Yield Normally Encountered cubic metres/day	Special Conditions Necessary for Yields in the Upper End of the Range Shown
Granite (Dartmoor)	2	0.5–7	
Other igneous and hard metamorphic rocks	3.5	1.5–9	
Lower and Upper Devonian	6	1–15	
Middle Devonian	6	1–225	Well fissured limestone
Pilton Shale	9	2.5–250	Limestone horizons with river recharge contribution
Carboniferous:			
Crackington Formation	6	1.5–15	
Bude Formation	6	1.5–20	
Permian:			
Breccio-conglomerate	10	2.5–100	
Sandstones	30	10–3000	Well sorted sandstone (for example dune bedded)
Marl/mudstone	5	2.5–100	Sandstone horizons.
Triassic:			
Pebble beds and sandstone	30	10–5000	Saturated thickness > 100 m
Marl/mudstone	5	2.5–100	Sandy horizons
Cretaceous:			
Upper Greensand	8	2–30	> 1000 from major springs
Chalk	15	0.5–1000	
Tertiary Quaternary	15	2–1000	Sand and gravels with river recharge

Table 13.3 GROUNDWATER QUALITY DATA FROM WELLS IN THE MAIN WATER-BEARING STRATA IN DEVON

Geology/Lithology	Location	Conductivity at 20 C µs	pH	Free Carbon Dioxide CO₂ mg/l	Total Hardness as CaCO₃ mg/l	Calcium Ca mg/l	Magnesium Mg mg/l	Sodium Na mg/l	Potassium K mg/l	Carbonate CO₃ mg/l	Sulphate SO₄ mg/l	Chloride Cl mg/l	Nitrate NO₃ mg/l
Middle Devonian:													
Limestone	Stoneycombe	490	7.4	15	242	85	7.2	16	2.3	115	32	25	29
Permian:													
Breccio-conglomerate	Sampford Courtenay	330	7.9	3.5	145	34	14	17	1.8	78	22	10	19
Sandstone	Dawlish	274	6.65	24	99	23	11	15	6.0	33.0	15	30	32
Marl/mudstone	Plymtree	560	6.9	20	280	94	11	38	16	80	55	64	140*
Triassic:													
Pebble beds	Colaton Raleigh	200	5.5	90	48	12	4.6	15	5.8	9.6	20	21	27
Sandstone	Greatwell	380	7.7	7.0	181	51	13	13	3.3	101	18	21	3.5
Marl/mudstone	Buckerell	850	8.0	—	578	90	86	16	4.5	182	242	31	14
Cretaceous:													
Greensand	Churchstanton	70	5.8	—	25	4.8	3.2	8.0	0.8	4.8	4.3	13	17
Greensand + chalk	Beer	500	7.5	16	252	95	3.4	15	1.5	130	23	20	17
Tertiary:													
Confined sands	Newton Abbot	615	6.9	32	140	44	7.3	84	3.9	78	19	127	<0.4
Pleistocene:													
Granite-derived river gravels	Taw Marsh	45	5.6	—	11	2.0	1.4	6.0	0.8	3.6	2.4	10	1.5
River gravels	Totnes	155	6.7	14	54	18	2.4	12	1.3	22	13	15	6.0

*Note: Anomalously high nitrate level at this site reflects local agricultural pollution.

Triassic

This, the major aquifer in the county, comprises the Budleigh Salterton Pebble Beds together with the overlying Otter Sandstone. These beds attain a maximum thickness of over 160 m. The outcrop extends over an area of 120 sq km, from Budleigh Salterton northwards along the lower Otter Valley through Ottery St Mary and on into Somerset. The specific yield of the aquifer is high, generally between 10 per cent and 20 per cent, and it has been estimated that over two hundred billion litres (50,000 million gallons) of water are stored beneath the area of outcrop in the Otter valley alone. To put this in perspective, it is over 10 times greater than the volume of water stored in Wimbleball Lake. Pumping-test analyses suggest that its field permeability ranges between 1 m and 5 m per day, and that it is highest in the lower levels of the aquifer. The transmissivity of the aquifer is in the range 50–500 cubic metres per day per metre, depending on its local saturated thickness. Groundwater quality differs in the two horizons, being soft, acid and corrosive in the pebble beds, moderately hard and more neutral in character in the overlying sandstone. High levels of iron and manganese occur at some localities, particularly where groundwater from the pebble beds has been drawn up into overlying sandstone. An account of the hydrogeology of this aquifer was given by Sherrell (1970).

Major production boreholes are generally drilled with a 70 cm diameter to depths of between 50 m and 140 m, usually yielding between one and three million litres per day for a drawdown of between 10 m and 30 m. Two boreholes at Dotton which give even higher yields are known to derive some of their supply by inducing recharge to the aquifer from the River Otter in their vicinity.

Considerable resource development has already occurred in the southern half of the aquifer, with over 18 major boreholes and a number of shallow spring sources utilised at present for public supply in the vicinities of Kersbrook, Colaton Raleigh, Dotton, Harpford, Greatwell and even in the confined zone of the aquifer at Sidford. Total abstraction for public supply presently exceeds 20 million litres per day during peak periods in the summer, and development is likely to continue into the next century. It appears that the recharge resources in the Otter Valley alone are adequate to allow at least double the present yield to be developed, but the method of development will require careful management to ensure that the aquifer is not over-exploited to the detriment of river flows and spring discharges in the area. Extensive studies are already being carried out on this aspect by the South West Water Authority. Exploration has also been initiated further north, in the Culmstock area.

Permian

The Permian aquifer comprises the sandstone and breccia formations, which undergo significant lateral change over the area of their outcrop. To the southwest of the Exe estuary the aquifer consists of dune sands, overlain and underlain by breccias. Boreholes in the Dawlish–Starcross area indicate permeabilities of approximately 4 m per day for the dune sands of this district. In a northerly direction these dune sands die out and are replaced by a sequence of more highly cemented breccias and sands. Although there are a few fairly high-yielding wells in these areas, little is known of the aquifer characteristics. The specific yield of the dune-sand horizons is probably similar to that encountered in the Triassic: however, storage resources are considerably lower due to a slightly more restricted outcrop and thickness. Recharge resources have been estimated to exceed 50 million litres per day over the 74 sq km of outcrop between Dawlish and Silverton. East of Exeter the aquifer is confined beneath the Littleham Mudstones ('Lower marls').

The area north of Exeter, lying at the western end of the Crediton trough, was investigated by the former east Devon Water Board and two major boreholes were commissioned to augment flows to the River Exe; this enabled surface abstraction at Exeter to continue at a high level under low-flow conditions. The yields from these were four million and three million litres per day respectively, although a degree of induced river recharge occurs. Some years ago the decision was made to support supplies to Tiverton and Exeter by augmentation of river flows in the Exe by releases from a reservoir in its headwaters at Wimbleball, which was completed in 1979. As a result further investigation of the potential for public supply of groundwater sources in this aquifer is geared to its possible use in the Crediton area.

Small scale development has taken place in the Dawlish Sands to the southwest of the Exe Estuary to augment supplies to the Dawlish–Teignmouth area. A borehole similar in construction to those in the Otter valley yields over two million litres per day for less than 8 m drawdown.

The breccia horizons are considerably less productive, being more indurated in nature and have a considerably smaller storage coefficient. Transmissivities are lower, but more variable because of local fissuring, and this affects well yields. The few adit and borehole sources of any magnitude in these horizons yield between 250,000 and one million litres per day, and are used for public and industrial supplies in the area. However, the likelihood of yields in the upper part of this range being obtained from exploratory sources is highly speculative.

Cretaceous

The considerable expanse of permeable Cretaceous rocks which caps the hills in east Devon has a major significance for water supplies in that area. Though Upper Greensand strata are dominant, a few outliers of Chalk also occur, where local folding or faulting have brought these beds below the general level of erosion in the uplands. Where both formations occur they form a single aquifer unit, since groundwater drains freely from the Chalk into the underlying deposit, but in most of this area only the Greensand strata remain. Given the transgressive nature of the Cretaceous unconformity, it is somewhat surprising to find that the beds above it constitute a distinct aquifer. However, they are underlain by mudstones, marls or clays wherever they occur in the county east of Exeter, as post-Cretaceous erosion has stripped them from those places where they once covered permeable strata such as the Triassic sandstones and conglomerates.

The high rainfall that occurs on these east Devon hills and the large outcrop area of the Cretaceous aquifer combine to ensure that the groundwater resources available from natural recharge are large. However, the deeply dissected and near-horizontal nature of these beds also means that they are efficiently drained by a large number of springs, so that there is little chance of large-scale exploitation except by tapping major springs. This has been carried out at a number of localities, notably at Newcombe, Sheldon, Churchstanton, Wilmington and Pinhay. The average yield of the first four of these sources is about one million litres per day. The last-named spring emerges on the coast near Lyme Regis, draining strata which have been thrown down by a major fault trending north-south. The fall-off in yield from this spring in dry weather is extremely low; for instance, it was still yielding over two million litres per day at the height of the drought in August 1976. This gives a good indication that the coefficient of storage of this aquifer is high.

The few boreholes which have been sunk in Cretaceous rocks have generally proved disappointing. Well sorted sands are encountered in the lower horizons of the aquifer and these are often the only ones that are fully saturated; being unconsolidated, they are a cause of 'running sand' conditions during drilling, and migration of the finer material limits the permeability. Yields are usually restricted to about a thousand litres per hour in order to prevent fine material from being drawn into the borehole and the pump mechanism.

Two shafts with adits penetrate saturated chalk in the centre of the syncline at Beer. The yield, in excess of half a million litres per day, is far more respectable, reflecting the higher permeability of these strata where they are still saturated.

The bacteriological quality of the groundwater in the Cretaceous aquifer is generally good, but it is often subject to contamination as it emerges from springs and mixes freely with polluted surface water draining from the higher ground. An interesting variation is found in the hardness of the water in different areas, closely reflecting the variation in the degree of calcareous content in the aquifer. The entirely non-calcareous Blackdown facies of the Upper Greensand, occurring northwest of a line from Sidmouth to Yarcombe, yields a soft aggressive water, while the calcareous Greensand strata further east yield a moderately hard water, hardest where the Chalk is also present.

Although not extensively developed for public supply, the aquifer supports numerous private supplies, mainly from tapped springs or wells situated along the edges of the outcrop. These provide a most useful source of water at a high topographic level for the isolated farms and communities in this region.

Further development of the water resources of the Cretaceous aquifer for public supply is only likely if it proves viable and economic in relation to the alternative. At present the economics appear to favour the piping-in of water from the Triassic aquifer further west, and from present evidence the development of major spring sources near outlying demand centres seems to be the only valid option worthy of detailed consideration.

The character of the east Devon rivers differs markedly from those in the rest of the county, mainly as a result of the high level of base-flow provided by the Cretaceous aquifer. In summer, river levels fall slowly in contrast to those draining the Palaeozoic rocks, and the Dartmoor granite in particular, which have a very small base-flow component. For example, the flow in the River Otter was higher than that in the River Taw during the latter part of the drought in 1976 which, when it is realised that the catchment area of the River Otter is one eighth that of the River Taw and its average flow is one sixth, shows just how important this base-flow component can be in 'low flow' conditions.

Drift

Wells and boreholes tapping superficial deposits yield good supplies at a number of localities. However, on detailed investigation, it is invariably found that the groundwater resources are not large and that high yields are derived at the expense of nearby river flows.

The public supply boreholes at Taw Marsh on Dartmoor are a good example of this, They have been sunk into the coarse-grained river terrace deposits that partially fill an upland valley on the northern slopes of the granite massif. Although they can yield over five million litres a day between them, most of their yield is derived at the expense of flow in the River Taw, which meanders through these deposits, and use of these sources is restricted because of a requirement to maintain the flow in that river for fisheries interests.

At Totnes two radial collector wells and a number of shallow boreholes have been sunk into river gravels flanking the River Dart. Twenty million litres per day can be pumped from these sources to public supply, but the water is derived from the river by the same mechanisms of induced recharge through these superficial deposits, as outlined above.

The blown sands which fringe the Taw–Torridge Estuary near Barnstaple have produced a number of useful local supplies with yields of up to 200,000 litres per day from certain boreholes. However, the risk of pollution is high, and saline intrusion from drawn-in sea water would probably limit any attempt at major development of the limited resources that they contain.

Other Formations

The Middle Devonian limestones around Torquay, Plymouth and Newton Abbot are noteworthy because yields of over 250,000 litres per day have been obtained from a number of boreholes drilled there. It has been mentioned above that the resources retained in storage are limited because the rock is too permeable, draining quickly after recharge; however, in certain areas where rivers traverse the outcrop, continuous recharge occurs to this formation as a result of leakage through the river bed, and this water travels through the limestone to emerge at springs elsewhere. Boreholes tapping strata where such a mechanism is operating 'capture' part of this through flow and, as a result, are able to support high yields. The Compton stream is known to lose over two million litres a day where it traverses outcrop limestone in the Stoneycombe area, and tracer tests have proved that this water travels through the limestone to emerge at springs in a neighbouring catchment at Kingskerswell, over a kilometre away. Two boreholes at Stoneycombe have each yielded over 250,000 litres per day.

Jurassic strata occurring in the far east of the area include sands of the Middle and Upper Lias, particularly the Bridport and Yeovil sands, which provide reasonable supplies to a number of isolated farms in the area. The water is generally hard an often ferruginous, and sand invasion can become a problem in boreholes. A number of reliable springs arise at favourable points along the outcrop, but the faulted nature of the strata makes them susceptible to pollution from agricultural activities.

High yields are sometimes quoted from boreholes tapping permeable horizons in the Oligocene deposits of the Bovey Basin. However, if continuous

Table 13.4   LICENSED ABSTRACTIONS IN DEVON DURING 1979
(Quantities in million cubic metres)

Type of Source	Use	Water Supply		Agricultural		Industrial	Total
		Public	Other	Non-Spray	Spray		
Surface water	No.	46	64	59	159	122	450
	quantity	171.8†	0.2	0.2	0.9	122.0*	295.1
Ground Water	No.	73	496	4265	58	76	4968
	quantity	41.6	0.6	8.2	0.2	3.5	54.1
Total	No.	119	560	4324	217	198	5418
	quantity	213.4	0.8	8.4	1.1	125.5	349.2

Note: †Includes Wimbleball Lake abstractions by Wessex Water Authority.
*Includes major leat abstractions at Morwellham and Mary Tavy.

pumping is attempted, the high initial yields almost invariably fall off rapidly to a much lower level at sites where the water is derived from confined groundwater stored within the permeable horizons themselves. As these become dewatered the yield drops, since minimal leakage occurs through the surrounding clay strata.

No major groundwater resources exist in the other rocks found in the county although, as has been mentioned above, small local supplies of a few thousand litres per day are derived from a multitude of tapped springs, wells and boreholes in almost all of them. Further data on the nature of the groundwater and typical well yields in these strata are presented in *Tables 13.2 and 13.3.*

The Palaeozoic and igneous rocks which comprise the bulk of the bedrock in the county are far more amenable to the retention of surface water by dams constructed across suitable valleys than to the development of their limited groundwater resources where major water supplies are needed. Thus the public water supply system in central and west Devon is reliant on surface-water abstractions by numerous upland reservoirs, in direct contrast to the exploitation of the groundwater resources of the major aquifers in east Devon for public supply to that area. The data on licensed sources in *Table 13.4* reflect this, with two-thirds of the total licensed abstractions in the county being for surface water sources. However, it should perhaps be remembered that hundreds, if not thousands, of farms and isolated households throughout the county rely on small private sources tapping groundwater from all these formations. The fact that there are about ten times as many licensed groundwater sources reinforces this point.

There are a number of reservoirs operating in the county, many on Dartmoor where the high rainfall makes them particularly productive, but the most recently completed scheme is on Exmoor, at Wimbleball. This impounds a tributary valley to the River Exe, and helps control flooding in that river as well as providing a more regulated flow during the summer months, enabling surface water abstractions to be secured for public supply to both Tiverton and Exeter.

References

M. T. ALI, 1975 'Environmental implications of infillings in the Upper Greensand of the Beer district, south Devon', *Proceedings of the Geologists Association* 85, 519–532.

M. T. ALI, 1976 'The significance of a mid-Cretaceous cobble conglomerate, Beer district, south Devon', *Geological Magazine* 113, 151–158.

P. ANDREIEFF, P. BOUYSSE, D. CURRY, B. N. FLETCHER, D. HAMILTON, C. MONCIARDINI and A. J. SMITH, 1975 'The stratigraphy of post-Palaeozoic sequences in part of the western Channel', *Philosophical Transactions of the Royal Society* A279, 79–97.

J. T. ANDREWS, D. Q. BOWEN and C. KIDSON, 1979 'Amino acid ratios and the correlation of raised beach deposits in South West England and Wales', *Nature* 281, 556–558.

E. A. N. ARBER, 1911 *The coast scenery of north Devon*, Dent, London.

M. A. ARBER, 1977 'A brickfield yielding elephant remains at Barnstaple, north Devon', *Quaternary Newsletter* 21, 19–21.

M. G. AUDLEY-CHARLES, 1970a 'Stratigraphical correlation of the Triassic rocks of the British Isles', *Quarterly Journal of the Geological Society* 126, 19–48.

M. G. AUDLEY-CHARLES, 1970b 'Triassic palaeogeography of the British Isles', *Quarterly Journal of the Geological Society* 126, 49–90.

J. P. N. BADHAM, 1982 'Strike-slip orogens—an explanation for the Hercynides', *Journal of the Geological Society* 139, forthcoming.

J. P. N. BADHAM and C. HALLS, 1975 'Microplate tectonics, oblique collisions and evolution of the Hercynian orogenic systems', *Geology* 3, 373–376.

H. W. BAILEY, 1975 'A preliminary microfaunal investigation of the Lower Senonian at Beer, southeast Devon', *Proceedings of the Ussher Society* 3, 280–285.

W. G. V. BALCHIN, 1952 'The erosion surfaces of Exmoor and adjacent areas', *Geographical Journal* 118, 453–476.

C. F. BARCLAY, 1931 'Some notes on the west Devon mining district', *Transactions of the Royal Geological Society of Cornwall* 16, 157–176.

K. E. BEER and P. J. FENNING, 1976 *Geophysical anomalies and mineralisation at Sourton Tors, Okehampton, Devon*, Institute of Geological Sciences Report 76/1, London.

K. E. BEER and T. K. BALL, 1977 'Barytes mineralisation in the Teign valley', *Transactions of the Institution of Mining and Metallurgy* 86, B91–B92.

W. A. BERGGREN, L. H. BURCKLE, M. B. CITA, H. B. S. COOKE, B. M. FUNNELL, S. GARTNER, J. D. HAYS, J. P. KENNETT, N. D. OPDYKE, L. PASTOURET, N. J. SHACKLETON and Y. TAKAYANAGI, 1980 'Towards a Quaternary time scale', *Quaternary Research* 13, 277–302.

F. G. H. BLYTH, 1957 'The Lustleigh fault in northeast Dartmoor', *Geological Magazine* 94, 291–296.

F. G. H. BLYTH, 1962 'The structure of the northeastern tract of the Dartmoor granite', *Quarterly Journal of the Geological Society* 118, 435–453.

F. G. H. BLYTH and D. J. SHEARMAN, 1962 'A conglomeratic grit at Knowle Wood, near Woolley, south Devon', *Geological Magazine* 99, 30–32.

M. H. P. BOTT, A. A. DAY and D. MASSON-SMITH, 1958 'The geological interpretation of gravity and magnetic surveys in Devon and Cornwall', *Philosophical Transactions of the Royal Society* A251, 161–191.

M. H. P. BOTT and P. SCOTT, 1966 'Recent geophysical studies in South West England', in *Present views of some aspects of the geology of Cornwall and Devon*, edited by K. F. G. Hosking and G. J. Shrimpton. 150th Anniversary Volume (for 1964) Royal Geological Society of Cornwall, Penzance, 25–44.

M. H. P. BOTT, A. P. HOLDER, R. E. LONG and A. L. LUCAS, 1970 'Crustal structure beneath the granites of South West England', in *Mechanisms of igneous intrusion*, edited by G. Newall and N. Rast. Geological Journal Special Issue 2, Liverpool, 93–102.

D. Q. BOWEN, 1977 'The coast of Wales', in *The Quaternary history of the Irish Sea*, edited by C. Kidson and M. J. Tooley. Seel House, Liverpool, 223–256.

D. Q. BOWEN, 1978 *Quaternary Geology: a stratigraphic framework for multidisciplinary work*, Pergamon, Oxford.

C. J. R. BRAITHWAITE, 1966 'The petrology of the Middle Devonian limestones in south Devon, England', *Journal of Sedimentary Petrology* 36, 176–192.

A. BRAMMALL, 1926a 'Gold and silver in the Dartmoor granite', *Mineralogical Magazine* 21, 14–20.

A. BRAMMALL, 1926b 'The Dartmoor granite' *Proceedings of the Geologists Association* 37, 251–277.

A. BRAMMALL and H. F. HARWOOD, 1923 'The Dartmoor granite: its mineralogy, structure and petrology', *Mineralogical Magazine* 20, 39–53.

A. BRAMMALL and H. F. HARWOOD, 1932 'The Dartmoor granites: their genetic relationships', *Quarterly Journal of the Geological Society* 88, 171–237.

C. M. BRISTOW, 1968 'The derivation of the Tertiary sediments in the Petrockstowe Basin, north Devon', *Proceedings of the Ussher Society* 2, 29–35.

C. M. BRISTOW, 1977 'A review of evidence for the origin of the kaolin deposits in South West England', *Proceedings of the Eighth International Kaolin Symposium and Meeting on Alunite, Madrid-Rome* K2, 1–19.

C. M. BRISTOW and D. E. HUGHES, 1971 'A Tertiary thrust fault on the southern margin of the Bovey Basin', *Geological Magazine* 108, 61–67.

A. V. BROMLEY, 1979 'Ophiolitic origin of the Lizard Complex', *Camborne School of Mines Journal* 79, 25–38.

M. BROOKS and M. S. THOMPSON, 1973 'The geological interpretation of a gravity survey of the Bristol Channel', *Journal of the Geological Society* 129, 245–274.

M. BROOKS and D. G. JAMES, 1975 'The geological results of seismic refraction surveys in the Bristol Channel, 1970–1973', *Journal of the Geological Society* 131, 163–182.

M. BROOKS, M. BAYERLY and D. J. LLEWELLYN, 1977 'A new geological model to explain the gravity gradient across Exmoor, north Devon', *Journal of the Geological Society* 133, 385–393.

A. P. BROWN, 1977 'Late Devensian and Flandrian vegetational history of Bodmin Moor, Cornwall', *Philosophical Transactions of the Royal Society* B276, 251–320.

D. BRUNSDEN, 1963 'Denudation chronology of the River Dart', *Transactions of the Institute of British Geographers* 32, 49–63.

D. BRUNSDEN, 1964 'The origin of decomposed granite on Dartmoor'. In: *Dartmoor Essays*, edited by I. G. Simmons. Devonshire Association, Exeter, 97–116.

D. BRUNSDEN, J. C. DOORNKAMP, C. P. GREEN and D. M. C. JONES, 1976 'Tertiary and Cretaceous sediments in solution pipes in the Devonian limestone of south Devon, England', *Geological Magazine* 113, 441–447.

N. E. BUTCHER and F. HODSON, 1960 'A review of the Carboniferous goniatite Zones in Devon and Cornwall', *Palaeontology* 3, 75–81.

J. CAMERON, 1951 'The geology of Hemerdon wolfram mine, Devon', *Transactions of the Institution of Mining and Metallurgy* 61, 1–14.

D. J. CARTER and M. B. HART, 1977 'Aspects of mid-Cretaceous stratigraphical micropalaeontology', *Bulletin of the British Museum (Natural History), Geology* 29, 1–135.

C. CARUS-WILSON, 1913 'Cupiferous sandstones at Exmouth', *Nature* 91, 530.

C. J. CASELDINE and D. J. MAGUIRE, 1981 'A review of the prehistoric and historic environment on Dartmoor', *Proceedings of the Devon Archaeological Society* 39, 1–16.

M. E. J. CHANDLER, 1957 'The Oligocene flora of the Bovey Tracey lake basin, Devonshire', *Bulletin of the British Museum (Natural History)* 3, 71–123.

M. E. J. CHANDLER, 1964 *The Lower Tertiary Floras of southern England, Part 4*, British Museum (Natural History), London.

D. M. CHURCHILL and J. J. WYMER, 1965 'The kitchen midden site at Westward Ho!, Devon, England: ecology, age and relation to changes in land and sea level', *Proceedings of the Prehistoric Society* 31, 74–84.

R. H. CLARKE, 1970 'Quaternary sediments off southeast Devon', *Quarterly Journal of the Geological Society* 125, 277–318.

A. W. CLAYDEN, 1906 *The history of Devonshire scenery*, Chatto and Windus, London.

J. A. CLEGG, M. ALMOND and P. H. S. STUBBS, 1954 'Remnant magnetisation of some sedimentary rocks in Britain', *Philosophical Magazine* 45, 583–598.

J. H. COLLINS, 1912 'Observations on the west of England mining region', *Transactions of the Royal Geological Society of Cornwall* 14.

J. J. CONYBEARE, 1823 'On the geology of Devon and Cornwall', *Annals of Philosophy* 5, 184 and 6, 35.

W. D. CONYBEARE and W. J. PHILLIPS, 1822 *Outlines of the geology of England and Wales*, William Phillips, London.

J. D. CORNWELL, 1967 'The palaeomagnetism of the Exeter lavas, Devonshire', *Geophysical Journal of the Royal Astronomical Society* 12, 181–196.

J. D. CORNWELL, 1971 'Geophysics of the Bristol Channel area', *Proceedings of the Geological Society* 1664, 286–289.

M. E. COSGROVE, 1972 'The geochemistry of the potassium-rich Permian volcanic rocks of Devonshire, England', *Contributions to Mineralogy and Petrology* 36, 155–170.

M. E. COSGROVE, 1975 'Clay mineral suites in the post-Armorican formations of South West England', *Proceedings of the Ussher Society* 3, 243.

M. E. COSGROVE and M. H. ELLIOTT, 1976 'Supra-batholithic volcanism of the South West England granites', *Proceedings of the Ussher Society* 3, 391–401.

D. CURRY, D. HAMILTON and A. J. SMITH, 1970 *Geological and shallow subsurface geophysical investigations in the Western Approaches to the English Channel*, Institute of Geological Sciences Report 70/3, London.

D. CURRY, D. HAMILTON and A. J. SMITH, 1971 'Geological evolution of the western English Channel and its relation to the nearby continental margin', in *The geology of the east Atlantic margin*, edited by F. M. Delany. Institute of Geological Sciences Report 70/14, London, 129–142.

D. DALZELL and E. M. DURRANCE, 1980 'The evolution of the Valley of Rocks, north Devon', *Transactions of the Institute of British Geographers* 5, 66–79.

J. DANGERFIELD and J. R. HAWKES, 1969 'Unroofing of the Dartmoor granite and possible consequences with regard to mineralisation', *Proceedings of the Ussher Society* 2, 122–131.

J. DANGERFIELD and J. R. HAWKES, 1981 'The Variscan granites of South West England: Additional Information', *Proceedings of the Ussher Society* 5, 116–120.

A. G. DARNLEY, T. H. ENGLISH, E. SPRAKE, E. R. PREECE and D. AVERY, 1964 'Ages of uraninite and coffinite from South West England', *Mineralogical Magazine* 34, 159–176.

W. T. DEAN, D. T. DONOVAN and M. K. HOWARTH, 1961 'The Liassic ammonite zones and subzones of the northwest European province', *Bulletin of the British Museum (Natural History), Geology* 4, 437–505.

W. R. DEARMAN, 1959 'The structure of the Culm Measures at Meldon, near Okehampton, north Devon', *Quarterly Journal of the Geological Society* 115, 65–106.

W. R. DEARMAN, 1963 'Wrench faulting in Cornwall and south Devon', *Proceedings of the Geologists Association* 74, 265–287.

W. R. DEARMAN, 1970 'Some aspects of the tectonic evolution of South West England', *Proceedings of the Geologists Association* 81, 483–491.

W. R. DEARMAN, 1971 'A general view of the structure of Cornubia', *Proceedings of the Ussher Society* 2, 220–236.

W. R. DEARMAN and N. E. BUTCHER, 1959 'The geology of the Devonian and Carboniferous rocks of the northwest border of the Dartmoor granite, Devonshire', *Proceedings of the Geologists Association* 70, 51–92.

W. R. DEARMAN and E. C. FRESHNEY, 1966 'Repeated folding at Boscastle, north Cornwall, England', *Proceedings of the Geologists Association* 77, 199–215.

H. T. DE LA BECHE, 1839 *Report on the geology of Cornwall, Devon and west Somerset*, Memoirs of the Geological Survey of Great Britain, London.

H. DEWEY, 1923 *Copper ores of Cornwall and Devon*, Memoirs of the Geological Survey of Great Britain, Mineral Resources Report 27, London.

H. DEWEY and C. N. BROMEHEAD, 1916 *Tungsten and manganese ores*, Memoir of the Geological Survey of Great Britain, Mineral Resources 1, London.

D. L. DINELEY, 1961 'The Devonian System in south Devonshire', *Field Studies* 1, 121–140.

D. L. DINELEY, 1963 'Contortions in the Bovey Beds (Oligocene), South West England', *Biuletyn Peryglacjalny* 12, 151–160.

D. L. DINELEY, 1966 'The Dartmouth Beds of Bigbury Bay, south Devon', *Quarterly Journal of the Geological Society* 122, 187–217.

H. G. DINES, 1956 *The metalliferous mining region of South West England*, Memoirs of the Geological Survey of Great Britain, London.

M. H. DODSON and L. E. LONG, 1962 'Age of Lundy granite, Bristol Channel', *Nature* 195, 975–976.

M. H. DODSON and D. C. REX, 1971 'Potassium-argon ages of slates and phyllites from South West England', *Quarterly Journal of the Geological Society* 126, 465–499.

A. T. J. DOLLAR, 1942 'The Lundy Complex: its petrology and tectonics', *Quarterly Journal of the Geological Society* 97, 39–77.

D. T. DONOVAN and A. H. STRIDE, 1975 'Three drowned coast lines of probable Late Tertiary age around Devon and Cornwall', *Marine Geology* 19, M35–M40.

J. C. DOORNKAMP, 1974 'Tropical weathering and the ultra-microscopic characteristics of regolith quartz on Dartmoor', *Geografiska Annaler* 56A, 73–82.

P. V. O. DRUMMOND, 1970 'The mid Dorset swell. Evidence of Albian–Cenomanian movements in Wessex', *Proceedings of the Geologists Association* 81, 679–714.

E. M. DURRANCE, 1969 'The buried channels of the Exe', *Geological Magazine* 106, 174–189.

E. M. DURRANCE, 1971 'The buried channel of the Teign estuary', *Proceedings of the Ussher Society* 2, 299–306.

E. M. DURRANCE, 1974 'Gradients of buried channels in Devon', *Proceedings of the Ussher Society* 3, 111–119.

E. M. DURRANCE and R. J. O. HAMBLIN, 1969 'The Cretaceous structure of Great Haldon, Devon', *Bulletin of the Geological Survey of Great Britain* 30, 71–88.

E. M. DURRANCE and M. C. GEORGE, 1976 'Metatyuyamunite from the uraniferous–vanadiferous nodules in the Permian marls and sandstones of Budleigh Salterton, Devon', *Proceedings of the Ussher Society* 3, 435–440.

E. M. DURRANCE, R. E. MEADS, R. R. B. BALLARD and J. N. WALSH, 1978 'The oxidation state of iron in the Littleham Mudstone Formation of the New Red Sandstone Series (Permian–Triassic) of southeast Devon, England', *Bulletin of the Geological Society of America* 89, 1231–1240.

M. J. EDEN and C. P. GREEN, 1971 'Some aspects of granite weathering and tor formation on Dartmoor, England', *Geografiska Annaler* 53A, 92–99.

E. A. EDMONDS, 1972a *The Pleistocene history of the Barnstaple area*, Institute of Geological Sciences Report 72/2, London.

E. A. EDMONDS, 1972b 'Drainage chronology of the South Molton area, Devonshire', *Bulletin of the Geological Survey of Great Britain* 42, 99–104.

E. A. EDMONDS, 1974 *Classification of Carboniferous rocks of South West England*, Institute of Geological Sciences Report 74/13, London.

E. A. EDMONDS, J. E. WRIGHT, K. E. BEER, J. R. HAWKES, M. WILLIAMS, E. C. FRESHNEY and P. J. FENNING, 1968 *The geology of the country around Okehampton*, Memoirs of the Geological Survey of Great Britain, London.

E. A. EDMONDS, M. C. McKEOWN and M. WILLIAMS, 1975 *British Regional Geology: South West England*, Institute of Geological Sciences, London.

E. A. EDMONDS, B. J. WILLIAMS and R. T. TAYLOR, 1979 *Geology of Bideford and Lundy Island*, Memoirs of the Geological Survey of Great Britain, London.

R. A. EDWARDS, 1970 *The geology of the Bovey Basin*, PhD Thesis, University of Exeter.

R. A. EDWARDS, 1973 'The Aller Gravels: Lower Tertiary braided river deposits in south Devon', *Proceedings of the Ussher Society* 2, 608–616.

R. A. EDWARDS, 1976 'Tertiary sediments and structure of the Bovey Basin, south Devon', *Proceedings of the Geologists Association* 87, 1–26.

M. A. H. EL SHARKAWI and W. R. DEARMAN, 1966 'Tin-bearing skarns from the northwest border of the Dartmoor granite, Devonshire, England', *Economic Geology* 61, 362–369.

C. S. EXLEY, 1959 'Magmatic alteration and differentiation in the St Austell granite', *Quarterly Journal of the Geological Society* 114, 197–230.

C. S. EXLEY and M. STONE, 1966 'The granitic rocks of South West England', in *Present views of some aspects of the geology of Cornwall and Devon*, edited by K. F. G. Hosking and G. J. Shrimpton. 150th Anniversary Volume (for 1964) Royal Geological Society of Cornwall, Penzance, 131–184.

M. J. R. FASHAM, 1971 'A gravity survey of the Bovey Tracey Basin, Devon', *Geological Magazine* 108, 119–130.

B. N. FLETCHER, 1975 'A new Tertiary basin east of Lundy Island', *Journal of the Geological Society* 131, 223–225.

P. A. FLOYD, 1972 'Geochemistry, origin and tectonic environment of the basic and acidic rocks of Cornubia, England', *Proceedings of the Geologists Association* 83, 385–404.

P. A. FLOYD, G. J. LEES and A. PARKER, 1976 'A preliminary geochemical twist to the Lizard's new tale', *Proceedings of the Ussher Society* 3, 414–423.

E. C. FRESHNEY, 1965 'Low-angle faulting in the Boscastle area', *Proceedings of the Ussher Society* 1, 175–179.

E. C. FRESHNEY, 1970 'Cyclical sedimentation in the Petrockstowe Basin', *Proceedings of the Ussher Society* 2, 179–189.

E. C. FRESHNEY and R. T. TAYLOR, 1971 'The structure of mid Devon and north Cornwall', *Proceedings of the Ussher Society* 2, 241–248.

E. C. FRESHNEY and R. T. TAYLOR, 1972 'The Upper Carboniferous stratigraphy of north Cornwall and west Devon', *Proceedings of the Ussher Society* 2, 464–471.

E. C. FRESHNEY, M. C. McKEOWN and M. WILLIAMS, 1972 *Geology of the coast between Tintagel and Bude*, Memoirs of the Geological Survey of Great Britain, London.

E. C. FRESHNEY, K. E. BEER and J. E. WRIGHT, 1979 *Geology of the country around Chulmleigh*, Memoirs of the Geological Survey of Great Britain, London.

T. N. GEORGE, 1958 'Lower Carboniferous palaeogeography of the British Isles', *Proceedings of the Yorkshire Geological Society* 31, 227–318.

T. N. GEORGE, G. A. L. JOHNSON, M. MITCHELL, J. E. PRENTICE, W. H. C. RAMSBOTTOM, G. D. SEVASTOPULO and R. B. WILSON, 1976 *A correlation of Dinantian rocks in the British Isles*, Geological Society Special Report 7, London.

A. J. W. GERRARD, 1974 'The geomorphological importance of jointing in the Dartmoor granite', *Institute of British Geographers Special Publication* 7, 39–51.

A. J. W. GERRARD, 1978 'Tors and granite landforms of Dartmoor and eastern Bodmin Moor', *Proceedings of the Ussher Society* 4, 204–210.

J. GILLULY, 1965 *Volcanism, tectonism and plutonism in the western United States*, Geological Society of America Special Paper 80, 1–69.

R. A. C. GODWIN-AUSTEN, 1842 'On the geology of the southeast of Devonshire', *Transactions of the Geological Society* 6, 433–439.

R. GOLDRING, 1962 'The bathyal lull: Upper Devonian and Lower Carboniferous sedimentation in the Variscan geosyncline', in *Some aspects of the Variscan fold belt*, edited by K. Coe, University Press, Manchester, 75–91.

R. GOLDRING, 1970 *The stratigraphy about the Devonian-Carboniferous boundary in the Barnstaple area of north Devon*, Sixth International Congress on Carboniferous Stratigraphy (1967), Sheffield, 807–816.

R. GOLDRING, 1971 *Shallow water sedimentation as illustrated in the Upper Devonian Baggy Beds*, Memoir of the Geological Society 5, London.

R. GOLDRING, 1978 'Baggy Sandstones', in *A Field Guide to Selected Areas of the Devonian of South West England*, edited by C. T. Scrutton, Palaeontological Association, London, 21–25.

A. J. GOODAY, 1975 'Ostracod ages from the Upper Devonian purple and green slates around Plymouth' *Proceedings of the Ussher Society* 3, 55–62.

T. M. GOULDSTONE, 1975 *Some aspects of the Quaternary history of the Bovey Basin, Devonshire*, MSc Thesis, University of Exeter.

C. P. GREEN, 1974 'Pleistocene gravels of the River Axe in southwestern England, and their bearing on the southern limit of glaciation in Britain', *Geological Magazine* 111, 213–220.

C. P. GREEN and M. J. EDEN, 1973 'Slope deposits on the weathered Dartmoor granite, England', *Zeitschrift für Geomorphologie*, Supplementband 18, 26–37.

J. F. N. GREEN, 1949 'The history of the River Dart, Devon', *Proceedings of the Geologists Association* 60, 105–124.

B. GREENWOOD, 1972 'Modern analogues and the evaluation of a Pleistocene sedimentary sequence', *Transactions of the Institute of British Geographers* 56, 145–169.

K. J. GREGORY, 1971 'Drainage density changes in South-West England', in: *Exeter Essays in Geography*, edited by K. J. Gregory and W. L. D. Ravenhill. University of Exeter, Exeter, 33–53.

A. W. GROVES, 1931 'The unroofing of the Dartmoor granite and the distribution of its detritus in the sediments of southern England', *Quarterly Journal of the Geological Society* 87, 62–96.

J. R. HAILS, 1975 'Sediment distribution and Quaternary history of Start Bay', *Journal of the Geological Society* 131, 19–35.

A. HALLAM, 1960 'The White Lias of the Devon coast', *Proceedings of the Geologists Association* 71, 47–60.

A. HALLAM, 1964 'Origin of the limestone-shale rhythm in the Blue Lias of England: a composite theory', *Journal of Geology* 72, 157–169.

R. J. O. HAMBLIN, 1973a 'The Haldon Gravels of south Devon', *Proceedings of the Geologists Association* 84, 459–476.

R. J. O. HAMBLIN, 1973b 'The clay mineralogy of the Haldon Gravels', *Clay Minerals* 10, 87–97.

R. J. O. HAMBLIN, 1974 'On the correlation of the Haldon and Aller Gravels', *Proceedings of the Ussher Society* 3, 103–110.

R. J. O. HAMBLIN and C. J. WOOD, 1976 'The Cretaceous (Albian–Cenomanian) stratigraphy of the Haldon Hills, south Devon', *Newsletters on Stratigraphy* 4, 135–149.

W. HAMILTON and W. B. MYERS, 1967 *The nature of batholiths*, United States Geological Survey Professional Paper 554-C, 1–30.

J. M. HANCOCK, 1969 'Transgression of the Cretaceous sea in South West England'. *Proceedings of the Ussher Society* 2, 61–83.

J. M. HANCOCK, 1976 'The petrology of the Chalk', *Proceedings of the Geologists Association* 86, 499–535.

J. M. HANCOCK and E. G. KAUFFMAN, 1979 'The great transgressions of the late Cretaceous', *Journal of the Geological Society* 136, 175–186.

R. K. HARRISON, 1975 'Concretionary concentrations of the rarer elements in Permo-Triassic red beds of South West England', *Bulletin of the Geological Survey of Great Britain* 52, 1–26.

T. R. HARROD, J. A. CATT and A. H. WEIR, 1973 'Loess in Devon', *Proceedings of the Ussher Society* 2, 554–564.

M. B. HART, 1971 'Micropalaeontological evidence of mid-Cenomanian flexuring in South West England', *Proceedings of the Ussher Society* 2, 315–325.

M. B. HART, 1973 'Some observations on the Chert Beds (Upper Greensand) of South West England', *Proceedings of the Ussher Society* 2, 599–608.

M. B. HART, 1975 'Microfaunal analysis of the Membury Chalk succession', *Proceedings of the Ussher Society* 3, 271–279.

M. B. HART and P. P. E. WEAVER, 1977 'Turonian microbiostratigraphy of Beer, southeast Devon', *Proceedings of the Ussher Society* 4, 86–93.

M. B. HART and H. W. BAILEY, 1979 'The distribution of planktonic Foraminiferida in the mid Cretaceous of North West Europe', *Aspekte der Kreide Europas*, Series A, 6, 527–542.

J. R. HAWKES, 1974 'Volcanism and metallogenesis: the tin province of South West England', *Bulletin Volcanologique* 38, 1125–1146.

J. R. HAWKES, R. R. HARDING and D. P. F. DARBYSHIRE, 1975 'Petrology and Rb:Sr age of the Brannel, South Crofty and Wherry elvan dykes, Cornwall', *Bulletin of the Geological Survey of Great Britain* 52, 27–42.

J. R. HAWKES and J. DANGERFIELD, 1978 'The Variscan granites of South West England', *Proceedings of the Ussher Society* 4, 158–171.

E. M. L. HENDRIKS, 1937 'Rock successions and structure in south Cornwall', *Quarterly Journal of the Geological Society* 93, 322–367.

E. M. L. HENDRIKS, 1939 'The Start-Dodman-Lizard boundary zone in relation to the Alpine structure of Cornwall', *Geological Magazine* 76, 385–401.

E. M. L. HENDRIKS, 1951 'Geological succession and structure in western south Devonshire', *Transactions of the Royal Geological Society of Cornwall* 18, 255–295.

E. M. L. HENDRIKS, 1959 'A summary of present views on the structure of Cornwall and Devon', *Geological Magazine* 96, 253–257.

S. HENLEY, 1972 'Petrogenesis of quartz porphyry dykes in South West England', *Nature* 235, 95–96.

S. HENLEY, 1974 'Geochemistry and petrogenesis of elvan dykes in the Perranporth area, Cornwall', *Proceedings of the Ussher Society* 3, 136–145.

M. R. HENSON, 1971 *The Permo-Triassic rocks of south Devon*, PhD Thesis, University of Exeter.

M. R. HENSON, 1973 'Clay minerals from the Lower New Red Sandstone of south Devon', *Proceedings of the Geologists Association* 84, 429–445.

W. J. HENWOOD, 1843 'On the metalliferous deposits of Cornwall and Devon', *Transactions of the Royal Geological Society of Cornwall* 5.

D. E. HIGHLEY, 1975 'Ball Clay'. *Institute of Geological Sciences Mineral Dossier* 11, London.

D. N. HINCKLEY, 1962 'Variability in crystallinity values among kaolin deposits of the coastal plain of Georgia and South Carolina', *Proceedings of the Eleventh Conference on Clays and Clay Minerals*.

D. M. HOBSON, 1976a 'A structural section between Plymouth and Bolt Tail, south Devon', *Proceedings of the Geologists Association* 87, 27–43.

D. M. HOBSON, 1976b 'The structure of the Dartmouth antiform', *Proceedings of the Ussher Society* 3, 320–332.

D. M. HOBSON, 1977 'Polyphase folds from the Start complex', *Proceedings of the Ussher Society* 4, 102–110.

D. M. HOBSON and D. J. SANDERSON, 1975 'Major early folds at the southern margin of the Culm synclinorium', *Journal of the Geological Society* 131, 337–352.

J. T. HOLLIN, 1977 'Thames interglacial sites, Ipswichian sea levels and Antarctic surges', *Boreas* 6, 33–52.

F. J. W. HOLWILL, 1962 'The succession of limestones within the Ilfracombe Beds (Devonian) of north Devon', *Proceedings of the Geologists Association* 73, 281–293.

K. F. G. HOSKING and G. J. SHRIMPTON, 1966 *Present views of some aspects of the geology of Cornwall and Devon*, 150th Anniversary Volume (for 1964) Royal Geological Society of Cornwall, Penzance.

M. R. HOUSE, 1957a 'Facies and time in Devonian tropical areas', *Proceedings of the Yorkshire Geological Society* 40, 233–288.

M. R. HOUSE, 1957b 'Faunas and time in the marine Devonian', *Proceedings of the Yorkshire Geological Society* 40, 459–490.

M. R. HOUSE and E. B. SELWOOD, 1966 'Palaeozoic palaeontology in Devon and Cornwall', in *Present views of some aspects of the geology of Cornwall and Devon*, edited by K. F. G. Hosking and G. J. Shrimpton, 150th Anniversary Volume (for 1964) Royal Geological Society of Cornwall, Penzance, 48–86.

M. R. HOUSE, J. B. RICHARDSON, W. G. CHALONER, J. R. L. ALLEN, C. H. HOLLAND and T. S. WESTOLL, 1977 *A correlation of Devonian rocks of the British Isles*, Geological Society Special Report 7, London.

A. R. HUNT, 1894 'Four theories of the age and origin of the Dartmoor granites', *Geological Magazine* 1, 97–108.

P. F. HUTCHINS, 1958 'Devonian limestone pebbles in central Devon', *Geological Magazine* 95, 119–124.

P. F. HUTCHINS 1963 'The Lower New Red Sandstone of the Crediton valley', *Geological Magazine* 100, 107–128.

P. R. INESON, J. G. MITCHELL and F. J. ROTTENBURY, 1977 'Potassium-argon isotopic age determinations from some north Devon mineral deposits', *Proceedings of the Ussher Society* 4, 12–23.

K. P. ISAAC, 1979 'Tertiary silcretes of the Sidmouth area, east Devon', *Proceedings of the Ussher Society* 4, 341–354.

K. P. ISAAC, 1981 'The Hercynian geology of Lydford Gorge, northwest Dartmoor, and its regional significance', *Proceedings of the Ussher Society* 5, 147–152.

K. P. ISAAC, P. J. TURNER and I. J. STEWART, 1982 'The evolution of the Hercynides of central South West England', *Journal of the Geological Society* 139, forthcoming.

C. V. JEANS, R. J. MERRIMAN and J. G. MITCHELL, 1977 'Origin of the Middle Jurassic and Lower Cretaceous Fuller's Earths in England', *Clay Minerals* 12, 11–44.

A. K. H. JENKIN, 1974 *Mines of Devon. Volume 1: The southern area*, David and Charles, Newton Abbot.

C. A. JENKINS and A. VINCENT, 1981 'Periglacial features in the Bovey Basin, South Devon', *Proceedings of the Ussher Society* 5, 200–205.

A. J. JUKES-BROWNE and W. HILL, 1900 *The Cretaceous rocks of Britain: Gault and Upper Greensand*, Memoirs of the Geological Survey of Great Britain, London.

A. J. JUKES-BROWNE and W. HILL, 1903 *The Cretaceous rocks of Britain: The Lower and Middle Chalk of England*, Memoirs of the Geological Survey of Great Britain, London.

A. J. JUKES-BROWNE and W. HILL, 1904 *The Cretaceous rocks of Britain: The Upper Chalk of England*, Memoirs of the Geological Survey of Great Britain, London.

N. C. KELLAND, 1975 'Submarine geology of Start Bay determined by continuous seismic profiling and core sampling', *Journal of the Geological Society* 131, 7–17.

G. A. KELLAWAY, J. H. REDDING, E. R. SHEPHARD-THORN and J–P. DESTOMBES, 1975 'The Quaternary history of the English Channel', *Philosophical Transactions of the Royal Society* A279, 189–218.

W. J. KENNEDY, 1970 'A correlation of the uppermost Albian and the Cenomanian of South West England', *Proceedings of the Geologists Association* 81, 613–677.

W. J. KENNEDY and R. E. GARRISON, 1975 'Morphology and genesis of nodular chalks and hardgrounds in the Upper Cretaceous of southern England', *Sedimentology* 22, 311–386.

C. KIDSON, 1962 'Denudation chronology of the River Exe', *Transactions of the Institute of British Geographers* 31, 43–66.

C. KIDSON, 1971 'The Quaternary history of the coasts of South West England, with special reference to the Bristol Channel coast', in *Exeter Essays in Geography*, edited by K. J. Gregory and W. L. D. Ravenhill. University of Exeter, Exeter, 1–22.

C. KIDSON, 1977 'The coast of South West England', in *The Quaternary History of the Irish Sea*, edited by C. Kidson and M. J. Tooley. Seel House, Liverpool, 257–298.

C. KIDSON and R. WOOD, 1974 'The Pleistocene stratigraphy of Barnstaple Bay', *Proceedings of the Geologists Association* 85, 223–238.

C. KIDSON and A. HEYWORTH, 1978 'Holocene eustatic sea level change', *Nature* 273, 748–750.

A. W. G. KINGSBURY, 1966 'Some minerals of special interest in South West England', in *Present views of some aspects of the geology of Cornwall and Devon*, edited by K. F. G. Hosking and G. J. Shrimpton. 150th Anniversary Volume (for 1964) Royal Geological Society of Cornwall, Penzance, 247–266.

G. DE V. KLEIN, 1962 'Sedimentary structures in the Keuper Marl (Upper Triassic)', *Geological Magazine* 99, 137–144.

D. C. KNILL, 1969 'The Permian igneous rocks of Devon', *Bulletin of the Geological Survey of Great Britain* 29, 115–138.

W. KREBS and H. WACHENDORF, 1973 'Proterozoic–Palaeozoic geosynclinal and orogenic evolution of central Europe', *Bulletin of the Geological Society of America* 84, 2611–2630.

P. H. KRINSLEY, P. F. FRIEND and R. KLIMENTIDIS, 1976 'Eolian transport textures on the surfaces of sand grains of early Triassic age', *Bulletin of the Geological Society of America* 87, 130–132.

D. J. C. LAMING, 1954 *Sedimentary processes in the formation of the New Red Sandstone of south Devonshire*, PhD Thesis, University of London.

D. J. C. LAMING, 1958 'Fossil winds', in *Polar wandering and continental drift–a symposium*, Journal of the Alberta Society of Petroleum Geologists 6, Calgary, 179–183.

D. J. C. LAMING, 1965 'Age of the New Red Sandstone in south Devonshire', *Nature* 207, 624–625.

D. J. C. LAMING, 1966 'Imbrications, palaeocurrents and other sedimentary features in the Lower New Red Sandstone, Devonshire, England', *Journal of Sedimentary Petrology* 36, 940–959.

D. J. C. LAMING, 1968 'New Red Sandstone stratigraphy in Devon and west Somerset', *Proceedings of the Ussher Society* 2, 23–25.

W. D. LANG, 1914 'The geology of the Charmouth cliffs, beach and foreshore', *Proceedings of the Geologists Association* 25, 293–360.

W. D. LANG, 1917 'The *ibex* Zone at Charmouth and its relation to the zones near it', *Proceedings of the Geologists Association* 28, 31–36.

W. D. LANG, 1924 'The Blue Lias of the Devon and Dorset coasts', *Proceedings of the Geologists Association* 35, 169–185.

W. D. LANG, 1932 'The Lower Lias of Charmouth and the Vale of Marshwood', *Proceedings of the Geologists Association* 43, 97–126.

W. D. LANG, 1936 'The Green Ammonite Beds of the Dorset Lias', *Quarterly Journal of the Geological Society* 92, 423–437 and 485–487.

G. R. LEWIS, 1908 *The Stannaries*, Reprinted 1965, Bradford Barton, Truro.

D. L. LINTON, 1955 'The problem of tors', *Geographical Journal* 121, 480–487.

S. McCOURT, 1975 'A geochemical division of Namurian shales of Devon and Cornwall and its significance', *Proceedings of the Ussher Society* 3, 228–233.

P. B. MACFARLANE, 1955 'Survey of two drowned river valleys in Devon', *Geological Magazine* 92, 419–429.

P. A. MADGETT and R. A. MADGETT, 1974 'A giant erratic on Baggy Point, North Devon', *Quaternary Newsletter* 14, 1–2.

B. MARSHALL, 1965 *The Start Boundary problem*, PhD Thesis, University of Bristol.

W. G. MATON, 1797 *Observations relative chiefly to the natural history, picturesque scenery and antiquities of the western counties of England, made in the years 1794 and 1797*. Salisbury.

S. C. MATTHEWS, 1975 'Exmoor thrust? Variscan front?', *Proceedings of the Ussher Society* 3, 82–94.

S. C. MATTHEWS, 1977a 'Carboniferous successions in Germany and in South West England', *Proceedings of the Ussher Society* 4, 67–74.

S. C. MATTHEWS, 1977b 'The Variscan fold belt in South West England', *Neues Jahrbuch fur Geologie und Palaeontologie, Abhandlungen* 154, 94–127.

S. C. MATTHEWS, 1981 'A cross section through South West England', *Geologie en Mijnbouw* 60, 145–148.

S. C. MATTHEWS and J. M. THOMAS, 1974 'Lower Carboniferous conodont faunas from northeast Devonshire', *Palaeontology* 17, 371–385.

G. MAW, 1864 'On a supposed deposit of boulder clay in North Devon', *Quarterly Journal of the Geological Society* 20, 445–451.

D. F. MAYHEW, 1976 'Comments on the British glacial–interglacial sequence', *Quaternary Newsletter* 19, 8–9.

J. R. MEREFIELD, 1981 'Caesium from former Dartmoor volcanism: its incorporation in New Red sediments of South West England', *Journal of the Geological Society* 138, 145–152.

G. V. MIDDLETON, 1960 'Splitic rocks in southeast Devon', *Geological Magazine* 97, 193–207.

J. A. MILLER and F. J. FITCH, 1962 'Age of the Lundy granites', *Nature* 195, 553–555.

J. A. MILLER and P. A. MOHR, 1964 'Potassium-argon measurements on the granites and some associated rocks from South West England', *Geological Journal* 4, 105–126.

J. A. MILLER, K. SHIBATA and M. MUNRO, 1961 'The potassium-argon age of the lava of Killerton Park near Exeter', *Geophysical Journal of the Royal Astronomical Society* 6, 394–396.

J. MILLES, 1750 *Devonshire Manuscripts 8*, Parochial Collections, Bodleian Library, Oxford.

G. F. MITCHELL, 1960 'The Pleistocene history of the Irish Sea', *Advancement of Science* 68, 313–325.

G. F. MITCHELL, 1972 'The Pleistocene history of the Irish Sea: second approximation', *Scientific Proceedings of the Royal Dublin Society* A4, 181–199.

G. F. MITCHELL, L. F. PENNY, F. W. SHOTTON and R. G. WEST, 1973 *A correlation of Quaternary deposits in the British Isles*, Geological Society Special Report 4, London.

S. MOORBATH, 1962 'Lead isotope abundance studies on mineral occurrences in the British Isles and their geological significance', *Philosophical Transactions of the Royal Society* A254, 295–360.

C. R. MOREY, 1976 'The natural history of Slapton Ley Nature Reserve IX. The morphology and history of the lake basins', *Field Studies* 4, 353–368.

N–A. MÖRNER, 1969 'Eustatic and climatic changes during the last 15,000 years', *Geologie en Mijnbouw* 48, 389–399.

D. N. MOTTERSHEAD, 1971 'Coastal head deposits between Start Point and Hope Cove, Devon', *Field Studies* 3, 433–453.

D. N. MOTTERSHEAD, 1977 'Devon valley of rugged rocks', *Geographical Magazine* 49, 711–714.

R. I. MURCHISON, 1854 *Siluria*, Murray, London.

A. E. MUSSETT, P. DAGLEY and M. ECKFORD, 1976 'The British Tertiary igneous province: palaeomagnetism and ages of dykes, Lundy Island, Bristol Channel', *Geophysical Journal of the Royal Astronomical Society* 46, 595–603.

D. NAYLOR and S. N. MOUNTENEY, 1975 *Geology of the northwest European continental shelf: Volume 1*, Graham Trotman Dudley, London.

M. J. ORCHARD, 1975 'Famennian conodonts and cavity infills in the Plymouth limestone (South Devon)', *Proceedings of the Ussher Society* 3, 49–54.

M. J. ORCHARD, 1977 'Plymouth–Tamar' in *A correlation of Devonian rocks of the British Isles*, Geological Society Special Report 7, London.

A. R. ORME, 1960 'The raised beaches and strandlines of South Devon', *Field Studies* 1, 109–130.

A. R. ORME, 1964 'The geomorphology of southern Dartmoor and the adjacent area', in *Dartmoor Essays*, edited by I. G. Simmons, Devonshire Association, Exeter, 31–72.

C. S. ORWIN, 1929 *The reclamation of Exmoor Forest*, Oxford University Press.

T. R. OWEN, 1971 'The structural evolution of the Bristol Channel', *Proceedings of the Geological Society* 1664, 289–294.

J. PALMER and R. A. NEILSON, 1962 'The origin of granite tors on Dartmoor, Devonshire', *Proceedings of the Yorkshire Geological Society* 33, 315–340.

E. PAPROTH, 1969 *Die parallelisierung von Kohlenkalk und Kulm*, Sixth International Congress on Carboniferous Stratigraphy (1967), Sheffield, 279–291.

R. L. PATON, 1974 'Capitosauroid labyrinthodonts from the Trias of England', *Palaeontology* 17, 253–289.

S. R. PATTISON, 1865 'A day in the north Devon mining district', *Transactions of the Royal Geological Society of Cornwall* 7, 223–227.

D. A. B. PEARSON, 1970 'Problems of Rhaetian stratigraphy with special reference to the lower boundary of the Stage', *Quarterly Journal of the Geological Society* 126, 125–150.

W. PENGELLY, 1863 'On the chronological value of the New Red Sandstone System of Devonshire', *Transactions of the Devonshire Association* 1, 31–43.

J. W. PERKINS, 1971 *Geology explained in south and east Devon*, David and Charles, Newton Abbot.

J. W. PERKINS, 1972 *Geology explained in Dartmoor and the Tamar valley*, David and Charles, Newton Abbot.

W. J. PHILLIPS, 1841 *Figures and descriptions of the Palaeozoic fossils of Devon, Cornwall and west Somerset*, Longman, Brown, Green and Longman, London.

R. POLWHELE, 1793 *The History of Devonshire (Three volumes; 1793, 1797, 1806)*, Reprinted 1977 with an introduction by A. L. Rowse, Kohler and Coombes, Dorking.

B. POMEROL and M. P. AUBRY, 1977 'Relation between west European Chalks and the opening of the North Atlantic', *Journal of Sedimentary Petrology* 47, 1027–1035.

J. E. PRENTICE, 1959 'Dinantian, Namurian and Westphalian rocks of the district southwest of Barnstaple, north Devon', *Quarterly Journal of the Geological Society* 115, 261–289.

P. F. RAWSON, D. CURRY, F. C. DILLEY, J. M. HANCOCK, W. J. KENNEDY, J. W. NEALE, C. J. WOOD and B. C. WORSSAM, 1978 *A correlation of the Cretaceous rocks of the British Isles*, Geological Society Special Report 9, London.

C. REID, G. BARROW, R. L. SHERLOCK, D. A. MACALISTER, H. DEWEY and C. N. BROMEHEAD, 1912 *The geology of Dartmoor*, Memoirs of the Geological Survey of Great Britain, London.

L. RICHARDSON, 1906 'On the Rhaetic and contiguous deposits of Devon and Dorset', *Proceeddings of the Geologists Association* 19, 401–409.

L. RICHARDSON, 1911 'The Rhaetic and contiguous deposits of west, mid and part of east Somerset', *Quarterly Journal of the Geological Society* 67, 1–74.

D. RICHTER, 1965 'Stratigraphy, igneous rocks and structural development of the Torquay area', *Transactions of the Devonshire Association* 97, 57–70.

D. RICHTER, 1966 'On the New Red Sandstone neptunian dykes of the Tor Bay area', *Proceedings of the Geologists Association* 77, 173–186.

D. RICHTER, 1967 'Sedimentology and facies of the Meadfoot Beds (Lower Devonian) in southeast Devon (England)', *Geologische Rundschau* 56, 543–561.

D. RICHTER, 1969 'Structure and metamorphism of the Devonian rocks south of Torquay, southeast Devon (England)', *Geologische Mitteilungen* 9, 109–173.

J. M. RIDGEWAY, 1974 'A problematical trace fossil from the New Red Sandstone of south Devon', *Proceedings of the Geologists Association* 85, 511–518.

T. RISDON, 1811 *Survey of the County of Devon*, Rees and Curtis, London.

D. ROBINSON, 1981 'Metamorphic rocks of an intermediate facies series juxtaposed at the Start Boundary, South West England', *Geological Magazine* 118, 297–301.

I. ROGERS and B. SIMPSON, 1937 'The flint gravel deposits of Orleigh Court, Buckland Brewer, north Devon', *Geological Magazine* 74, 309–316.

F. J. ROTTENBURY, 1974 *Geology, mineralogy and mining history of the metalliferous mining areas of Exmoor*, PhD Thesis, University of Leeds.

F. J. ROTTENBURY and R. D. YOUELL, 1974 'New evidence on the mineral workings of Exmoor', *Proceedings of the Ussher Society* 3, 180–181.

P. M. G. RUSSELL, 1979 'Manganese mining', *Devon and Cornwall Notes and Queries* 31, 205–213.

P. M. SADLER, 1973 'An interpretation of new stratigraphic evidence from south Cornwall', *Proceedings of the Ussher Society* 2, 535–550.

P. M. SADLER, 1975 'An appraisal of the "Lizard–Dodman–Start Thrust" concept', *Proceedings of the Ussher Society* 3, 71–81.

D. J. SANDERSON and W. R. DEARMAN, 1973 'Structural zones of the Variscan fold belt in South West England, their location and development', *Journal of the Geological Society* 129, 527–536.

H. SCHMIDT, 1926 'Schwellen und Beckenfazies im osrheinischen Palaozoikum', *Zeitschrift der Deutschen Geologischen Gesellschaft* 77, 226–234.

C. J. SCHMITZ, 1973 *The Teign valley lead mines*, Northern Cave and Mine Research Society Occasional Publication 6, Sheffield.

R. C. SCRIVENER and K. E. BEER, 1971 'Cassiterite in the Aller Gravels near Newton Abbot', *Proceedings of the Ussher Society* 2, 326–329.

R. C. SCRIVENER and J. H. WALBEOFFE-WILSON, 1975 'Alluvial tin at Colston, Buckfastleigh', *Proceedings of the Ussher Society* 3, 237–242.

R. C. SCRIVENER, B. V. COOPER and O. A. BAKER, 1977 'Some notes on the mineralisation of the Dartmoor granite', *Proceedings of the Ussher Society* 4, 7–10.

J. B. SCRIVENOR, 1948 'The New Red Sandstone of south Devonshire', *Geological Magazine* 85, 317–332.

C. T. SCRUTTON, 1977 'Facies variations in the Devonian limestones of eastern south Devon', *Geological Magazine* 114, 165–193.

A. SEDGWICK, 1852 'On the slate rocks of Devon and Cornwall', *Quarterly Journal of the Geological Society* 8, 1–19.

A. SEDGWICK and R. I. MURCHISON, 1839 'Classification of the older stratified rocks of Devonshire and Cornwall', *London and Edinburgh Philosophical Magazine and Journal of Science* 14, 241–260 and 354.

A. SEDGWICK and R. I. MURCHISON, 1840 'On the physical structure of Devonshire, and on the subdivisions and geological relations of its older stratified deposits', *Transactions of the Geological Society* 2, 633–704.

E. B. SELWOOD, 1971a 'Structures along the southern margin of the Culm synclinorium, northwest of Dartmoor', *Proceedings of the Ussher Society* 2, 237–240.

E. B. SELWOOD, 1971b 'Successions at the Devonian-Carboniferous boundary between Boscastle and Dartmoor', *Proceedings of the Ussher Society* 2, 275–285.

E. B. SELWOOD, 1974 'The age of the Upper Palaeozoic volcanics between Bodmin Moor and Dartmoor', *Proceedings of the Ussher Society* 3, 63–70.

E. B. SELWOOD and S. McCOURT, 1973 'The Bridford thrust', *Proceedings of the Ussher Society* 2, 529–535.

E. B. SELWOOD, R. A. EDWARDS, S. SIMPSON, J. A. CHESHER, R. J. O. HAMBLIN, M. R. HENSON, B. W. RIDDOLLS and R. A. WATERS, 1982 *Geology of the country around Newton Abbot*, Memoirs of the Geological Survey of Great Britain, London, forthcoming.

W. G. SHANNON, 1927 'The petrography and correlation of the Permian rocks of the Torquay promontory', *Proceedings of the Geologists Association* 38, 133–144.

D. J. SHEARMAN, 1967 'On Tertiary fault movements in north Devonshire', *Proceedings of the Geologists Association* 78, 555–566.

D. SHELLEY, 1966 'The significance of granophyric and myrmekitic textures in the Lundy granites', *Mineralogical Magazine* 35, 678–692.

S. M. F. SHEPPARD, 1977 'The Cornubian batholith, South West England; D/H and $^{18}O/^{16}O$ studies of kaolinite and other alteration minerals', *Journal of the Geological Society* 133, 573–591.

F. W. SHERRELL, 1970 'Some aspects of the Triassic aquifer in east Devon and west Somerset', *Quarterly Journal of Engineering Geology* 2, 255–286.

A. H. SHORTER, W. L. D. RAVENHILL and K. J. GREGORY, 1969 *Regions of the British Isles: South West England*, Nelson, London.

F. W. SHOTTON, 1956 'Some aspects of the New Red desert in Britain', *Geological Journal* 1, 456–465.

I. G. SIMMONS, 1964 'An ecological history of Dartmoor', in *Dartmoor Essays*, edited by I. G. Simmons, Devonshire Association, Exeter, 191–215.

S. SIMPSON, 1951 'Some solved and unsolved problems of the stratigraphy of the marine Devonian in Great Britain', *Abhandlungen der Senckenbergischen naturforschenden Gesellschaft* 485, 53–66.

S. SIMPSON, 1953 'The development of the Lyn drainage system and its relation to the origin of the coast between Combe Martin and Porlock', *Proceedings of the Geologists Association* 64, 14–23.

S. SIMPSON, 1957 'On the trace fossil Chondrites', *Quarterly Journal of the Geological Society* 112, 475–500.

S. SIMPSON, 1964 'The Lynton Beds of north Devon', *Proceedings of the Ussher Society* 1, 121–122.

S. SIMPSON, 1969 'Geology' in *Exeter and its Region*, edited by F. Barlow. University of Exeter, Exeter, 5–26.

S. SIMPSON, 1970 'The structural geology of Cornwall', *Proceedings of the Geological Society* 1662, 1–3.

S. SIMPSON, 1971 'The Variscan structure of south Devon', *Proceedings of the Ussher Society* 2, 249–252.

J. B. SISSONS, 1979 'The Loch Lomond Stadial in the British Isles', *Nature* 280, 199–203.

J. B. SISSONS, 1981 'British shore platforms and ice sheets', *Nature* 291, 473–475.

A. J. SMITH and D. CURRY, 1975 'The structure and geological evolution of the English Channel', *Philosophical Transactions of the Royal Society* A279, 3–20.

D. B. SMITH, R. G. W. BRUNSTROM, P. I. MANNING, S. SIMPSON and F. W. SHOTTON, 1974 *A correlation of Permian rocks in the British Isles*, Geological Society Special Report 5, London.

W. E. SMITH, 1957 'The Cenomanian limestone of the Beer district, south Devon', *Proceedings of the Geologists Association* 68, 115–133.

W. E. SMITH, 1961a 'The Cenomanian deposits of southeast Devonshire: the Cenomanian limestones and contiguous deposits west of Beer', *Proceedings of the Geologists Association* 72, 91–133.

W. E. SMITH, 1961b 'The detrital mineralogy of the Cretaceous rocks of South West England with particular reference to the Cenomanian', *Proceedings of the Geologists Association* 72, 303–331.

W. E. SMITH, 1965 'The Cenomanian deposits of southeast Devonshire: the Cenomanian limestone east of Seaton', *Proceedings of the Geologists Association* 76, 121–136.

W. W. SMITH, 1878 'On the occurrence of metallic ores with garnet rock', *Transactions of the Royal Geological Society of Cornwall* 9, 38–45.

D. K. SMYTHE, 1973 'Structure of the Devonian limestone at Brixham', *Proceedings of the Ussher Society* 2, 617–625.

SOUTH WEST WATER AUTHORITY, 1976 *Hydrometric Report Part 3: Ground-water*, South West Water Authority, Exeter.

SOUTH WEST WATER AUTHORITY, 1977 *Hydrometric Report Part 3: Ground-water*, South West Water Authority, Exeter.

SOUTH WEST WATER AUTHORITY, 1978 *Survey of existing water use and management*, South West Water Authority, Exeter.

N. STEPHENS, 1966 'Some Pleistocene deposits in north Devon', *Biuletyn Peryglacjalny* 15, 103–114.

N. STEPHENS, 1970 'The lower Severn valley' and 'The West Country and southern Ireland', in *The glaciations of Wales and adjoining regions*, edited by C. A. Lewis, Longman, London, 107–124 and 267–314.

N. STEPHENS and F. M. SYNGE, 1966 'Pleistocene shorelines', in *Essays in Geomorphology*, edited by G. H. Dury, Heinemann, London, 1–51.

I. J. STEWART, 1981a 'Late Devonian and Lower Carboniferious conodont faunas from north Cornwall, and their stratigraphical significance', *Proceedings of the Ussher Society* 5, 179–185.

I. J. STEWART, 1981b 'The Trekelland thrust', *Proceedings of the Ussher Society* 5, 163–167.

M. STONE, 1968 'A study of the Praa Sands elvan and its bearing on the origin of elvans', *Proceedings of the Ussher Society* 2, 37–42.

A. STRAW, 1974 *Quaternary Research Association Field Handbook, Easter Meeting 1974*, Exeter.

A. J. SUTCLIFFE, 1969 'Pleistocene faunas of Devon', in *Exeter and its Region*, edited by F. Barlow, University of Exeter, Exeter, 66–70.

A. J. SUTCLIFFE, 1976 'The British glacial–interglacial sequence: a reply', *Quaternary Newsletter* 18, 1–7.

D. H. TARLING, 1979 'Palaeomagnetic reconstructions and the Variscan Orogeny', *Proceedings of the Ussher Society* 4, 233–261.

C. W. TAYLOR, 1956 'Erratics of the Saunton and Fremington areas', *Transactions of the Devonshire Association* 88, 52–64.

J. J. H. TEALL, 1902 'Petrology of the Exeter traps', in *The geology of the country around Exeter*, W. A. E. Ussher, 1902, Memoirs of the Geological Survey of Great Britain, London.

E. TERRELL, 1920 'The Hemerdon wolfram–tin mine', *Mining Magazine* 22, 75–87.

J. M. THOMAS, 1963a 'The Culm Measures succession in northeast Devon and northwest Somerset', *Proceedings of the Ussher Society* 1, 63–64.

J. M. THOMAS, 1963b 'Sedimentation in the Lower Culm Measures round Westleigh, northeast Devon', *Proceedings of the Ussher Society* 1, 71–72.

W. G. TIDMARSH, 1932 'Permian lavas of Devon', *Quarterly Journal of the Geological Society* 88, 712–773.

C. E. TILLEY, 1923 'The petrology of the metamorphosed rocks of the Start area', *Quarterly Journal of the Geological Society* 79, 172–204.

G. R. TRESISE, 1960 'Aspects of the lithology of the Wessex Upper Greensand', *Proceedings of the Geologists Association* 71, 316–339.

G. R. TRESISE, 1961 'The nature and origin of chert in the Upper Greensand', *Proceedings of the Geologists Association* 72, 333–356.

M. E. TUCKER, 1974 'Sedimentology of Palaeozoic pelagic limestones: the Devonian Griotte (southern France) and Cephalopodenkalk (Germany)', *Special Publications of the International Association of Sedimentologists* 1, 71–92.

I. P. TUNBRIDGE and A. WHITTAKER, 1978 'Lynton Beds and Hangman Sandstone Group', in *A field guide to selected areas of the Devonian of South West England*, edited by C. T. Scrutton, Palaeontological Association, London, 8–13.

C. TURNER, 1975 'Der Einfluss grosser Mammalier auf die interglaziale Vegetation', *Quartärpaleontologie* 1, 13–19.

F. J. TURNER, 1968 *Metamorphic Petrology*, McGraw–Hill, New York.

P. J. TURNER, 1981 'Aspects of the structure of the Chillaton area, southwest Devonshire', *Proceedings of the Ussher Society* 5, 153–162.

W. A. E. USSHER, 1888 'The granite of Dartmoor. Parts 1 and 2', *Transactions of the Devonshire Association* 20, 141–157.

W. A. E. USSHER, 1904 *The geology of the country around Kingsbridge and Salcombe*, Memoirs of the Geological Survey of Great Britain, London.

W. A. E. USSHER, 1906 *Geology of the country between Wellington and Chard*, Memoirs of the Geological Survey of Great Britain, London.

W. A. E. USSHER, 1907 *The geology of the country around Plymouth and Liskeard*, Memoirs of the Geological Survey of Great Britain, London.

W. A. E. USSHER, 1912 *The geology of the country around Ivybridge and Modbury*, Memoirs of the Geological Survey of Great Britain, London.

W. A. E. USSHER, 1913 *The geology of the country around Newton Abbot*, Memoirs of the Geological Survey of Great Britain, London.

E. K. USTIYEV, 1970 'Relations between volcanism and plutonism at different stages of the tectonomagmatic cycle'. in *Mechanisms of igneous intrusion*, edited by G. Newall and N. Rast, Geological Journal Special Issue 2, Liverpool, 1–22.

C. VANCOUVER, 1808 *General view of the Agriculture of the County of Devon*, Phillips, London.

J. E. VAN HINTE, 1976 'A Cretaceous time scale', *Bulletin of the American Association of Petroleum Geologists* 60, 498–516.

A. VAUGHAN, 1904 'Note on the Lower Culm of north Devon', *Geological Magazine* 1, 530–533.

W. VICARY and J. W. SALTER, 1864 'On the pebble bed of Budleigh Salterton with a note on the fossils', *Quarterly Journal of the Geological Society* 20, 283–302.

A. VINCENT, 1974 *Sedimentary environments of the Bovey Basin*, MPhil Thesis, University of Surrey.

P. G. L. VIPAN, 1959 'Lead and zinc mining in South West England', in *The future of non-ferrous mining in Great Britain and Ireland*, Institution of Mining and Metallurgy, London, 337–351.

A. D. WALKER, 1969 'The reptile fauna of the 'Lower Keuper' sandstone', *Geological Magazine* 106, 470–476.

G. WARRINGTON, 1971 'Palynology of the New Red Sandstone sequence of the south Devon coast', *Proceedings of the Ussher Society* 2, 307–314.

G. WARRINGTON, 1976 'British Triassic Palaeontology', *Proceedings of the Ussher Society* 3, 341–353.

G. WARRINGTON and B. OWENS, 1977 *Micropalaeontological biostratigraphy of off-shore samples from South West Britain*, Institute of Geological Sciences Report 77/7, London.

R. A. WATERS, 1970 'The Variscan structure of eastern Dartmoor', *Proceedings of the Ussher Society* 2, 191–197.

R. S. WATERS, 1960a 'Erosion surfaces on Dartmoor and adjacent areas', *Transactions of the Royal Geological Society of Cornwall, Proceedings of the Second Conference*, 28–29.

R. S. WATERS, 1960b 'The bearing of superficial deposits on the age and origin of the upland plain of east Devon, west Dorset and south Somerset', *Transactions of the Institute of British Geographers* 28, 89–97.

R. S. WATERS, 1961 'Involutions and ice-wedges in Devon', *Nature* 189, 389–390.

R. S. WATERS, 1962 'Altiplantation terraces and slope development in Vest-Spitsbergen and South West England', *Biuletyn Peryglacjalny* 11, 89–101.

R. S. WATERS, 1964 'The Pleistocene legacy to the geomorphology of Dartmoor', in *Dartmoor Essays*, edited by I. G. Simmons, Devonshire Association, Exeter, 73–96.

R. S. WATERS, 1965 'The geomorphological significance of Pleistocene frost action in South West England', in *Essays in Geography for Austin Miller*, edited by J. B. WHITTOW and P. D. Wood, Reading University, Reading, 39–57.

R. S. WATERS, 1971 'The significance of Quaternary events for the landform of South West England', in *Exeter Essays in Geography*, edited by K. J. Gregory and W. L. D. Ravenhill, University of Exeter, Exeter, 23–31.

B. D. WEBBY, 1965a 'The stratigraphy and structure of the Devonian rocks in the Brendon Hills, west Somerset', *Proceedings of the Geologists Association* 76, 39–60.

B. D. WEBBY, 1965b 'The Middle Devonian marine transgression in north Devon and west Somerset', *Geological Magazine* 102, 478–488.

B. D. WEBBY, 1966a 'The stratigraphy and structure of the Devonian rocks in the Quantock Hills, west Somerset', *Proceedings of the Geologists Association* 76, 321–343.

B. D. WEBBY, 1966b 'Middle-Upper Devonian palaeogeography of north Devon and west Somerset' *Palaeogeography, Palaeoclimatology, Palaeoecology* 2, 27–46.

R. G. WEST, 1977 *Pleistocene Geology and Biology*, Longman, London.

W. WHITAKER, 1869 'On the succession of beds in the 'New Red' on the south coast of Devon, and on the locality of a new specimen of Hyperodapedon', *Quarterly Journal of the Geological Society* 25, 152–158.

A. WHITTAKER, 1975 'A postulated post-Hercynian rift valley system in southern Britain', *Geological Magazine* 112, 137–149.

A. WHITTAKER, 1976 'The distribution of the ammonite *Psiloceras planorbis* in South West Britain', *Proceedings of the Ussher Society* 3, 360.

A. WHITTAKER, 1978a 'Ilfracombe Slate' in *A field guide to selected areas of the Devonian of South West England*, edited by C. T. Scrutton, Palaeontological Association, London, 13–18.

A. WHITTAKER, 1978b 'Discussion of the gravity gradient across Exmoor, north Devon', *Journal of the Geological Society* 135, 353–354.

A. WHITTAKER and R. C. SCRIVENER, 1978 *Institute of Geological Sciences boreholes in 1976*, Institute of Geological Sciences Report 77/10, London, 12–13.

A. WOOD, 1976 'Successive regressions and transgressions in the Neogene', *Marine Geology* 22, M23–M29.

H. B. WOODWARD and W. A. E. USSHER, 1911 *The geology of the country near Sidmouth and Lyme Regis*, Memoirs of the Geological Survey of Great Britain, London.

F. E. ZEUNER, 1959 *The Pleistocene Period*, Hutchinson, London.

P. A. ZIEGLER, 1975 'North Sea Basin history in the tectonic framework of North Western Europe', in *Petroleum and the Continental Shelf of North West Europe; Volume 1, Geology*, edited by A. W. Woodland, Applied Science, London, 131–149.

W. H. ZIEGLER, 1975 'Outline of the geological history of the North Sea', in *Petroleum and the Continental Shelf of North West Europe; Volume 1, Geology*, edited by A. W. Woodland, Applied Science, London, 165–190.

J. D. A. ZIJDERVELD, 1967 'The natural remanent magnetisations of the Exeter volcanic traps', *Tectonophysics* 4, 121–153.

Index